PHYSICAL GEOLOGY
A PROCESS APPROACH

PHYSICAL GEOLOGY

A PROCESS APPROACH

DAVID ALT
University of Montana

WADSWORTH PUBLISHING COMPANY   Belmont, California

A Division of Wadsworth, Inc.

Cover photograph: Marble Gorge in the Grand Canyon, looking upstream from President Harding Rapids. Photograph by David Cross.

Geology editor: Thomas P. Nerney
Designer: Janet Bollow
Copy editor: Stuart Kenter
Technical illustrators: Ayxa Art Studio; Barbara Hack

Printed in the United States of America

2 3 4 5 6 7 8 9 10—86 85 84 83 82

Library of Congress Cataloging in Publication Data

Alt, David D.
  Physical geology.

  Includes bibliographies and index.
    1. Physical geology   I. Title.
QE28.2.A47   551   81–16291
ISBN 0–534–01034–2   AACR2

CONTENTS

EIGHT: GROUND WATER   135

NINE: COASTAL PROCESSES   155

TEN: WIND   171

PREFACE

Physical Geology: A Process Approach is addressed primarily to students taking an introductory course in physical geology, especially to the large number of those who have no initial plans to pursue a major in this area. For their sake, the text conveys a sense of geology, as well as the way geologists think and work, with a minimum of technical jargon.

Like all sciences, geology consists of a body of factual observation embedded in a connecting tissue of theory that varies greatly in strength from one region of the subject to another. I have tried to sketch both the observations and the ideas in a way that will show their relationship but not obscure the distinctions between them, to maintain a delicate balance between describing natural phenomena on the one hand and presenting abstract concepts on the other. I hope that students who read this book will not only learn some basic ideas about the earth but will also appreciate the kinds of observations behind these ideas, and thus perhaps sense something of the distinctive style and tradition of geologic reasoning.

Physical Geology: A Process Approach is organized into major sections dealing with earth materials, surface processes, and internal processes. Unlike many other texts, however, this book does not contain separate chapters covering economic mineral deposits and environmental problems. Those topics are integrated into the sections dealing with the processes that create the mineral deposits or pose the hazards. Although the section on earth materials includes a chapter devoted to the description and classification of rocks, the discussion of the origin of rocks is woven into the chapters about the processes that create them.

Except for the section on earth materials, which naturally precedes the rest of the book, and the chapter on plate tectonics, which is a natural prelude to the ensuing chapters on internal processes, the sections and chapters are independent enough to be read in almost any order. There are many reasonable ways to organize a course in elementary physical geology and this book may be easily adapted to most of them.

The suggested readings at the ends of the chapters emphasize widely available books and also include a few more technical papers and collections of papers. Each list contains a variety of titles, ranging from some that are quite general and nontechnical to others that are considerably more demanding.

Introductory texts are necessarily eclectic. No author of such a book can possibly claim much credit for the content, which must be drawn from the material of the entire discipline and the work of innumerable professional colleagues. It is impossible for me to properly acknowledge my sources of information and ideas because I acquired much of the material for this book many years before the urge to write surfaced and then finally overcame my saner impulses. All of my own teachers and close colleagues contributed, as have the authors of many professional papers, my fellow students in graduate and undergraduate school, and my own students during the years of my teaching career.

I hope the many left unmentioned will forgive me for singling out D. W. Hyndman to whom I owe special thanks, as well as G. R. Thompson, D. Winston, R. M. Weidman, D. M. Fountain, A. I. Qamar, I. Lange, J. P. Wehrenberg, T. Bateridge, and F. N. Blanchard, all of whom will recognize certain of their own distinctive contributions. I also thank the numerous reviewers who read all or part of the manuscript at various stages in its preparation and greatly influenced the final product. They include Kennard B. Bork, Denison University; Robert Boutilier, Bridgewater State College; F. W. Cambray, Michigan State University; E. Julius Dasch, Oregon State University; Richard A. Davis, Jr., University of South Florida; W. R. Farrand, The University of Michigan; Paul D. Fullagar, The University of North Carolina at Chapel Hill; Jon S. Galehouse, San Francisco State University; Lawrence W. Knight, William Rainey Harper College; Edward G. Lidiak, University of Pittsburgh; Gary L. Peterson, San Diego State University; Robert F. Schmalz, Pennsylvania State University; Charles R. Singler, Youngstown State University; James E. Slosson, Los Angeles Valley College; David L. Southwick, University of Minnesota, Twin Cities. My thanks also to Shirley Petterson and Jan Dobson, who typed the final draft of the manuscript. And special appreciation goes to my wife Sandy, who, with her usual forebearance, patiently endured all the strains and commotions that authorship seems to cause.

This book of course represents a creative partnership between author and publisher. Tom Nerney, Joan Garbutt, Jean Francois Vilain, and Jeremiah Lyons all provided editorial guidance and support during various stages of the writing.

David Alt

PHYSICAL GEOLOGY
A PROCESS APPROACH

CHAPTER ONE

We like to imagine that other planets akin to our own may exist somewhere in the unmeasured vastness of space. We do not, however, know of them and may not survive as a species to learn about them. This photograph of the Yosemite Valley was taken almost a century ago, before the Valley was paved. (J. K. Hillers, U.S. Geological Survey)

A SINGULAR PLANET

It is irresistibly tempting to suppose that the human race is not alone, that somewhere in the vast emptiness of space there must be other worlds like ours, where clouds float in blue summer skies, children play, and screen doors slam. But such other worlds, if they do exist, would be impossibly remote. None of us who now wonder about them, perhaps not even our species, can hope to survive to know them. The only other planets we may ever see are those in our own solar system, and they are strange worlds indeed.

THE TERRESTRIAL PLANETS AND SATELLITES

The inner members of the solar system, Mercury, Venus, Earth, and Mars, are called the **terrestrial** planets because they consist largely of rocks. Mercury, Venus, and Mars are in various other ways directly comparable to the earth. Beyond Mars, there is an outer group of giant planets, including Jupiter, Saturn, Uranus, and Neptune that consists largely of gases and liquids. These outer planets are not directly comparable to the earth and provide little insight into geology. However, our moon, the four larger satellites of Jupiter, and one large satellite of Saturn also consist largely or entirely of rock. For this reason, and also because these bodies are large enough, they may be considered along with the group of terrestrial planets, even though they orbit other planets instead of the sun.

Mercury and the Moon

The nearest planet to the sun, Mercury, and our own moon resemble each other closely. Both are considerably smaller than the earth. Each, therefore, has a gravitational field much too weak to retain either surface water or an atmosphere. Their surfaces consist mainly of vast expanses of cratered terrain, such as that shown in Figure 1–1, an orbital view of the moon.

Virtually all of those craters, large and small alike, formed when meteorites exploded upon hitting the planet. A meteorite speeding through space at cosmic velocity, generally, about 25 to 50 kilometers per second, possesses tremendous energy of motion. If it strikes a body such as Mercury or the moon, all that energy of motion is suddenly converted into heat, enough to vaporize the meteorite itself, as well as accelerate the atoms of vapor to extremely high velocity. The result is a violent explosion without parallel in human experience—a meteorite traveling at cosmic velocity contains more energy per gram of its weight than a hydrogen bomb. Every meteorite striking the surface of a planet or satellite blasts a crater roughly proportional in size to its own mass. The earth's atmosphere shields its surface from small meteorites but not from large ones.

Most of the small and medium craters on Mercury and the moon are circular in outline and approximately hemispherical in form. Many contain central peaks that were formed when the rocks beneath the crater rebounded from

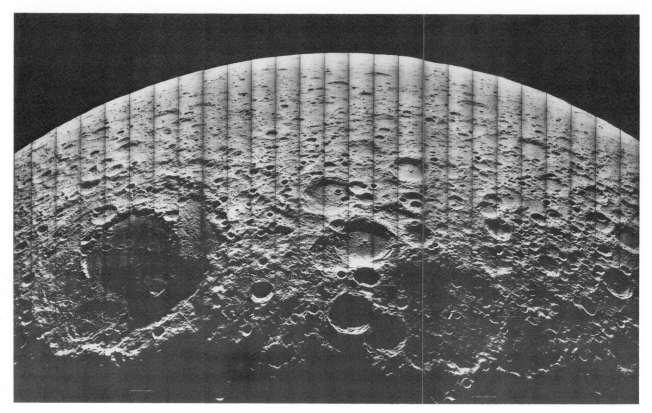

FIGURE 1-1

The moon's surface, like that of Mercury, is mainly a chaotic landscape of circular craters formed by meteorites of all sizes that exploded upon impact. Although most of the craters are well over 2 billion years old, they remain almost intact. Few surface or internal processes exist to modify them significantly. (NASA)

the shock of the explosion. The larger craters, also circular in outline, have flat floors, the surfaces of molten lava lakes that flooded the crater after it formed. Figure 1–2 shows one of those lava plains pocked by later impacts of smaller meteorites. Evidently, the moon and Mercury, like the earth, are so hot within that the rocks at depth would melt if they were not under the great pressure exerted by the weight of those above. The larger meteorite craters relieved this pressure on the rocks at depth, permitting them to partially melt and then fill the crater with lava.

Ages of moon rocks reveal that most of the major cratering occurred early in the moon's history and had essen-

tially ended by about 2 billion years ago. Most geologists assume that meteorite cratering was a major process going on throughout the solar system during the first couple of billions of years of its existence. If so, then the earth must then have had a cratered surface similar to what we now see on the moon and on Mercury. However, the earth's extremely active internal and surface processes have almost completely erased all signs of early meteorite cratering. A few geologists, though, speculate that the "greenstone" belts discussed in Chapter 20 may be remnants of an early cratered terrain. Nevertheless, the earth does contain a number of younger meteorite craters, of

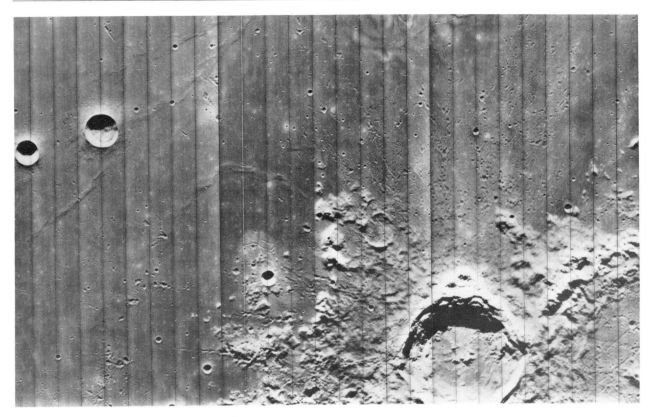

FIGURE 1-2

A level plain of lava on the moon pocked by several small meteorites that struck after the lava had cooled. The big crater on the lower right, created by a larger meteorite, is surrounded by a lumpy debris blanket. (NASA)

which the 1.7 billion-year-old Sudbury complex in eastern Ontario is probably the best known. Meteor Crater in Arizona (Figure 1–3) is about 30,000 years old, probably the youngest sizeable impact crater on the earth. Cratering is no longer a major process.

In contrast, even ancient craters on the moon and Mercury remain relatively fresh because those bodies lack wind, rain, streams, and other familiar agents of erosion. Lunar scientists estimate that the astronauts' footprints on the moon may well last 100 million years or more before gradual lunar erosion—mostly impacts from minute meteorites—finally obliterate them. Not much happens to mark the passage of time on the airless, waterless, and lifeless surfaces of the moon and Mercury.

Mercury and the moon appear to be as dead internally as on their surfaces, probably because both are too small to generate enough heat to drive much internal motion. Orbital photographs of these bodies, especially pictures of Mercury, do, however, show a few long straight cliffs. These structures were evidently formed by vertical movements along fractures similar to the displacements that raise some mountain ranges on the earth. Details depict a few enigmatic features that may be small volcanoes. Neither Mercury nor the moon show surface evidence of other

FIGURE 1-3

Meteor Crater, Arizona, a hole blasted by a large meteorite that exploded upon impact. The highway on the left gives an impression of the crater's size, which is small by lunar standards. (W. B. Hamilton, U.S. Geological Survey)

significant internal activity. Neither one possesses the great linear mountain ranges wrinkling the earth's surface, revealing our planet's interior motion. Moreover, explosion craters formed by impacting meteorites several billions of years ago on both the moon and Mercury remain perfectly circular today. They show no evidence of deformation by any process except later impacts of still more meteorites.

Venus

The second planet from the sun, Venus, is remarkably similar to the earth in many important respects. It is, in fact, often regarded as the earth's twin. Venus is virtually the same size as the earth, has the same density, mass, and gravitational field, and possesses a dense atmosphere.

The similarities between Venus and the earth are so numerous and striking that astronomers wondered for years if an earthlike surface might be hidden beneath the thick clouds that continuously shroud Venus. They speculated about the possibility of oceans and rivers existing on Venus, perhaps steaming jungles, perhaps even intelligent beings with whom we could communicate. The American and Soviet Venus probes of the early 1970s abruptly ended such speculation.

The unmanned Soviet *Venera* spacecraft that penetrated the atmosphere of Venus and landed on its surface radioed back the information that the planet is a dreadful place, totally unsuited for life. The clouds consist largely of sulfuric acid mist and surface temperatures are hot enough to melt lead. Evidently, the exceedingly dense atmosphere of Venus, which consists mainly of carbon dioxide, traps heat at the surface of the planet. The earth might have developed along this line also had green plants not appeared early in geologic time and effectively removed most of the carbon dioxide from our atmosphere.

We know very little about the surface of Venus because its dense cloud cover shields it from our telescopes. One of the Soviet *Venera* spacecraft that landed on the surface survived the heat long enough to transmit photographs of a landscape densely littered with large rocks, which appear curiously rounded, as though they might somehow be weathering.

For broader views of the landscape of Venus we must rely on the extremely coarse imagery obtained by reflecting radar signals from the earth and more recently from the Venus orbiter, *Pioneer*. The relatively poor images now available show Venus looking quite unlike Mercury and the moon. Nor does its surface appear to resemble the earth's, except in having large high and low regions that may be comparable to continents and ocean basins. Most geologists expect that more detailed views of the surface of Venus will yield evidence of internal processes at least partially comparable to those operating within the earth.

Mars

The fourth planet from the sun, Mars, is larger than Mercury and the moon but smaller than the earth and Venus. Its gravitational field is strong enough to retain an extremely thin atmosphere, which consists mostly of carbon dioxide and some water. Clouds of ice crystals have been observed on Mars, ice frost has been photographed on its surface, and we presume that ice may exist in the Martian soil. However, the atmospheric pressure on Mars is too low to permit water in liquid form to exist on the surface of the planet; any ice that melts must become water vapor.

Although the Martian atmosphere is extremely thin, howling winds blow across the bleak surface and raise clouds of dust that sometimes obscure the entire planet. The first Martian orbiter, one of the *Mariner* spacecraft, arrived during such a planetary dust storm. For a while the scientists directing the mission feared that they would not be able to obtain photographs. However, the air did finally clear, and the detailed pictures of the Martian surface reveal vast fields of dunes. They also show a striking absence of small impact craters, which evidently have been eroded away by the wind.

Ground photographs from landed spacecraft show an undulating rocky surface liberally furnished with small dunes and drifts of sand or dust. These formations have exactly the same cleanly sculptured surfaces that such winddriven landforms develop on the earth. Were it not for the total absence of plants, photographs of the Martian surface might seem to have been taken in many of the more extreme desert regions of the earth.

One of the *Viking* landers was equipped with various mechanized experiments designed to scoop up soil samples and test them for signs of life. The results were peculiar and difficult to explain. At first many scientists thought the findings might constitute evidence of life. However, it now seems far more probable that the results reflected the existence of some unusual soil minerals. In any case, the experiments did *not* demonstrate the existence of life on Mars. Neither, though, did they conclusively demonstrate that there are no living organisms on the planet.

Some orbital photographs of Mars depict landforms that closely resemble small stream drainage systems. If that is indeed what these formations are, then we must conclude that, at some time in its past, Mars had surface water—rain fell from the clouds in its atmosphere. Such rainfall could not happen unless the surface temperatures were much warmer than those now prevailing on Mars, and unless the atmospheric pressure were much greater. At the moment, no satisfactory explanation has been advanced for those features that look as if they might be streams. We currently have no solid theory to account for the conditions that could have permitted their formation.

Except for evidence of wind erosion, such as the scarcity of small craters, about half the Martian surface looks quite a bit like the landscapes of Mercury and the moon. That half of Mars demonstrates little internal activity. Its other hemisphere, which has relatively few craters, does show some evidence of internal activity. For example, a long valley exists there that somewhat resembles the oceanic rift valleys on the earth. Also, this hemisphere contains quite a few volcanoes of the type that appear in Hawaii. One, called Olympus Mons, is the largest volcano discovered so far in the solar system, and is considerably larger than any on earth. Apparently, Mars is just barely a large enough planet to accumulate sufficient internal heat to drive some of the processes that stir much more vigorously within the earth.

Io

The gaseous giant planet Jupiter has four large satellites that were discovered with a primitive telescope by Galileo in 1610. The *Pioneer 10* and *11* spacecraft finally made detailed imagery of Jupiter available in the late 1970s. Then the two *Voyager* spacecraft returned detailed imagery of Jupiter's four largest satellites, which surprised nearly everyone by revealing that they are utterly unlike Mercury and the moon. Neither do they resemble each other. The most astounding images came from the innermost of the four, Io, which is approximately the size of the moon.

The *Voyager* spacecraft photographed seven actively erupting volcanoes on Io and revealed that its surface contains no craters. Io therefore appears to be fairly young. Peculiar yellow and orange colors dominate the surface, making it look like a pizza and utterly unlike any volcanic region on earth. There is evidence that Io's volcanoes erupt sulfur.

Clearly, Io must have abundant internal heat despite its being no larger than the moon, which has very little. It is difficult to imagine that Io could be generating all that heat solely through decay of radioactive elements. A more likely explanation would be that Io gains heat from tidal motions induced by its close proximity to Jupiter. Continual tidal flexing of the rocks could heat them in the same way that an underinflated tire becomes overheated through excessive flexing.

Europa

The second large satellite out from Jupiter, Europa, has a relatively smooth surface nearly devoid of craters. However, it is completely covered with a dense pattern of cracks that make pictures of the planet resemble photographs of a cracked hard-boiled egg. Europa is slightly smaller than our moon and much less dense so it seems likely that it must consist in part of ice. In fact, it appears that the entire planet is covered with a thick shell of ice.

No one knows how the thick skin of ice on Europa acquired its network pattern of cracks. However, Europa, like Io, is close enough to Jupiter to experience large tidal movements. Perhaps the tidal flexing adds enough heat to the ice to keep it mobile enough to erase craters on the surface. In fact, some scientists now suspect that liquid water may exist beneath that thick shell of ice, and that the cracked surface may reflect movements akin to those in the Arctic pack ice on earth. Europa is a baffling puzzle.

Ganymede and Callisto

The next two large satellites out from Jupiter, Ganymede and Callisto, are both considerably larger than the moon but smaller than Mars. Like Europa, Ganymede and Callisto have relatively low densities, a fact suggesting that they contain large amounts of water, again in the form of ice. Both appear to be covered by a thick shell of ice, a solidly frozen ocean covering each entirely.

Unlike Europa, however, both Ganymede and Callisto are heavily pocked by impact craters that appear in the *Voyager* photographs as startling white spots. It is difficult at first to imagine that those craters could possibly date from the earlier years of the solar system. Ice just seems too unstable to retain its form for such a long time. How-

ever, the surface temperatures of Ganymede and Callisto are close to the absolute zero of space and neither satellite is near enough to Jupiter to gain much heat from tidal flexing. At such extremely low temperatures, ice is as stable and enduring as almost any rock.

Titan

The planet Saturn has a swarm of small satellites and one large one—Titan—that can reasonably be included in the "terrestrial" category. Titan is especially interesting to scientists because its atmosphere is evident through telescopic observation. A reasonably close look at Titan was obtained when *Voyager I* passed Saturn. The views were disappointing.

As had been feared, Titan's atmosphere completely obscures its surface—its landscapes cannot be seen. However, the spacecraft did obtain data showing that Titan's atmosphere is mostly nitrogen mixed with a few percent of methane and much smaller quantities of other gases. Like Ganymede and Callisto, Titan has an extremely low density, slightly less than twice that of water, and must therefore consist largely of ice. Because Titan is far from the sun, its temperatures are extremely low. Thus, it seems likely that methane and perhaps nitrogen are condensing as liquids on its surface.

THE EARTH

Of all the known planets, only the earth has the combination of an atmosphere, liquid surface water, and an actively moving interior.

Solar energy moves the atmosphere and liquid surface water to drive wind, rain, waves, streams, and other surface geologic agents. Such agents then convert rocks into soil, transport the debris, and deposit it as sediment, shaping our complex landscapes. In their constantly changing forms, then, earth landscapes reflect the workings of the geologic processes involved. However, if it were not for the earth's *internal* processes, the surface geologic processes would long ago have reduced the continents to featureless plains, filled the ocean basins with sediment, and displaced sea water over the entire surface of the planet.

Heat derived from decay of radioactive elements within the earth drives its internal processes, which perpetually renew the earth's surface. These processes replace old ocean floor with new, and return sediment eroded from the continents back to the continents in the form of new ranges of mountains. If the earth were smaller, it could not retain enough air and water to maintain its surface processes. Nor could it generate enough interior heat to drive its inner movements. But being the size it is, the earth possesses interesting geologic processes. Let us now consider the composition of the earth.

The Earth's Interior

One of the commoner kinds of meteorites consists of a dark rock called **peridotite** (see Chapter 3). Many of these stony meteorites contain blobs of a metallic alloy of iron and nickel strewn through them. Most geologists envision the earth forming as an aggregation of just such a mix of stony and metallic matter heating up as the young planet grew larger.

The growing earth eventually became hot enough to make the material within it mobile. At that juncture, the heavy metallic alloy sank to the center to form a **core** with a radius of 3,488 kilometers (km). The earth's core is almost exactly the same size as its moon. Meanwhile, the stony material rose to form an outer shell composed of peridotite. We call this outer shell the **mantle.** The mantle is about 2,900 km thick and is by far the largest part of the earth. Figure 1–4 shows a schematic cross-sectional view of the planet.

The outer part of the mantle (see Figure 1–5) is directly responsible for most of the internal processes affecting the

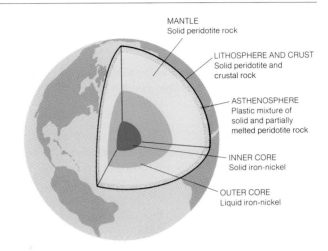

FIGURE 1–4

A cross-section showing the earth's inaccessible interior as it is now interpreted on the basis of various evidence.

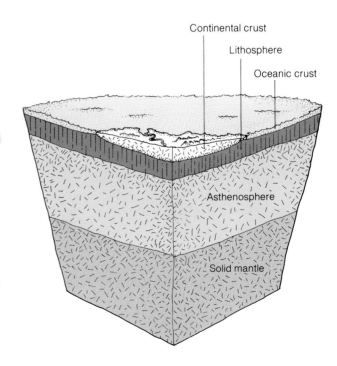

FIGURE 1–5 (right)

The mantle has an outer rind about 100 km thick called the **lithosphere**, which is relatively cold and therefore rigid. The lithosphere grades downward into another zone called the **asthenosphere** in which the rock is partially melted and therefore very weak and mobile. At a depth of about 250 km the asthenosphere grades downward into solid rock. The earth's crust is a thin skin on the lithosphere. In oceanic regions, the crust consists of lava flows; on land, it consists of a floating slab of lighter rocks beneath the continents.

earth's surface. The outer rind of the mantle is called the **lithosphere.** This section is about 100 km thick in most regions. It consists of relatively cool and therefore relatively rigid rock. Immediately below the lithosphere is the **asthenosphere,** a zone of extremely weak and probably partially melted rock. This zone extends to a depth of about 250 km.

The lithosphere consists of a mosaic of seven large and perhaps a dozen small segments called **plates.** These plates fit together like the bones in a skull to completely tile the earth's surface. All of the lithospheric plates move at rates of as much as several centimeters per year by gliding on the weak rocks of the asthenosphere beneath. The discovery of lithospheric plates is relatively new in geology. The theory dealing with their existence and motion is called **plate tectonics.** This modern theory has completely changed our view of the earth's internal processes, and to some extent our understanding of the surface processes.

FIGURE 1–6

New lithosphere, and new oceanic crust composed of volcanic lava, flows from where two plates pull away from each other. Where two plates collide, one sinks into the mantle and disappears as it gets hot. It thus loses its identity as cold and rigid lithosphere. Much of the volcanic rock that formed its oceanic crust melts. This melted rock returns to the surface through a chain of volcanoes along the edge of the plate that remains at the surface.

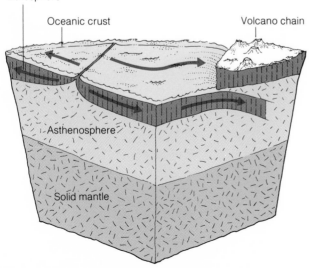

The outermost skin of the lithosphere is called the earth's **crust.** The crust constitutes only a minuscule part of the entire planet, proportionately less than the skin on a large apple. Nevertheless, it is quite important to us because it is the only part of our planet we can actually see.

The **oceanic crust** (Figure 1–6) covers about 60 percent of the earth's surface. It consists mostly of lava flows and averages about 10 km thick. The earth constantly generates new oceanic crust and destroys the old. Thus, no part of the ocean floor has any great antiquity in the geologic sense of the word. Most of the oceanic crust is flooded by seawater to an average depth of about 4 km. That makes the oceans comparable to the size of the earth as a film of water is to a wet basketball. The **continental crust** consists of rafts of lighter rocks averaging about 40 km thick embedded in the upper surfaces of some lithospheric plates. Unlike the oceanic crust, the continental crust is a permanent part of the earth's surface. Some portions of the continents date back to the early stages of the planet's evolution.

The weak rocks of the asthenosphere permit the lithosphere and crust to sink wherever a heavy load is placed on them, and to rise as they are unloaded. Precise measurements of the earth's gravity field show that most parts of its surface are in almost perfect floating equilibrium, a condition geologists refer to as **isostasy.** Many observations have shown that the lithosphere sinks beneath such loads as large glaciers, volcanoes, thick deposits of sediment, and even newly filled reservoirs behind large dams. It rises in response to unloading as glaciers melt, or as erosion strips soil from a landscape.

An Active Planet

Modest vertical and large horizontal movements of the lithosphere combine to keep the earth's surface dynamically active. Old ocean basins vanish as new ones open and continents move from place to place. Sediment dumped into the oceans returns to the continents in the form of new mountain ranges. The intricate interplay of internal and surface geologic processes constantly changes the earth's surface, leaving no part static. Each moment in our planet's history is unique, unlike any that have gone before or will come after. All of this activity is the province of geology.

SUMMARY

The earth is unique among known planets in having liquid surface water, an atmosphere, and active internal processes. Other planets and satellites resemble the earth in having one or more of these attributes. None, however, have the same combination of size and distance from the sun that make the earth the active and liveable planet we know.

The earth contains a metallic core about the size of the moon surrounded by a thick rocky shell called the mantle. The relatively cold and rigid outer rind of the mantle—the lithosphere—consists of a mosaic of segments called plates. These plates move independently on the weaker and partially melted rocks of the asthenosphere beneath. Plate movements constantly renew the earth's surface.

They create new ocean basins while destroying old ones and return eroded sediments to the continents.

The only part of the earth we can observe directly is the thin outer crust of the lithosphere. The oceanic crust is extremely thin and is constantly renewed. The continental crust, on the other hand, is much thicker and is a permanent part of the earth's surface. The crust and lithosphere float isostatically on the deeper rocks of the asthenosphere, sinking where they are loaded and rising where they are unloaded.

Geology is the study of the processes that create and change the earth's rocks and landscapes: (1) the surface processes that depend on air, water, and solar energy for their operation; and (2) the internal processes that depend upon the earth's interior heat, which is derived from radioactive decay.

KEY TERMS

asthenosphere	isostasy	peridotite
continental crust	lithosphere	plates
core	mantle	plate tectonics
crust	oceanic crust	terrestrial

REVIEW QUESTIONS

1. Why do meteorites explode upon impact even though they are composed of rock and metal, which are not ordinarily explosive substances? *high velocity*

2. Mercury is larger and denser than the moon and therefore has the stronger gravitational field. What effect would you expect that difference to have on the diameters and on the shapes of the debris blankets surrounding meteorite impact craters on those two bodies?

3. Why does the earth have an atmosphere whereas the moon does not? *earth· internal action by heat moon· too small for heat*

4. Mars is just barely large enough to sustain a low level of surface and internal activity. Explain why both might have been more intense if the planet had been larger.

5. If Mars had been farther from the sun, how might that have affected the intensity of its surface processes? *slow down*

6. If Europa were somehow moved from its present orbit around Jupiter to one around the earth, how might that affect its mantle of ice? *may melt, if not the mantle will move more + erode the surface*

7. Do you think it likely that future, more detailed investigation will reveal stream valleys on Venus? If any were to appear, how would you suggest that they be interpreted?

SUGGESTED READINGS

Calder, N. *The Restless Earth.* New York: Viking Press, 1974.

A general review of the earth written for a nontechnical audience.

French, Bevan M. *The Moon Book.* Harmondsworth, Middlesex, England: Penguin Books Ltd., 1977.

A thorough review of our knowledge of the moon, ranging from the historical background to the results of recent research.

Lowman, Paul D. *Lunar Panorama.* Zurich, Switzerland: Weltflugbild, 1969.

A collection of orbiter photographs of the moon accompanied by brief and extremely informative narrative.

Mutch, T. A., R. E. Arridson, J. W. Head, K. L. Jones, and R. S. Saunders. *The Geology of Mars.* Princeton, N.J.: Princeton University Press, 1976.

A pleasantly written, beautifully illustrated, and comprehensive review of our knowledge of Mars.

York, D. *Planet Earth.* New York: McGraw-Hill Book Co., 1975.

A good general review of the earth.

CHAPTER TWO

A group of quartz crystals. (W. T. Schaller,
U.S. Geological Survey)

MINERALS

Minerals are the basic materials of the earth. They may be broadly defined as naturally occurring chemical elements or compounds classified by their compositions and **crystal structures,** the orderly internal arrangement of atoms. Although generally sound, this definition blurs around the edges. Petroleum, for instance, is not a mineral even though it is a mineral resource; a few minerals such as opal lack a crystal structure; and at least one mineral, native mercury, is a liquid. However defined, minerals are important: they are sources of basic raw materials, they provide scientific information about the earth, and some are objects of great beauty.

ELEMENTS, BONDS, AND CRYSTALS

Atoms and Ions

Just as all other substances, minerals are composed of atoms. All atoms consist of an extremely minute **nucleus** surrounded by a vastly larger "cloud" of electrons. If an atomic nucleus were expanded to the size of the sun, its cloud of electrons would be larger than the solar system. Atomic nuclei consist of a tight cluster of **protons,** which carry a positive electrical charge, and **neutrons,** which weigh as much as a proton, but carry no charge. Each electron carries a negative charge and their number in the electron cloud must equal the number of protons in the nucleus in order that the positive and negative electrical charges balance.

Appendix C shows the naturally occurring chemical elements arranged in a table according to their atomic numbers. The **atomic weights** of the elements generally exceed their atomic numbers because the nuclei of all atoms, except those of hydrogen, contain neutrons as well as protons. Many elements have different kinds of atoms called **isotopes,** which contain different numbers of neutrons in their nuclei. The isotopes of an element vary in atomic weight but not in atomic number.

The various isotopes of the elements have little effect on minerals because they do not differ in chemical behavior. Nevertheless, some isotopes are geologically important because their nuclei turn into nuclei of other elements through **radioactive decay.** It is radioactive decay that provides the earth with the continuing supply of internal heat necessary to drive its internal processes. Radioactive decay is also the basis for the "natural clocks" that enable geologists to measure the ages of many minerals and rocks.

The atoms of most elements tend to gain or lose electrons. When they do either, they become electrically charged **ions.** If they lose electrons to acquire a positive charge, atoms are called **cations;** if they become negatively charged by gaining electrons, they are called **anions.**

Chemical Bonds

The number of electrons surrounding the nucleus of an atom must balance the number of protons within that nucleus to maintain electrical neutrality. Therefore, oppositely charged ions must bond together to form chemical compounds in which the positive and negative charges balance. They do that in several ways.

Chemical bonds differ greatly in the extent to which the joined ions share electrons. When **ionic bonds** form, each ion retains its own discrete cloud of electrons. Ionic bonding occurs to some extent in many minerals but is dominant in relatively few. Ionically bonded substances tend to be water soluble and therefore occur as minerals only in rather special geologic situations. Ions linked together through **covalent bonds** completely share one or more of their electrons. Covalent bonding is dominant in most of the abundant minerals even though many also contain some ionic bonds. In **metallically bonded** substances, some electrons circulate freely without being attached to any particular atom. As the name suggests, all metallically bonded substances, including many minerals, do indeed have a metallic appearance.

Crystals

All crystalline solids consist of bonded atoms arranged in three-dimensional patterns that can be continued indefinitely like those of tiled floors. Crystals must consist of positive and negative ions bonded to each other in ways that not only balance their electrical charges but also fit geometrically. These requirements present a problem because the ions of different elements vary greatly in size. There is a limit to the number of two-dimensional shapes that can fit together to continuously tile a floor. It is impossible, for example, to completely cover a floor with five-sided tiles. In like manner, there is a limit to the number of

three-dimensional shapes that can be continuously stacked to form a solid crystal. Bricks, say, could not fit together in a wall if they were made with seven sides instead of six.

We can, in principle, slice a crystal down into **unit cells,** the smallest assemblages of atoms and parts of atoms that embody a crystal's composition and form. Suppose such slicing were done. We could then imagine reassembling the crystal by stacking its unit cells, just as a haystack can be constructed by piling individual bales so they fit neatly together. The form of the resulting crystal would reflect the shape of its unit cells in the same way that the rectangular shape of our completed haystack reflects the rectangular form of each single bale in it. If bales of hay had hexagonal instead of rectangular outlines, haystacks would of course be hexagonal in form.

Many years ago, crystallographers found that different crystals of the same mineral resemble each other in certain basic ways although they may vary greatly in superficial appearance. Among other things, they found that the angles between equivalent crystal faces remain constant even though the forms of the crystals differ. They also discovered that different crystals of the same mineral shared basic elements of symmetry even though their forms differed. Figure 2–1 illustrates those observations.

From such simple observations, the early crystallographers proceeded to an elegant theoretical analysis of the number of basic forms crystals might assume without violating those basic constraints. Many years later, when atoms were discovered, the reasoning of crystallographers was extended to show that there is a similarly limited number of regular arrays, called **lattices,** in which atoms can be arranged to form a crystal. Figure 2–2 shows an example of one simple crystal lattice that occurs in a large number of minerals.

Polymorphs and Substitution

Many pairs of minerals share the same composition but differ in their crystal structures. Others consist of different kinds of atoms arranged in similar crystal lattices. For these reasons it is necessary to define the mineral species in terms of both chemical composition and crystal structure.

Everyone, for instance, should be familiar with diamond,

FIGURE 2-1
Different crystals of the same mineral may vary in shape as their faces vary in relative size but the angles between corresponding faces remain constant.

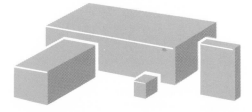

the hardest substance known, and also with graphite, the soft mineral used to make pencil leads. Both are composed of carbon so they have the same chemical composition. Yet their crystal structures and other properties are quite different. Diamond and graphite are an example of what we call a **polymorphic** pair, a single chemical substance that occurs in different crystal forms.

The converse phenomenon—in which pairs of minerals have different chemical compositions but similar crystal structures—is called **isomorphism.** Examples of isomorphism are easily found because the total number of possible crystal lattices is so limited. To illustrate: Halite, which is the mineral form of common salt, NaCl, and the lead sulfide mineral galena, PbS, have the same crystal lattices. They differ, however, in composition, type of chemical bonding, and every other respect except the cubic forms of their crystals.

Isomorphism commonly occurs to a more limited extent as a few atoms of one element substitute for a few of another in a crystal lattice. For example, many crystals of galena contain small amounts of silver, which substitutes

for lead in the crystal lattice. If atoms of different elements are about the same size and form similar chemical bonds, they can substitute for each other in crystals just as different colored bricks of the same size can be incorporated into a wall. Therefore, most mineral species vary in composition. Many, in fact, blend imperceptibly into each other. Such blendings merely represent the final results of the continual process of variation in crystal composition.

Isomorphous substitution commonly involves atoms of several elements that substitute in different positions in the crystal lattice. This phenomenon makes the mineral in question vary widely. Complex substitutions make cumbersome chemical formulas, which helps to explain why geologists retain their long tradition of referring to minerals by name. It also helps explain why different specimens of many minerals vary greatly in appearance. Identifying these accurately requires considerable experience. The author painfully recalls an examination in undergraduate mineralogy in which all of the 20 specimens to be identified were examples of the same mineral, and no two looked quite alike.

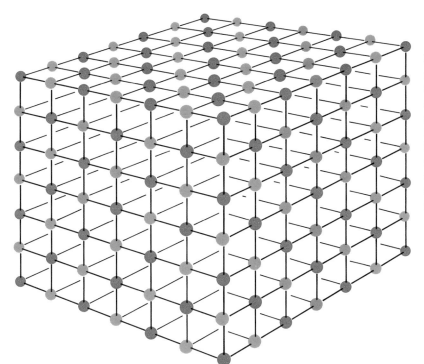

FIGURE 2-2

An exploded view of a sodium chloride or halite crystal lattice showing the sodium and chloride ions arranged at alternate corners of cubes, which repeat indefinitely to make a crystal. If this drawing were to scale, the smaller sodium and larger chloride ions would appear to touch, as though they were softballs and basketballs packed neatly into a box.

The Limiting Factors

Approximately 2,200 species of minerals have been recognized and described. That number may seem large, but it is actually surprisingly small when we consider that about 90 chemical elements are known to exist in nature. Furthermore, the majority of the known mineral species are rare. Many occur in just a few places and some in just one place. The number of abundant **rock-forming** minerals that comprise a substantial part of the earth's mantle and crust is actually quite small, hardly two dozen. The problem is not to explain why so many mineral species exist but rather to understand why there are so few.

Of course, many of the possible combinations of the naturally occurring elements are gases or liquids that cannot form minerals. Many others dissolve so readily in water that they never occur in nature. But the main reason for the small number of abundant minerals is simple: Only about 10 of the naturally occurring chemical elements, some of which are shown in Figure 2–3, exist in large enough quantities to comprise as much as 1 percent of the earth's mantle and crust. Therefore, all of the common rock-forming minerals must consist mostly of these few elements. Many of the less common elements appear only as trace impurities substituting for atoms of the more common elements. Many others form separate minerals of their own only where they are locally concentrated.

Oxygen is by far the most abundant element in the earth's mantle and crust. Among atoms of the most abundant elements, those of oxygen are alone in their tendency to gain extra electrons to form anions. Therefore, atoms of the other more abundant elements generally bond to oxygen atoms. Most of the rock-forming minerals are based on such combinations. Furthermore, atoms of oxygen are much larger than those of most of the other abundant elements. As a result, the crystal structures of the common

FIGURE 2–3

(a) The approximate elemental composition of the continental crust calculated in terms of weight percents. (b) The approximate elemental composition of the continental crust calculated in terms of volume percents. Oxygen, the most abundant element, has large atoms that account for most of the volume of the common rocks.

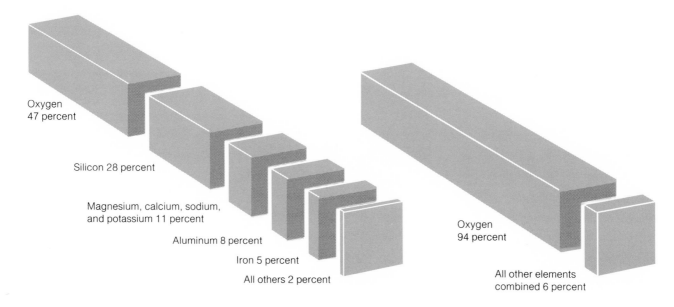

Oxygen
47 percent

Silicon 28 percent

Magnesium, calcium, sodium,
and potassium 11 percent

Aluminum 8 percent

Iron 5 percent

All others 2 percent

Oxygen
94 percent

All other elements
combined 6 percent

(a)   (b)

rock-forming minerals essentially consist of closely packed oxygen atoms, with the smaller atoms of the other elements tucked into the spaces between them. These latter bind the structure together. The bulk of the volume of common rocks, of hills, and of mountains consists of oxygen. If all the other elements could somehow be sucked out of these formations, they would hardly shrink at all.

THE ROCK-FORMING MINERALS

After oxygen, silicon is the next most abundant element. Most of the abundant rock-forming minerals—the **silicates**—are based on combinations of silicon and oxygen, along with various other elements. Together, the silicates comprise about 99 percent of the earth's mantle and crust.

A much smaller but nevertheless important group of minerals is the **oxides,** which consist of simple combinations of various elements with oxygen and without silicon.

The **carbonates,** which include two important rock-forming minerals, have crystal structures controlled by carbon. Though not one of the more abundant elements itself, carbon plays a highly significant role in many geologic processes that also involve plants and animals.

Like carbon, sulfur is also not one of the abundant elements. It is, nonetheless, geologically significant. Sulfur atoms, like those of oxygen, can gain extra electrons and bond directly to cations. This capability allows the formation of a large class of minerals, the **sulfides,** many of which are valuable as ores. Sulfur atoms also have the capacity to lose electrons and bond to oxygen to form another group of minerals, the **sulfates,** some of which form large bodies of uncommonly interesting rock.

The Silicates

Silicon atoms are extremely small, even by atomic standards. They lose four electrons whereas the much larger oxygen atoms gain two electrons. Nearly every silicon ion in the earth's mantle and crust is surrounded by four covalently bonded oxygen ions, as though it were a golf ball nestled in the space between four clustered basketballs. This assemblage is called a **silicate group.** Each of the four oxygens shares one of its extra electrons with the silicon at the center of the cluster. The other remaining elec-

tron can be used to form bonds with other cations. It is these bonds that join silicate groups together to form crystals.

Figure 2–4 shows that the four oxygens surrounding the silicon are arranged as though they were at the corners of a tetrahedral pyramid. That arrangement explains why the assemblage is often called a **silicate tetrahedron.** All of the silicate minerals consist basically of silicate tetrahedra arranged in a variety of crystal structures and linked to each other by a variety of cations.

The earth contains only about three times as many oxygen atoms as silicon atoms. Thus, it is necessary for most silicon atoms to share oxygen atoms between them so that each can surround itself with its full complement of four. When that happens, one oxygen bonds to two different silicon atoms, sharing one of its extra electrons with each and thereby linking two silicate tetrahedra together. Each of the several ways in which silicate tetrahedra can be combined to share oxygen atoms leads to a distinctive type of crystal structure. Each type of structure defines a major group of silicate minerals.

The simplest silicates are those in which none of the oxygens is shared between different silicons. In this case

FIGURE 2–4

The silicate tetrahedron shown in an exploded view, left, and drawn more nearly to scale, right. The exploded view displays the small silicon atom that nestles within the cluster of four oxygen atoms and holds the assembly together. The scale drawing depicts the oxygen atoms packed tightly together, hiding the silicon atom within the little space between them.

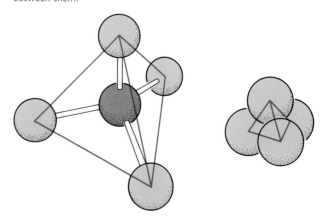

each silicate tetrahedron exists as an isolated unit within the crystal lattice. The structure is held together by doubly positive cations of such elements as iron, magnesium, and calcium, which can form bonds with two different oxygens belonging to adjacent tetrahedra, connecting them.

In the **olivine** minerals, cations of iron and magnesium substitute for each other in all proportions. Thus, the formula is written as $(Mg,Fe)_2SiO_4$. The comma inserted between the symbols indicates that the elements substitute for each other. The olivines generally form pale green crystals, which comprise an important part of the earth's mantle. They are also abundant in the oceanic crust although we rarely see them in continental rocks. The **garnets** are another important group of minerals based on isolated silicate tetrahedra. In their crystal lattices, cations of iron, magnesium, and calcium substitute for each other in some positions while those of aluminum and iron substitute in others. The result is a mineral group with a complex range of compositions that can be generally described by the formula $(Ca,Mg,Fe)_3(Al,Fe)_2(SiO_4)_3$. Garnets occur in a wide variety of rocks. Their bright colors, most commonly red, and elegantly formed 12-sided crystals make them attractively conspicuous.

The **single-chain silicates** form as silicate groups join at their corners. The joining occurs by sharing an oxygen atom to make long daisy chains, such as those shown diagrammatically in Figure 2–5. Crystals of the **pyroxene** minerals are the most abundant single-chain silicates, consisting essentially of such chains stacked side by side as though they were spaghetti noodles. They are linked to each other by doubly positive ions, most commonly those

of iron, magnesium, and calcium, which bond to their unshared oxygens. Pyroxenes typically grow into stubby prismatic crystals (Figure 2–6) generally having a dull surface luster. Such crystals may be black or some shade of green, depending upon their composition. The most common and abundant pyroxene is **augite,** a black mineral with a complex and somewhat variable composition. Augite can be generally described by the rather cumbersome formula $Ca(Mg,Fe,Al)(Si,Al)_2O_6$. Remember that the elements whose symbols are separated by commas substitute freely for each other.

Silicate double chains consist essentially of two single chains linked side by side through additional shared oxygens to make broad ribbons, such as that shown schematically in Figure 2–7. As in the single-chain silicates, the double chains stack like noodles. They are linked together by doubly positive ions that bond to the unshared oxygens. The most abundant double-chain silicates are the **amphiboles,** which tend to crystallize into long needles (Figure 2–8) that typically have a glossy surface luster. Amphibole species occur in a wide variety of compositions and colors. The most common and abundant is **hornblende,** a black mineral. Hornblende has one of those formulas that vividly illustrate why geologists refer to minerals by name: $Ca_2Na(Mg,Fe)_4(Al,Fe,Ti)_3Si_6O_{22}(O,OH)_2$.

The **layer silicates** consist basically of silicate groups that share three of their oxygen atoms to make a flat sheet. The remaining unshared oxygen atoms rise above the plane of the sheet as the projecting peaks of the tetrahedral pyramids. Figure 2–9 shows the general scheme. Positive ions link the projecting unshared oxygens together.

FIGURE 2–5
Silicate tetrahedra joined by shared oxygens into single chains.

FIGURE 2–6
The pyroxenes tend to crystallize into stubby prisms that cleave in two directions, intersecting at right angles.

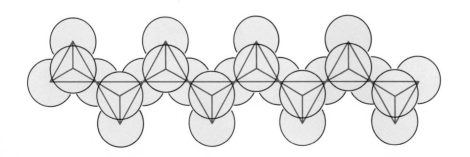

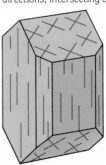

FIGURE 2-8

The amphiboles crystallize into long prisms that cleave in two directions, intersecting at oblique angles.

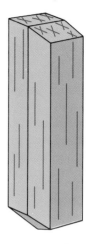

FIGURE 2-7

A silicate double chain. The oxygen at the apex of each tetrahedron is unshared, as is one additional oxygen in alternate tetrahedra. Doubly positive ions of iron, calcium, or magnesium form bonds with the extra electron remaining on the unshared oxygens to link adjacent double chains together into an amphibole structure.

FIGURE 2-9

A silicate sheet. One oxygen at the apex of each tetrahedron remains unshared and capable of bonding with positive ions. Various cations link the sheets together face to face, making layers that stack like sheets of paper to form crystals. All the minerals so constructed cleave easily in one direction because the layers are weakly bonded together.

They join the silicate sheets face to face, forming double layers that stack on each other like sheets of paper to form crystals. Because the bonding between the stacked layers is loose, the crystals tend to split easily into thin flakes.

The most abundant layer silicates are the **micas.** The most common micas are biotite, a black mineral with the formula $K(Mg,Fe)_3(AlSi_3O_{10})(OH)_2$; and muscovite, a colorless mineral with the formula $KAl_2(AlSi_3O_{10})(OH)_2$.

Other abundant groups of layer silicates include the **chlorites** and the **serpentines.** These groups resemble each other in that, in both, magnesium plays the major role in binding the silicate sheets together. They differ mainly in that the chlorites also contain iron whereas the serpentines do not. Both chlorites and serpentines tend to be greenish. They are the minerals mostly responsible for green colors in rocks.

The **clay** minerals are another important group of layer silicates that occur in a wide range of compositions. **Kaolinite** generally forms in the soils of humid regions and other environments in which water circulates freely. It is widely used in the manufacture of porcelain and other ceramic products. The **smectites** form in arid soils and other situations that limit the circulation of water. The smectites have the interesting property of absorbing water between their silicate layers. This characteristic makes smectite crystals expand as they get wet, just as if they were miniature accordions.

Oxygen sharing among silicon atoms reaches its logical end in the **framework silicates.** Here, all four oxygens in each silicate tetrahedron are shared with an adjacent tetrahedron. The crystal structure becomes a complexly interconnected web of silicate tetrahedra that no longer resemble pyramids except, perhaps, to the willing eye of faith. Nevertheless, the tetrahedra are there and each silicon atom nestles within its cluster of four oxygens.

The simplest framework silicate is **quartz,** SiO_2. In quartz, each silicon atom is bonded to four oxygen atoms and each oxygen is shared between two silicons. Quartz forms under a wide range of conditions. It is the most widely distributed and most commonly seen of all minerals although not the most abundant in a quantitative sense. There are dozens of varieties of quartz, including material as diverse as ordinary sand grains and the semiprecious gemstones agate and amethyst. Entire volumes have been written on the mineralogy of quartz.

The most abundant minerals in the continental crust are the **feldspars.** In this group of framework silicates, aluminum atoms substitute for some of the silicon atoms in the tetrahedral groups. Aluminum atoms are almost as small as those of silicon. They fit nicely into the space between closely packed oxygen atoms. However, the aluminum atom loses only three electrons instead of the four that silicon loses. Thus, the crystal must incorporate additional positive ions to maintain the balance between positive and negative electrical charges.

The two most abundant feldspars are **orthoclase,** $KAlSi_3O_8$, and **plagioclase,** which consists of the two compositions $NaAlSi_3O_8$ and $CaAl_2Si_2O_8$ mixed in all proportions. Both tend to form blocky crystals (see Figure 2–10). Also, both may be white although orthoclase commonly occurs in pale shades of salmon pink and plagioclase generally shows at least a faint touch of green.

The Carbonates

Carbon atoms are very small and, like silicon atoms, they lose four electrons. However, they share electrons with only three oxygen atoms to form a carbonate group, $CO_3^=$. In this group, the oxygens are arranged at the corners of a triangle with the carbon at its center. After the three oxygen atoms have shared four of their six extra electrons with the carbon atom, the two remaining electrons can form bonds with other positive ions. They appear in the formula above as a pair of minus signs. There are two abundant rock-forming carbonate minerals: **calcite,** $CaCO_3$, in which calcium ions link the carbonate groups together; and **dolomite,** $(Ca,Mg)CO_3$, in which the linking function is performed by calcium and magnesium ions in equal proportion.

The Sulfates

Sulfur atoms can lose as many as six electrons. They therefore shrink to a size even smaller than that of silicon atoms. When their electrons are lost, sulfur atoms surround themselves with four oxygens arranged at the corners of a tetrahedral pyramid to make a sulfate group, $SO_4^=$. The four oxygens, with six of their extra electrons shared with the sulfur atom at the center of the group, still retain two extra electrons that can bond to other positive

FIGURE 2-10

Blocky crystals of orthoclase feldspar collected from an igneous rock exposed in the Tobacco Root Mountains of Montana. These crystals are about the size of acorns.

ions. Doubly positive calcium ions commonly link the sulfate groups together, forming either **anhydrite,** $CaSO_4$, or **gypsum,** $CaSO_4(2H_2O)$. Both of these sulfates most commonly form in places where bodies of concentrated brine evaporate. Figure 2–11 shows some large crystals of gypsum. This mineral is familiar to most people as the main constituent of wallboard. It is also the substance that forms as plaster of paris is mixed with water and sets.

The Sulfides

Sulfur ranks sixteenth in the order of abundance of the elements in the earth's crust. After oxygen, sulfur is the most abundant element with the ability to form negatively charged ions. Sulfur atoms can gain one electron to form sulfide anions. Those anions, can, in turn, bond to a wide variety of cations to form the sulfide minerals, most of which have a distinctly metallic appearance. The iron sulfide **pyrite,** FeS_2, is the only member of this group common enough to be regarded as a rock-forming mineral. Figure 2–12 shows a large crystal of pyrite. Much smaller crystals

commonly occur widely disseminated through many kinds of rocks. Despite its valuable appearance, which has earned it the popular name "fool's gold," pyrite is virtually worthless. There are much better sources of both iron and sulfur. But a majority of the other sulfide minerals are valuable. They provide most of the world's supply of copper, nickel, zinc, lead, molybdenum, and silver, among other mineral commodities.

The Oxides

Oxide minerals form when positive ions bond to individual oxygen ions instead of to groups of them clustered about some other cation. Numerous oxide minerals, especially those of iron, copper, manganese, tin, and lead, make valuable ore deposits, though only the iron oxides are common rock-forming minerals.

Hematite, Fe_2O_3, is the common red iron oxide that occurs in a wide variety of rocks and soils. **Goethite** is similar to hematite chemically, except that it contains variable amounts of water, whereas hematite does not. Goethite

FIGURE 2–11

Gypsum, like mica, cleaves in one direction to form thin flakes. Unlike mica flakes, however, those of gypsum are soft enough to scratch with a fingernail. They break easily if bent.

FIGURE 2–12

The iron sulfide mineral pyrite forms pale yellow crystals with a brilliant metallic luster. This specimen is about the size of a walnut.

occurs in various rusty shades of yellow and brown. Both minerals make excellent pigments and are used extensively in paints. They also account for most of the many shades of red, yellow, and brown we see in rocks.

Black crystals of **magnetite,** Fe_3O_4, lightly pepper many kinds of rocks. These crystals are, as their name suggests, strongly magnetic. As rocks form, the magnetite crystals tend to align themselves with the earth's magnetic field. They remain locked in position, retaining a permanent record of the earth's magnetic field direction at the time and place where the rock they are in formed. Study of that magnetic memory has contributed significantly to our knowledge of the earth by showing that the continents and ocean floors move about the face of the planet.

RECOGNIZING MINERALS

Rocks are the natural habitat of minerals. Certain mineral species consistently occur in association with certain others to form the various kinds of rocks. Therefore, geologists in the field generally begin the process of identifying an unknown mineral by examining the rock in which it appears. They attempt to determine what other minerals the rock contains and thus narrow the probable identity of the unknown species to a few likely possibilities. Next, they scratch and probe the unknown mineral to see how hard it is and how it breaks. The specimen's color and crystal form is then examined through a lens. In most cases, this informed on-the-spot observation of the mineral's physical properties will reveal its identity. However, a final confirming of such field identification may require laboratory work. The mineral's optical properties may be studied under a microscope, for example, or its crystal structure may be determined through X-ray analysis. In more difficult cases, actual chemical analysis may be necessary.

Many diagnostic physical properties of minerals can be observed only with the aid of sophisticated instruments. Nonetheless, several of the more basic properties can reveal quite a bit if examined with nothing more elaborate than a pocketknife and magnifying lens. These can be quite effective diagnostic tools if the user has a perceptive eye and the proper training. Field geologists, in fact, identify most of the minerals they encounter using only those basic tools. Later laboratory analysis generally shows that the field identifications were either absolutely correct or very close. Fundamental observations include considering: how the mineral surface appears in reflected light; the mineral's color, transparency, hardness, and external crystal form; and the way the mineral breaks.

Luster

The term **luster** refers to the appearance of the mineral's surface in reflected light. On the basis of this property all minerals may be broadly divided into two large groups: those with a metallic or submetallic appearance, and those that are clearly nonmetallic.

A metallic or submetallic luster is evidence of some degree of metallic bonding in the crystal lattice. All of the native metals, such as gold, silver, or copper, have metallic lusters. So do most of the sulfide minerals and some of the oxides. All minerals with a metallic or submetallic luster are absolutely opaque, even in the form of thin flakes.

With the exception of pyrite, which is metallic, and of magnetite, which is submetallic, all of the common rock-forming minerals are nonmetallic. Geologists distinguish many different kinds of nonmetallic lusters. Some of those, such as "glassy," "pearly," "earthy," and "silky" are fairly obvious. The more subtle distinctions, such as that between a "waxy" and "greasy" luster, are best left to an experienced observer. Minerals with a nonmetallic luster may be transparent. All transmit at least some light although it may be necessary to look at the edge of a thin flake to verify that.

Color

A mineral's most obvious physical property is its color, which may or may not provide a useful guide to a mineral's identity. The metallic and submetallic minerals always do have specific diagnostic colors as do a few of the nonmetallic species. However, many of the nonmetallic minerals, including some of the commonest and most abundant species, vary greatly in color.

The colors of the metallic and submetallic minerals are distinctive enough to identify many almost beyond doubt.

For example, the common iron sulfide mineral pyrite has a pale yellow color that will turn to a rich gold if copper substitutes for part of the iron to make the closely related mineral **chalcopyrite.** Arsenic substituted for part of the iron will make yet another closely related mineral called **arsenopyrite,** which has an equally distinctive, extremely pale, color. Although closely related in composition and crystal structure, the three minerals can be confidently distinguished on the basis of their colors.

Some of the nonmetallic minerals, such as olivine, have clearly diagnostic colors. Most, however, do not. Many occur in a wide range of colors that may vary even within the confines of a single crystal. To cite an extreme example, some crystals of **tourmaline,** a fairly common silicate mineral, are pink at one end and green at the other. Geologists exercise extreme caution in using the colors of the nonmetallic minerals as a guide to their identity.

The color variety of nonmetallic minerals is due partly to variations in composition and partly to imperfections in their crystal lattices, which affect the way they transmit light. For example, the zinc sulfide mineral **sphalerite,** which is one of the few nonmetallic sulfides, varies in color from pale yellow through dark red to black as its iron content increases. Lattice imperfections arise from several causes. In many cases they form as decay of radioactive isotopes knocks atoms out of their proper places in the lattice. Unscrupulous dealers occasionally adjust the colors of gems, mainly diamonds and sapphires, simply by exposing them to strong X-rays. Color in minerals caused by radiation damage to the lattice generally disappears or changes if the specimen is heated enough to permit the atoms to snap back into their proper places. As that happens, the mineral glows.

Grinding a mineral specimen to a fine powder reveals its intrinsic color, which does not vary from specimen to specimen as does the color of crystals. With most minerals, the intrinsic color of the fine powder can be conveniently seen by using the specimen like a piece of chalk to draw a **streak** on an unglazed porcelain tile. A collection of specimens of the iron oxide mineral hematite, for instance, may display colors ranging from black through brown to red but all will produce the same maroon streak. Many geologists obtain streak colors by gouging the surface of the specimen with a sharp knife tip and then reducing the chips to a fine powder.

Hardness

The property of **hardness** in a mineral refers to its qualities as an abrasive. Hardness is a measure of what a mineral will or will not scratch; it is *not* a measure of how easily a mineral breaks. Minerals will scratch anything that has a hardness equal to or less than their own. They can be scratched by anything as hard or harder than themselves.

A formal hardness scale called the **Moh's Scale** (Figure 2–13) consists of 10 minerals ranked in order of increasing hardness from talc at the bottom to diamond at the top. However, most geologists check the hardness of minerals rather informally. They use fingernails, approximate hardness of 2; a copper penny, hardness of 3; or a knife blade, approximate hardness of 5.5. These handy items range through enough of the Moh's hardness scale to usefully bracket most of the abundant rock-forming minerals.

Cleavage and Fracture

Nearly all mineral grains have the orderly internal structures of crystals. Many, though, do not reflect that in their external forms. If mineral grains do exhibit a regular crystal form, that always provides a useful clue to their identity. Some minerals, such as garnet and pyrite, regularly occur in well-formed crystals whereas others, such as quartz, do so rarely.

Many minerals also express their orderly internal structures through the way they break. Many—by no means

FIGURE 2–13

The Moh's hardness scale.

Diamond	10	
Corundum	9	
Topaz	8	
Quartz	7	
Orthoclase	6	
Apatite	5	knife blade
Fluorite	4	
Calcite	3	copper penny
		fingernail
Gypsum	2	
Talc	1	

FIGURE 2-14

Common table salt, the mineral halite, cleaves readily in three directions intersecting at right angles to produce cubic fragments.

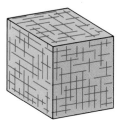

all—tend to break along regular plane surfaces that correspond to planes of weakness in their crystal lattices. This property is known as **cleavage.** Common table salt, for example, cleaves easily in three directions oriented at right angles to each other to make cubic fragments (see Figure 2–14). Nearly every grain of salt that comes out of the shaker is a pretty little cube. Other minerals cleave in one, two, four, or six directions that intersect at various characteristic angles. Mica, for example, cleaves in only one direction to produce flat flakes (Figure 2–15). Calcite cleaves in three directions that intersect at 72° angles to produce rhombohedral prisms (Figure 2–16).

Sometimes, a broken surface of a rock will show numerous mirror reflections when turned against the light. Such reflections very likely come from flat cleavage surfaces on mineral grains. Any doubt can be resolved by breaking one of the grains to see if it will cleave again along a parallel surface. In most cases the mineral must be carefully examined under a lens. It can then be determined whether the grains cleave in only one direction to produce flat flakes, or in two or more intersecting directions to produce stepped surfaces. Cleavages may also appear as fine sets of parallel lines or as hairline cracks.

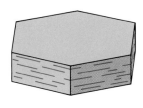

FIGURE 2-15

Geologists often refer to mica crystals as ''books'' because they cleave easily in one direction to form thin sheets that suggest pages.

FIGURE 2-16

Calcite crystals cleave in three directions intersecting at 72° angles to produce rhombohedral fragments which, in this case, do not resemble the form of the original crystal.

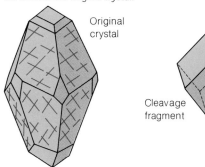

Original crystal

Cleavage fragment

Some minerals cleave easily to produce large flat faces. Others cleave only with difficulty to produce more ragged surfaces that contain many small cleavage faces. Still others do not cleave at all. Even members of this last category fracture in characteristic ways just as glass and bricks break to form distinctive surfaces. For example, neither quartz nor pyrite cleave but they break along smoothly curving surfaces resembling those of broken glass. Other minerals break along surfaces that may be smooth, rough, jagged, or splintery.

SUMMARY

Minerals are naturally occurring chemical elements or compounds defined in terms of both their compositions and crystal structures. Many distinctly different mineral species share the same composition or crystal structure but not both. Numerous minerals vary widely in composition as well as in appearance. This phenomenon exists because the atoms of various elements that resemble each other in size and chemical behavior substitute for each other in the crystal structure.

Oxygen is the most abundant element in the earth's mantle and crust. It is the only one of the more significantly abundant elements that forms negative ions. Therefore, atoms of the other most abundant elements, of which the majority are silicon, most commonly bond to oxygen atoms. Most of the rock-forming minerals are based on combinations of oxygen and silicon. Abundant rock-forming minerals are severely limited in number due to two factors: (1) the requirement that positive and negative ions must combine with each; and (2) that there is a limited number of abundant chemical elements.

In oxygen-silicon bonding, each silicon atom exists within a cluster of four oxygen atoms. The oxygen atoms are disposed as though they were at the corners of a tetrahedral pyramid. Silicon atoms share oxygens between them thus linking the silicate tetrahedra together in several different patterns. Each individual pattern defines a major class of silicate minerals.

Other groups of rock-forming minerals, which together comprise less than 1 percent of the earth's mantle and crust, include the carbonates, the sulfates, and the sulfides. The carbonates and sulfates are based on groups of oxygen atoms clustered around atoms of carbon or sulfur, respectively. The oxides consist of atoms of various elements bonded directly to atoms of oxygen instead of to groups of oxygen atoms. The sulfides consist of various atoms bonded directly to those of sulfur which, like oxygen, can form negatively charged ions.

Geologists identify mineral species in the field in several ways. They observe a number of simple physical properties, such as hardness and color. They also examine the rock where the mineral is found. They do so because certain mineral species exist in association with certain others. Thus, if one mineral in a rock is known, others there may be recognized through this association. Final verification depends upon more precise laboratory examination of physical properties. In many cases direct identification may be achieved through X-rays of the crystal structure or through sophisticated chemical analysis of the mineral's composition.

KEY TERMS

amphiboles	cleavage	ionic bonds	mica	pyrite	sphalerite
anhydrite	covalent bonds	ions	Moh's Scale	pyroxene	streak
arsenopyrite	crystal structure	isomorphism	neutrons	radioactive decay	sulfates
atomic numbers	feldspar	isotopes	nucleus	rock forming	sulfides
atomic weights	garnet	kaolinite	olivine	serpentine	tourmaline
augite	goethite	lattice	orthoclase	silicates	unit cells
calcite	gypsum	layer silicates	oxides	silicate double chains	
chalcopyrite	hardness	luster	plagioclase	silicate tetrahedra	
chlorites	hematite	magnetite	polymorphic	single-chain silicates	
clay	hornblende	metallic bond	proton	smectites	

REVIEW QUESTIONS

1. Would you expect different polymorphic mineral varieties of the same substance to have the same unit cell? *Yes*

2. The earth's mantle consists largely of pyroxene and olivine whereas the continental crust contains large amounts of feldspar and quartz. What does that difference in composition imply about the relative proportions of silicon and oxygen in the mantle and continental crust?

3. The minor element zirconium commonly occurs in the silicate mineral zircon and is rarely found as a trace constituent of other minerals. To judge from that observation, would you expect to find a major element that forms ions similar in size and chemical bonding behavior to those of zirconium?

4. The common minerals quartz and calcite often look a bit alike. How could you use a pocketknife to distinguish them? *Hardness test*

5. Calcite cleaves in three directions whereas quartz does not cleave. Suggest three observations that might distinguish between those minerals on the basis of that difference.

6. Why do many minerals form prism-shaped crystals with four or six sides, and none with five or seven sides?

SUGGESTED READINGS

Berry, L. G. and B. Mason. *Elements of Mineralogy.* San Francisco: W. H. Freeman & Co., 1968.

A standard introductory text on mineralogy.

Deer, W. A., R. A. Howie, and J. Sussman. *An Introduction to the Rock-Forming Minerals.* New York: John Wiley, 1966.

A comprehensive reference.

Ernst, W. G. *Earth Materials.* Englewood Cliffs, N.J.: Prentice-Hall, 1969.

An excellent brief introduction to mineralogy. Extremely readable.

Hamilton, W. R., A. R. Woolley, and A. C. Bishop. *The Larousse Guide to Minerals, Rocks and Fossils.* New York: Larousse & Co., 1977.

An unusually well done popular guide with excellent descriptions and numerous color photographs as well as clear line diagrams.

Hurlbut, C. S. *Minerals and Man.* New York: Random House, 1969.

A nontechnical discussion of minerals and their uses.

Vanders, I., and P. F. Kerr. *Mineral Recognition.* New York: John Wiley, 1967.

An excellent descriptive reference for the serious mineral collector. Full of attractive color pictures.

CHAPTER THREE

A sawed slice of quartz-plagioclase-hornblende gneiss showing tightly folded layers, probably the remnants of sedimentary beds. The specimen is about 10 cm long.

ROCKS

Geologists engage in the study of rocks—**petrology**—to determine what they are and how they formed. Rocks vary greatly in type and origin but geologists describe and classify all of them in terms of their mineral compositions and their **textures**. Textures depend upon the sizes and patterns of the mineral grains that make up the fabric of the rock. Most rocks also contain various **structures,** such as gas bubbles, layering, ripple marks, and folds, all of which help to reveal their identities and explain their origins.

Igneous rocks, those that crystallize from a melt, are the most fundamental type. All other kinds of rocks must ultimately derive from them in one way or another. In the beginning, all the earth's rocks must have been igneous and that would still be true if the earth did not have active internal and surface processes.

Sedimentary rocks form from sediments such as sand, mud, or gravel deposited on the earth's surface by such geologic agents of erosion and transportation as running water, wind, or glacial ice. Although they comprise an insignificant proportion of the earth's volume, most of us see sedimentary rocks more often than any other kind. They cover much of the earth's surface in much the same way that a coat of paint covers a board. Sedimentary rocks are economically important because they are the source of all the earth's fossil fuels as well as a large proportion of its other mineral resources. Furthermore, the sedimentary rocks contain most of the archives, as it were, of earth history, including the records of the origin and evolutionary development of living things.

Metamorphic rocks form from older igneous or sedimentary rocks that were recrystallized under various combinations of temperature and pressure deep within the earth's crust. In general metamorphism makes rocks more coarsely crystalline. It also replaces old mineral assemblages and rock structures with new ones that reflect the temperatures, pressures, and intense deformation that typically accompany recrystallization. Rocks may undergo all degrees, or *grades*, of metamorphism, ranging from slight recrystallization that leaves the original rock virtually unchanged to high-grade metamorphism that generally transforms the original rocks almost beyond recognition. Metamorphic rocks inform geologists about the pressure and temperature environments that exist deep beneath the earth's surface.

This chapter will briefly introduce the major types of rocks, primarily from a descriptive point of view. These rocks will be considered in greater detail in later chapters dealing with the processes that create them.

IGNEOUS ROCKS

All igneous rocks crystallize from a hot melt called **magma** as long as it remains below the earth's surface, and **lava** when it erupts from a volcano. *Plutonic* igneous rocks de-

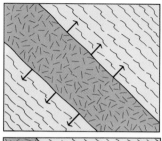

STEP 1: Intrusion of sill concordantly to banding in the gneiss.

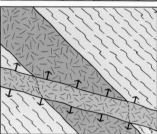

STEP 2: Intrusion of dike which crosses the sill and the banding in the gneiss. Note that the dike offsets the sill.

FIGURE 3-1

A small dike of pale granite cuts diagonally across a larger sill of similar granite sandwiched within the layers of a banded gneiss. The sill is about 1 m across.

rive from magmas that crystallize within the earth's crust to form either large masses called **plutons** or else smaller, sheetlike bodies. These smaller bodies are called **dikes** if they fill fractures that cut across the structures of the enclosing country rocks, and are called **sills** if they parallel those structures (Figure 3-1). In general, the plutonic rocks are composed of individual mineral grains large enough to be visible under a simple magnifying glass. The *volcanic* igneous rocks form from magmas that erupt as lava, either in the form of flows that pour across the ground surface or volcanic ash ejected into the sky. They are generally so fine-grained that many, or in some cases all, of the individual mineral grains become visible only with the aid of a microscope. Plutonic and volcanic igneous rocks differ so radically in appearance that it is difficult to believe that they are indeed closely related in origin and composition.

All igneous rocks fall within a spectrum of chemical compositions (Figure 3-2) that closely correlate with their mineral compositions, colors, and occurrences. The rocks at one end are rich in magnesium and iron, are composed largely of pyroxenes, olivine, and plagioclase feldspar, and are generally dark. These are called **mafic** rocks, a term coined to refer to their high content of magnesium and iron. The rocks at the opposite end of the spectrum are rich in silicon and aluminum, are composed largely of feldspar, quartz, and micas, and are generally pale. These are called **sialic** rocks, another artificial word coined from the first letters of ''silicon'' and ''aluminum.'' In general, the dark mafic rocks tend to be noticeably denser than the paler sialic rocks, and they melt at much higher temperatures. Most igneous rocks fall somewhere between the mafic and sialic extremes.

Mineral Sequence

Experimental work with synthetic magmas in addition to field study of igneous rocks reveal that their minerals often crystallize in two parallel sequences called **Bowen's reaction series.** One sequence follows the stages of progressively greater oxygen-sharing in the silicate minerals; the other follows the progressive substitution of sodium for calcium in the plagioclase feldspars (Figure 3-3). For example, the first minerals to appear during crystallization of

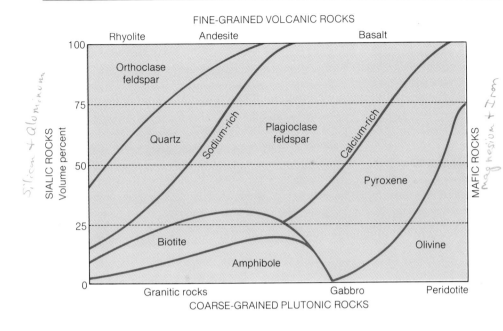

FIGURE 3–2

A simplified classification of the common igneous rocks. The verticals describe the mineral compositions of the plutonic rock types at the bottom and their volcanic equivalents at the top. The absence of boundary lines between the rock types emphasizes that they blend continuously into each other instead of fitting into natural pigeonholes.

FIGURE 3–3

The mineral reaction series. As a magma cools, the first minerals to crystallize at high temperature react with the remaining melt to form the mineral next down the sequence. Broad horizontal slices through the diagram describe mineral assemblages likely to occur in common igneous rocks. For example, olivine, pyroxene, and calcium-rich plagioclase are likely to occur in association with each other but not with orthoclase or quartz. Certain uncommon rocks provide exceptions.

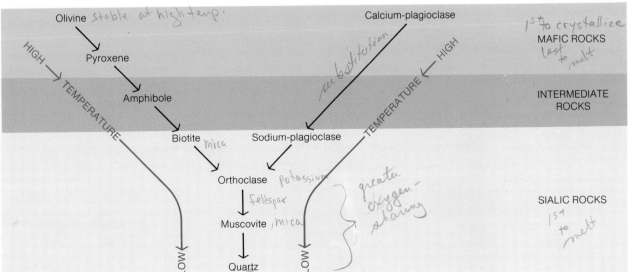

a mafic magma are olivine and calcium-rich plagioclase. As crystallization proceeds, these early minerals react with the remaining magma to form pyroxene and plagioclase with a smaller content of calcium. At that point the mafic magmas become completely crystalline and the progression stops. The more silica-rich sialic magmas start lower down in the sequence and finish with a rock composed of sodium-rich plagioclase, quartz, and orthoclase feldspar, which are the last minerals to appear.

If the reactions in the mineral crystallization sequence are completed, they obliterate the earlier minerals, leaving no trace of them in the final rock. Therefore, minerals remote from each other in the crystallization sequence are incompatible in the same rock because the earlier ones should have reacted with the magma and disappeared before the later ones began to form. For example, olivine should occur with pyroxene and calcium-rich plagioclase, never with sodium-rich plagioclase, orthoclase, or quartz. Therefore, knowing the mineral reaction series enables geologists to anticipate mineral associations in igneous rocks. In many cases mineral associations can be ascertained by simply looking at the color of the rock. Dark igneous rocks generally contain pyroxene, calcium-rich plagioclase, and perhaps olivine whereas the lighter-colored rocks are likely to contain orthoclase, sodium-rich plagioclase, hornblende, and quartz.

Grain Size

The textures of the igneous rocks depend primarily upon the sizes of their mineral grains or, to express the same idea in other terms, the number of mineral grains within a given volume of rock. The coarse-grained plutonic rocks contain relatively few large mineral grains within a given volume whereas the fine-grained volcanic rocks contain a great many small grains. This difference depends mainly upon the rate at which the rocks crystallize although other factors are also involved. Consider, for example, the situation within a magma as the first crystals begin to appear.

In earliest stages of growth minute crystals lead a precarious existence. At this time, they consist of only a few atoms that may dissolve back into the magma instead of collecting more atoms to grow larger. We may compare

their predicament to that of larval fish struggling to get started in a hostile ocean. The chances of survival are especially poor for tiny crystals that begin to grow when temperature and pressure are only marginally appropriate for their existence. Most that start under such circumstances vanish back into the magma. Only a lucky few survive to grow larger as changing temperature and pressure bring the magma into a condition in which they can stabilize.

Plutonic rocks form from magmas that cool extremely slowly because the enclosing rock provides excellent thermal insulation. Therefore, such magma lingers near the temperature at which each mineral species just starts to crystallize, which is also the temperature at which minute crystals are least likely to survive. Few of the crystals that start to grow actually do survive. Thus, plutonic rocks contain relatively few mineral grains within a given volume and are, therefore, coarse-grained. Unusually coarse-grained plutonic rocks called **pegmatites** (Figure 3–4) are composed of giant crystals that, on the average, range from a few centimeters to several meters. They apparently crystallize from extremely fluid magmas with high water contents in which the survival rate of beginning crystals is evidently low, and only a very few grow large. However, once a pegmatite crystal does become large enough to keep from dissolving back into the magma, the same fluidity that endangered its start now enhances its growth by permitting rapid diffusion of the necessary ions to its surface.

The volcanic rocks generally chill quickly after they erupt. They do not linger in the critical temperature range in which minute crystals form and then dissolve. Instead, they soon cool to temperatures at which all the tiny crystals that start to grow can succeed. This process forms a very fine-grained rock that contains large numbers of extremely small crystals within any given volume, far more than a similar volume of coarse-grained plutonic rock would contain.

Some sialic volcanic rocks cool without crystallizing to form glassy rocks called **obsidian** (Figure 3–5). These rocks are, in effect, extremely viscous melts, not crystalline rocks composed of aggregates of mineral grains. Obsidian differs chemically from crystalline sialic rocks in having an extremely low water content which evidently inhibits formation of minute crystals and causes the magma to solidify into noncrystalline glass. More mafic lavas may also form

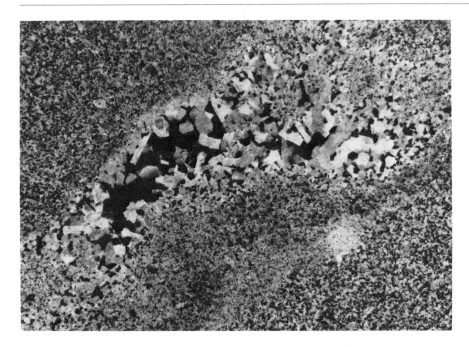

FIGURE 3-4

A patch of pegmatite in a polished slab of granite. The black mineral is biotite mica and the blocky white crystals are feldspar—both plagioclase and orthoclase. The larger feldspar crystals are about 2 cm across.
(D. W. Hyndman)

glass if they happen to cool extremely quickly. However, rapid chilling can not explain the glassiness of obsidian, which typically forms large masses much too thick to have cooled quickly (Figure 3-6).

Many igneous rocks, both plutonic and volcanic, contain isolated larger crystals called **phenocrysts,** which are scattered through a much finer-grained matrix, giving the rock a **porphyritic** texture (Figure 3-7). Porphyritic textures probably reflect a change in conditions within the crystallizing melt; that is, conditions increasing the chances that small crystals could survive after the larger ones had already reached considerable size. For example, a rapid drop in temperature would plunge the magma into a condition favoring survival of small crystals. That may well explain many porphyritic textures in volcanic rocks, which do indeed cool quickly after they erupt. However, postulating a magma that could cool quickly while it is within the earth's crust is difficult, so porphyritic textures in plutonic rocks probably record some other kind of event, such as a rapid drop in pressure or perhaps a loss of water.

FIGURE 3-5

A hand specimen of obsidian from Glass Butte, an obsidian mountain near Burns, Oregon. Streaks of glass mark a swirling flow structure developed while the extremely viscous magma was still slowly moving.

FIGURE 3-6

A large obsidian lava flow in the crater of Newberry volcano, Oregon. Such crudely corrugated surfaces typically develop on extremely viscous flowing masses. (Oregon Department of Geology and Mineral Industries)

FIGURE 3-7

Porphyritic texture in a volcanic rock as it appears through a microscope. The large crystals of feldspar set in a fine-grained matrix average about 3 mm across. (John Herrlin)

Plutonic Igneous Rocks

Recognizing plutonic igneous rocks is generally easy because most are composed of crystals large enough to see with the unaided eye or with a simple magnifier. Although there are exceptions, the mineral grains of plutonic igneous rocks are typically in random orientation so a specimen looks essentially the same from all directions and lacks any hint of a grain. Geologists describe such textures as **massive** (Figure 3–8). Plutonic igneous rocks typically weather into distinctive bouldery outcrops, which are easy to recognize even from a distance.

The most abundant plutonic igneous rock, indeed the most abundant earthly rock of any kind, is **peridotite,** an extremely mafic and dark-colored rock composed primarily of pyroxene and olivine accompanied in some cases by mica or garnet. Peridotites probably constitute the bulk of the earth's mantle as well as most of the moon and presumably the other terrestrial planets. Many stony meteorites consist of peridotite. Although it is abundant we rarely see peridotite because little of it is exposed at the earth's surface. One variety called kimberlite occurs locally in the earth's crust and commonly contains diamonds, which can form only under the extremely high pressure that prevails deep within the earth's mantle. The presence of diamonds in those peridotites clearly indicates that they must have indeed come from the mantle.

A few large masses of plutonic igneous rock are formed by **gabbro,** which is a dark mafic rock composed mostly of augite pyroxene and calcium-rich plagioclase accompanied in some cases by olivine. Some of those masses, such as the one at Sudbury in southeastern Ontario, appear to fill large meteorite explosion craters essentially similar to the craters on the moon. The magma that crystallizes to form gabbro appears to originate through partial melting of peridotite in the mantle, a process we will consider in Chapter 19. The relative rarity of gabbro is surprising in light of the great abundance of its eruptive equivalent, the volcanic rock **basalt,** which forms most of the oceanic crust as well as many volcanoes both on continents and in the ocean basins.

The common sialic rock **granite** comprises a majority of the continental crust and is therefore the plutonic rock we most often see. Granite is composed of orthoclase, sodium-rich plagioclase, and quartz in varying proportions

FIGURE 3–8

An example of massive texture.

accompanied in most cases by lesser amounts of amphibole or mica. Like most sialic rocks, granites are light-colored, generally either pale gray or pink. Granite typically occurs in enormous masses called **batholiths,** which generally consist of many smaller plutons packed tightly together. Batholiths are commonly exposed over areas measured in thousands of square kilometers.

Volcanic Igneous Rocks

The volcanic igneous rocks occur in a bewildering variety of forms depending mostly upon the circumstances of their eruption. Some erupt as lava flows that pour fluidly across the ground surface whereas others blow out of the vent in blasts of escaping steam to form fragments. In some cases, the volcanic fragments are hot enough to fuse together into solid rocks called **welded ash** (Figure 3–9) as they settle to the ground; in other cases, they form loose deposits of coarse **volcanic cinders** or fine **volcanic ash.** Wind and running water erode and redeposit many fragmental deposits thus blurring the distinction between volcanic and sedimentary rocks.

All volcanic rocks are too fine-grained to be easily identified by their constituent minerals. Even those with porphyritic textures have a fine-grained matrix or groundmass in which individual mineral grains become visible only with the aid of a microscope. Therefore, field identification of volcanic rocks is based primarily upon their colors, which depend upon their compositions. Final identification generally depends upon microscopic study in the laboratory or upon chemical analysis. We will concern ourselves here only with the field identification.

The most abundant and widely distributed volcanic rock is **basalt,** which forms most of the oceanic crust that covers about two-thirds of the earth's surface. Basalt also erupts from many volcanoes on the continents. It is a mafic rock easily recognizable by its flat black color and general lack of visible mineral grains. Microscopic examination reveals a mass of tiny grains of calcium-rich plagioclase and augite accompanied in many cases by small amounts of olivine. Basalt erupts as a fluid lava that spreads across the ground surface in thin flows that commonly cover large areas. However, if the rising magma absorbs some water before it erupts, it may generate enough steam to cough out fragments that settle to form deposits of black volcanic cinders and ash.

At the sialic end of the compositional spectrum of the volcanic rocks we find **rhyolite,** which differs from basalt in every important respect. Rhyolite occurs in a variety of pale shades of gray, pink, and yellow and commonly contains recognizable larger crystals of quartz or feldspar set in a fine-grained matrix. Although fluid, rhyolite magmas are much too viscous to spread out into thin lava flows. Instead, they commonly extrude quietly to make steep, dome-shaped mounds of lava or very thick lava flows, many of which cool without crystallizing to become obsidian. If rhyolite magmas reach the surface with a large steam content, they erupt in an explosive blast that tears the lava to shreds. The fragments may fuse together into solid masses of welded ash if they fall near the vent while still very hot, or they may drift long distances downwind before they finally settle to make sparkling white beds of rhyolitic volcanic ash.

The vast intermediate compositional range between the spectrum poles of mafic basalts and sialic rhyolites is occupied by rocks commonly called **andesite** by field geologists. Precisely subdividing andesites into more narrowly defined rock types generally requires laboratory analysis. Freshly erupted andesites occur in shades of gray ranging from almost black in those near the basaltic end of the spectrum to quite pale in those more nearly approaching rhyolite in composition. Andesite lavas at times erupt as flows and, at other times, as clouds of ash and other fragmental material. These lavas build the picturesque cones that most people associate with volcanoes. They form long volcanic chains, such as the Cascades and Aleutians in western North America.

SEDIMENTARY ROCKS

In studying sedimentary rocks geologists generally try to reconstruct as accurately as possible the environment in which the original sediments accumulated. Were they, for

FIGURE 3–9
These two sawed hand specimens of welded ash clearly show the flow structure that developed as the hot ash continued to move while it settled and fused into solid rock.

example, deposited in a lake, on the deep ocean floor, in a coastal lagoon, or on a beach? Such interpretations require patient study of numerous outcrops scattered over large areas and frequently involve drilling expensive holes to obtain samples of rocks not exposed at the surface. However, the results are generally well worth the effort and expense because they often lead to discovery of new mineral deposits and play a vital role in planning the orderly development of known deposits.

Interpretation of sedimentary rocks depends heavily upon study of their **primary sedimentary structures,** including features such as **mudcracks** or **ripple marks** (Figure 3–10) that formed in the original sedimentary environment and survive preserved in the rock. The most conspicuous and widespread of all primary sedimentary structures is **layering** or **stratification** (Figure 3–11). Stratification marks nearly all sedimentary rocks, making them easily recognizable even from a considerable distance. Other common primary sedimentary structures include various kinds of **cross-beds,** which are inclined layers contained within larger layers; **fossils,** which are the remains of animals or plants; many special kinds of lay-

FIGURE 3–10
Ripple marks in sedimentary rock.

FIGURE 3–11
Layering is the single most conspicuous characteristic of sedimentary rocks.

ering; and so forth. In later chapters we will consider a number of primary sedimentary structures in connection with the processes that create them.

Sedimentary materials can be divided into two broad categories: (1) the **clastic sediments,** which consist of debris derived from erosion of older rocks or soil; and (2) the **nonclastic sediments,** which consist of material produced within the sedimentary environment. The commonest clastic sedimentary materials are familiar as gravel, sand, silt, and clay. Mud is a mixture of silt and clay. Sea shells are the most familiar nonclastic sediment—they form in the sedimentary environment and do not consist of fragments eroded from older rocks. The most important nonclastic sediment, which few people ever see, is **calcareous mud.** Others include **peat,** an accumulation of semidecomposed plant matter, and deposits of minerals, such as gypsum or halite, that crystallize from evaporating salty waters.

FIGURE 3-12

A sawed hand specimen of conglomerate showing pebbles set in a matrix of sand now cemented into solid rock. The larger pebbles are about 2 cm in diameter.

Although geologists commonly apply different names to unconsolidated sediments and the rocks derived from them, they attach no great conceptual significance to the distinction. In some cases, the name of the rock is closely related to that of the sediment. For example, sand hardens into **sandstone,** mud into **mudstone,** and silt into **siltstone.** In other cases, the names of the rocks and the original sediments differ. For example, deposits of clay become **shale,** gravel becomes **conglomerate** (Figure 3-12), peat becomes **coal,** and calcareous mud becomes **limestone** or **dolomite.** Sediments formed by crystallization of soluble minerals from evaporating salty waters are called **evaporites.** If, as quite commonly happens, a rock contains a mixture of sedimentary materials, geologists simply describe the mixture in naming the rock. Thus, we hear of such rocks as "pebbly sandstones," "muddy sandstones," "sandy limestones," and "calcareous shales." Although several elaborate schemes of sedimentary rock nomenclature exist and are widely used in technical publications, most geologists find the common sense descriptive names suggested in quotes perfectly adequate for ordinary occasions.

Lithification

Most deposits of unconsolidated sediment, even the softest, eventually harden into solid rock, a transformation known as **lithification.** It is amazing to think of soft mud or loose sand becoming rigid rock. The change, however, is actually fairly simple. It involves the operation of several different processes that start to work as soon as the sediment accumulates and generally continue long after it has become rock. These processes are collectively known as **diagenesis.**

The simplest diagenetic process is **compaction,** the early stages of which are familiar to everyone who has ever squeezed water out of a handful of mud to convert it into a solid ball suitable for throwing. Freshly deposited sediments typically contain large amounts of water—more than 50 percent by volume in the case of many silts and clays. This water escapes as the weight of new layers of sediment compresses those beneath. In addition to expelling water, compaction brings the mineral grains into close physical contact so other diagenetic processes can weld them into a solid mass.

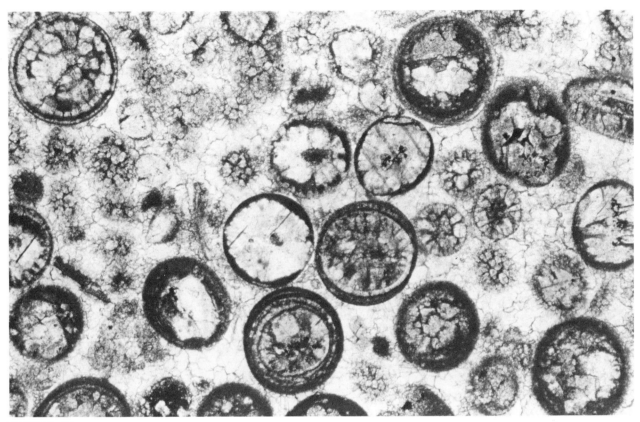

FIGURE 3–13

A photomicrograph of a thin-section of a limestone composed of small round bodies called **oolites** cemented together by crystalline calcite precipitated between them from circulating water. Oolites are spherical pellets of calcite that grow on shallow, wave-washed sea bottoms; those in this picture average about 1 mm in diameter. (John Herrlin)

Cementation is a basic diagenetic process that involves filling the void spaces between mineral grains with new mineral matter deposited from water circulating within the sediment (Figure 3–13). Cementation is primarily responsible for the lithification of many granular sediments, such as sand or gravel, and also helps consolidate deposits of sea shells into solid masses of limestone or dolomite. Geologists commonly include the identity of the cement in their descriptions of sedimentary rocks, using such expressions as ''calcite-cemented sandstone'' or ''quartz-cemented conglomerate.''

Mineral particles in sediments tend to dissolve in the water that fills the void spaces between them. The dissolved material then deposits elsewhere, both cementing and welding the grains together. This process is called **recrystallization** because it involves the dissolving of the original mineral grains and corresponding growth of new crystals. Recrystallization is familiar as the process that converts table salt and sugar into solid lumps during periods of wet weather. It plays a major role in the lithification of most limestones and many sandstones.

In some cases the diagenetic processes transform loose

sediments into solid rocks quickly and in others the change occurs slowly or not at all. Although older rocks commonly tend to be more thoroughly lithified than younger ones, no consistent relationship exists between the age of a sedimentary rock and its degree of lithification. All geologists have encountered examples of old sedimentary deposits that remain nearly unconsolidated as well as much younger ones already hardened into solid rock. In fact, there are samples of thoroughly lithified beach sands that contain broken fragments of modern beer bottles.

As the temperature passes beyond the general region of 200°C, the domain of diagenesis blends imperceptibly into that of low-grade metamorphism with no clear dividing line between them. Geologists make the distinction arbitrarily as certain minerals, such as chlorites, begin to appear.

METAMORPHIC ROCKS

Metamorphic rocks form as igneous or sedimentary rocks recrystallize at high temperature, destroying old mineral grains and rock textures while creating new mineral assemblages and distinctive metamorphic textures and structures. In most cases, high pressures and deforming stresses accompany the high temperatures and leave their distinctive imprints on the metamorphosed rocks. The colorful minerals and complex structures of many metamorphic rocks make them strikingly beautiful and some also contain valuable economic minerals, such as talc, asbestos, and garnet.

One of the basic problems geologists must solve in their study of metamorphic rocks concerns the question of what the parent rock was before metamorphism occurred. In many cases, the composition of the metamorphic rock answers that question immediately. Only a pure sandstone, for example, could turn into a metamorphic rock consisting almost entirely of quartz, and only a limestone could recrystallize into a rock composed of calcite. In other cases, enough primary structure, such as bedding or ripple marks, survives metamorphism to identify the parent rock. However, high-grade metamorphism generally reconstitutes the original rock so completely that little or no trace of primary structures remains. In such cases identifying the parent rock may prove difficult—it may even be impossible to determine whether the original rock was igneous or sedimentary.

Geologists who study metamorphic rocks also attempt to determine the temperature and pressure under which metamorphism occurred. They do that by comparing the mineral assemblages found in metamorphic rocks to synthetic assemblages grown experimentally in the laboratory under conditions of known temperature and pressure. The results provide information about conditions deep within the earth's crust, as well as being significant for interpreting the rock itself.

Field observation of metamorphic rocks shows that certain assemblages of minerals consistently occur together. Furthermore, rocks with identical chemical compositions crystallize into distinctly different mineral assemblages within well-defined areas that can be plotted on a map. In many cases, for example, it is possible to follow the same metamorphosed rock formation across the countryside and find one mineral assemblage passing into another within a distance of perhaps a few kilometers. Evidently such changes reflect metamorphism of the same formation under differing conditions of temperature and pressure. The problem is to determine precisely what those conditions were, a difficult feat because most common minerals form within rather wide ranges of temperature and pressure—if they did not, they could not be common.

In general, it is safe to assume that all the metamorphic rocks within a reasonably small area must have formed under essentially similar conditions of temperature and pressure. If several kinds of rocks with different chemical compositions exist within such a small area, they will contain among them a variety of minerals and mineral assemblages that have been studied experimentally and are known to be stable under slightly different conditions of temperature and pressure. Each mineral and assemblage of minerals restricts the possible range of temperatures and pressures under which the rocks could have formed in the same way that the sight of snow on the ground restricts our thinking about the possible temperatures outside. In most cases, the combination of restrictions imposed by all the minerals and mineral assemblages in the area defines a fairly narrow range of possible temperature and pressure conditions. So by applying knowledge of laboratory experiments to the observation of rocks, geologists have been able to identify the pressure and temperature ranges in which most kinds of metamorphic rocks form (Figure 3–14).

The One-Way Path

Most metamorphic rocks contain the mineral assemblage that formed under the highest temperature they experienced and show little evidence of having continued to recrystallize as they cooled. Evidently rocks recrystallize as the temperature rises but not as it falls. What magic does an increasing temperature have that a decreasing one lacks?

Many of the mineral reactions that occur with increasing temperature produce water, which presumably escapes from the rock as steam. If the same reactions are to proceed in the opposite direction as the temperature decreases after the peak of metamorphism has been reached, then water must enter the rock to replace that lost as the temperature rose. However, the reverse metamorphic reactions proceed with an increase in volume so the new minerals expand as they form and block the channels through which the water entered. Therefore, metamorphic rocks tend to seal themselves against entry of water. They thus preserve the mineral assemblage that formed at the highest temperature of metamorphism—except locally, where numerous fractures open the rock beyond its capacity to reseal itself. If that were not so, geologists would be hard pressed to find rocks containing direct evidence of the pressure and temperature conditions that prevail deep within the earth's crust.

Metamorphic Rock Types

Determining the temperature and pressure environment in which a metamorphic rock crystallized is an interpretation, not a description, which does not necessarily convey much information about the rock's mineral composition or appearance. Geologists also classify metamorphic rocks in terms of their mineral compositions and textures using a simple system of descriptive names. Metamorphic rock types depend partly upon the kind of parent rock that existed before metamorphism, partly upon the extent to which the original rock is recrystallized, and partly upon whether or not deformation accompanied metamorphism.

Quartz remains stable throughout the enormously wide

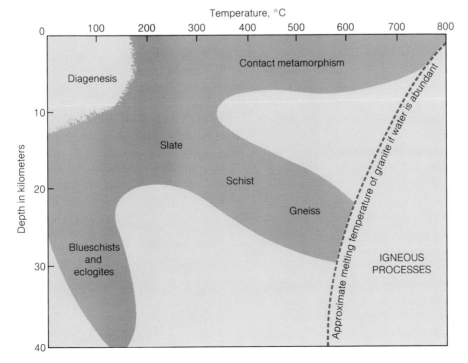

FIGURE 3-14

The processes of metamorphism operate in the pressure and temperature region between the domain of diagenesis at one extreme and that of igneous processes at the other. Most metamorphic rocks form under pressure and temperature conditions corresponding to the shaded areas of the diagram.

range of metamorphic temperatures and pressures so pure quartz sandstones pass through all grades of metamorphism essentially unchanged. Metamorphism merely welds the sand grains together into a brittle mass of solid quartz called **quartzite,** in which it is difficult to see the ghostly vestiges of the original sand grains. Calcite and dolomite likewise remain stable through most of the range of metamorphic temperatures and pressures, so they simply recrystallize into more coarsely granular rocks called **marble**. Limestone or dolomite become white marble if they are pure, colorful marble if impure. Sawed surfaces of many impure marbles reveal intricately swirling patterns that record the severe deformation the rock experienced during metamorphism. These patterns make impure marble desirable as ornamental building stone.

Muddy sandstones, mudstones, siltstones, and shales generally contain both quartz and clays as well as other minerals that react with each other during metamorphism. Such rocks pass through an interesting sequence of changes in both mineral composition and texture, with increasing intensity of metamorphism. In all those rocks the new minerals that form during metamorphism strongly tend to align themselves in preferred directions, in military formation as it were, to create distinctive textures quite unlike those of either igneous or sedimentary rocks. If the aligned mineral grains form flat crystals, such as mica flakes, the texture is called a **foliation.** If they form long slender crystals, such as needles of amphibole, it is called a **lineation.**

At relatively low temperatures, sedimentary rocks rich in clay metamorphose into **slate,** which superficially resembles the original sedimentary rock. Slate, however, is much harder and no longer splits into thin flakes parallel to the original sedimentary layering. Figure 3–15 shows how slates split into slabs along a direction parallel to that of an imaginary plane that would bisect the folds in the rock into mirror images. This property is called **slaty cleavage** and it reflects the development of an incipient foliation—minute flakes of mica have begun to grow in the rock and they are

FIGURE 3–15

These tightly folded layers of slate with slaty cleavage developed parallel to the plane of symmetry of the fold. (A. Keith, U.S. Geological Survey)

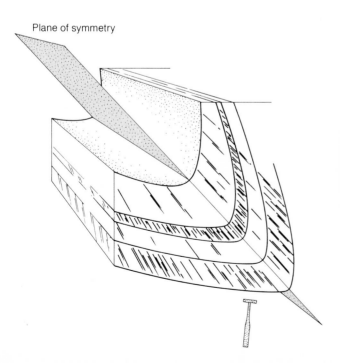

Plane of symmetry

aligned in a direction perpendicular to the stress that created the folds.

Slightly more intense metamorphism converts slate into **phyllite,** a rock in which the original clays have recrystallized into mica flakes. These flakes are just barely large enough to make the foliation easily visible and to give surfaces split along it a silken sheen. Still further increase in metamorphic temperatures converts the phyllite into **schist,** in which the aligned mica crystals give the rock a distinctly flaky character and spangle broken surfaces, making them sparkle in the sun. Some rocks form amphibole instead of mica and develop into schists with a splintery instead of a flaky character.

Still more intense metamorphism may convert schist into **gneiss** (Figure 3–16), a banded and streaky rock with distinct foliation or lineation but lacking enough mica or amphibole to confer a flaky or splintery character. Gneisses are typically complex rocks full of structures that clearly tell of intense deformation (Figure 3–17). Many form at temperatures near the upper limit of the metamorphic environment where they begin to melt to form sialic magmas. Such gneisses commonly consist of an intimate mixture of plutonic igneous and metamorphic rock stirred into patterns that strongly suggest marble cake. Since they are mixed igneous and metamorphic rocks, they are called **migmatites.** Beautifully cut and polished specimens of migmatite appear in nearly every cemetery as ornamental tombstones.

Unlike terms such as quartzite or marble that convey information about both mineral composition and texture in a single word, the expressions ''schist'' and ''gneiss'' refer only to rock texture. To complete the rock description, it is necessary to add a list of the minerals a rock contains, customarily in order of increasing abundance. Therefore, geologists commonly use expressions such as ''garnet-quartz-mica schist'' or ''hornblende-plagioclase gneiss'' to describe metamorphic rocks.

Most metamorphic rocks form as great masses of older rock are carried deep below the earth's surface where they are simultaneously heated and crumpled into tight folds. However, heat escaping from crystallizing igneous plutons may recrystallize the surrounding rocks, forming **contact metamorphic rocks** within narrow zones rarely much more than 100 meters wide. Contact metamorphic rocks generally form through the influence of simple heating without accompanying deformation and therefore typically

FIGURE 3–16

An outcrop of banded gneiss seen from a distance of a few meters.

FIGURE 3–17

A sawed slab of quartz-plagioclase-hornblende gneiss. This specimen shows its banded structure trending parallel to the foliation direction, and scattered large crystals of feldspar. A small dike of granite about 3 cm across slices through at the right hand side.

lack foliation. Instead, they have a distinctive spotted appearance and commonly look as though they might have developed a severe case of colorful freckles. Contact metamorphism creates spectacularly striking rocks where igneous plutons, especially granites, invade limestones. The chemical reactions between the molten magma and the limestone create a wide variety of minerals, including ore minerals in many cases, which commonly grow into large and well-formed crystals. The contact zones between granitic plutons and limestones provide choice opportunities for prospectors and mineral collectors.

So far, we have considered metamorphism chiefly as a matter of recrystallization at high temperature. However, there are some metamorphic rocks that evidently recrystallize under conditions of extremely high pressure and only moderately elevated temperature. They form in places where sedimentary and volcanic rocks are rapidly carried to great depths. There, they recrystallize and then return to shallower levels of the crust before they can reach high temperatures. Sedimentary rocks metamorphosed under such conditions become **blueschists** and basalts **eclogites.** Although neither kind of rock is common or abundant, their study played an important part in explaining the processes of plate tectonics.

SUMMARY

Although geologists describe and classify all rocks according to their mineral compositions and textures, the approach varies greatly with the type of rock.

There are two large categories of igneous rocks: the plutonic rocks that crystallize within the earth's crust, and the volcanic rocks that solidify on the surface. Most plutonic rocks are coarse-grained because relatively few crystals grow within them as they slowly cool whereas most volcanic rocks are fine-grained because they contain large numbers of crystals. Both kinds of igneous rocks may contain larger crystals set in a finer-grained matrix, a texture that apparently records some change in physical conditions while the rock was crystallizing.

The mineral compositions and colors of igneous rocks reflect their chemical makeup. The dark-colored mafic rocks contain relatively little silicon, aluminum, and potassium and relatively large amounts of iron, magnesium, and calcium. They consist mostly of pyroxenes, olivines, and calcium-rich plagioclase in varying proportions. The light-colored sialic rocks contain much larger quantities of silicon, aluminum, sodium, and potassium and relatively less iron, magnesium, and calcium. They consist mostly of sodium-rich plagioclase, orthoclase, and quartz along with lesser amounts of mica and amphibole.

Mafic melts are rather fluid, although they contain very little water, whereas sialic melts are typically quite viscous even though they often contain large amounts of water. Those properties greatly affect their manner of eruption from volcanoes and therefore the textures of the volcanic rocks they produce.

Sedimentary rocks consist of (1) various mixtures of clastic sediments, which are fragments of older rocks or soil; and (2) nonclastic sediments, which consist of materials that originate within the sedimentary environment. Description and classification of sedimentary rocks depends largely upon recognition of the original sedimentary materials. Study of these materials usually leads to reconstruction of the original sedimentary environment.

A complex of diagenetic processes operates within sediments after deposition to lithify them into solid rocks, and after lithification to further transform the rock. These processes include simple compaction, which usually causes expulsion of water; cementation; recrystallization and dolomitization; and many others.

At temperatures near 200°C, diagenesis becomes metamorphism, which further transforms the mineral compositions and textures of both sedimentary and igneous rocks. Progressive mineral reactions occur with increasing temperature, making it possible to infer the temperature—and to some extent the pressure—of metamorphism through study of mineral assemblages. The metamorphic grade of a rock depends upon its mineral composition, which reflects the temperature and pressure under which it recrystallized.

Most metamorphic rocks recrystallize under directed stress, which orients their mineral grains thus giving them a foliated texture. The textures of metamorphic rocks vary according to their mineral compositions and degree of foliation. Naming them involves listing their principal mineral constituents and appending an appropriately descriptive textural term.

At temperatures near 750°C—the exact value varying with both pressure and water content—metamorphic rocks begin to melt to generate sialic magmas.

KEY TERMS

andesite
basalt
batholith
blueschist
calcareous mud
clastic sediment
coal
compaction
conglomerate
contact metamorphic rock
cross bed
diagenesis
dike
dolomite
eclogite
evaporite
foliation
fossil
gabbro
gneiss
granite
igneous rock
lava
layering
limestone
lineation
lithification
mafic rock
magma
marble
metamorphic rock

migmatite
mudcrack
mudstone
nonclastic sediment
obsidian
peat
pegmatite
peridotite
petrology
phenocryst
phyllite
pluton
porphyritic
quartzite
recrystallization
rhyolite
ripple mark
sandstone
schist
sedimentary rock
shale
sialic rock
sill
siltstone
slate
slaty cleavage
stratification
structure
texture
volcanic ash
volcanic cinder
welded ash

REVIEW QUESTIONS

1. It is possible to find examples of granite, gneiss, and sandstone all composed of quartz, feldspar, and mica in similar proportions. What characteristics would distinguish rocks so similar in composition but so different in origin?

2. Most black igneous rocks owe their color to a large content of augite. Explain why you would or would not expect to find quartz or olivine in the rock.

3. What factors control the mineral compositions of metamorphic rocks?

4. How do clastic, nonclastic, and diagenetic minerals in sedimentary rocks differ in origin?

5. What characteristics of color and texture distinguish between mafic and sialic or plutonic and volcanic igneous rocks?

6. What factors seem to determine the grain size of igneous rocks?

SUGGESTED READINGS

Blatt, H., G. Middleton, and R. Murray. *Origin of Sedimentary Rocks.* Englewood Cliffs, N.J.: Prentice-Hall, 1972.
An excellent general text on the sedimentary rocks.

Hyndman, D. W. *Petrology of Igneous and Metamorphic Rocks.* New York: McGraw-Hill Book Co., 1972.
A definitive general text and reference on the igneous and metamorphic rocks.

Jackson, K. C. *Textbook of Lithology.* New York: McGraw-Hill Book Co., 1970.
A good introductory textbook on petrology.

Winkler, H. G. F. *Petrogenesis of Metamorphic Rocks.* 3d ed. New York: Springer-Verlag, 1974.
This standard text closely relates the results of experimental investigations and field observations.

CHAPTER FOUR

A spectacular angular unconformity in the Grand Canyon. Such exposures stimulated early geologists to begin thinking of an earth with a long and eventful history. (N. Carkhuff, U.S. Geological Survey)

GEOLOGIC TIME

The appreciation and finally the accurate measurement of the vast length of geologic time rank among the greatest and most disturbing intellectual achievements of Western culture. The new concept of geologic time span destroyed the old view of a relatively young and unchanging earth and replaced it with an endless historical perspective full of change. Within this perspective human presence proportionately dwindles to an event of the last fleeting moments. We find ourselves removed from the center of time, perched instead on the endpoint of a long historical line that recedes into a past too remote to imagine. And that presumably will extend into an equally endless and eventful future.

Two centuries ago most scientists agreed that the earth still existed essentially as it had been created only a few thousand years before, and had little history beyond that known to human experience. In those days scientists studied the earth mainly to amplify the biblical account of its creation and early history. During the nineteenth century the new science of geology demolished that approach by demonstrating that rocks carry the record of a long series of events that must have consumed vast stretches of time. Nineteenth-century geologists envisioned these spans, which were finally measured by their twentieth-century counterparts. During the first half of the nineteenth century, geologists worked out a general history of the earth in terms of a series of historical periods—the roll call of names shown in Figure 4–1—which is still in standard use today. Development of that geologic time scale depended upon systematically applying a set of basic principles, all of which seem obvious now even though several were nothing less than heresy when first articulated.

THE BASIC PRINCIPLES

Perhaps the most fundamental principle is that of **cross-cutting relationships**—the idea that an object must exist before anything can happen to it. We apply this concept automatically when, for example, we look at graffiti and recognize the newer inscriptions because they are scrawled across those written earlier. Similarly, each successive geologic event overprints a previous record in the rocks or landscape. Various geologic applications of this principle are illustrated in Figures 4–2, 4–3, and 4–4. Figure 4–5 shows how methodically applying the principle of cross-cutting relationships can sort a complex geologic situation into an orderly sequence of simple logical steps.

The **principle of superposition,** first clearly stated in 1669 by Nicolaus Steno, is basically an application of the principle of cross-cutting relationships. The principle of superposition holds that sedimentary layers become progressively younger from bottom to top in an undisturbed se-

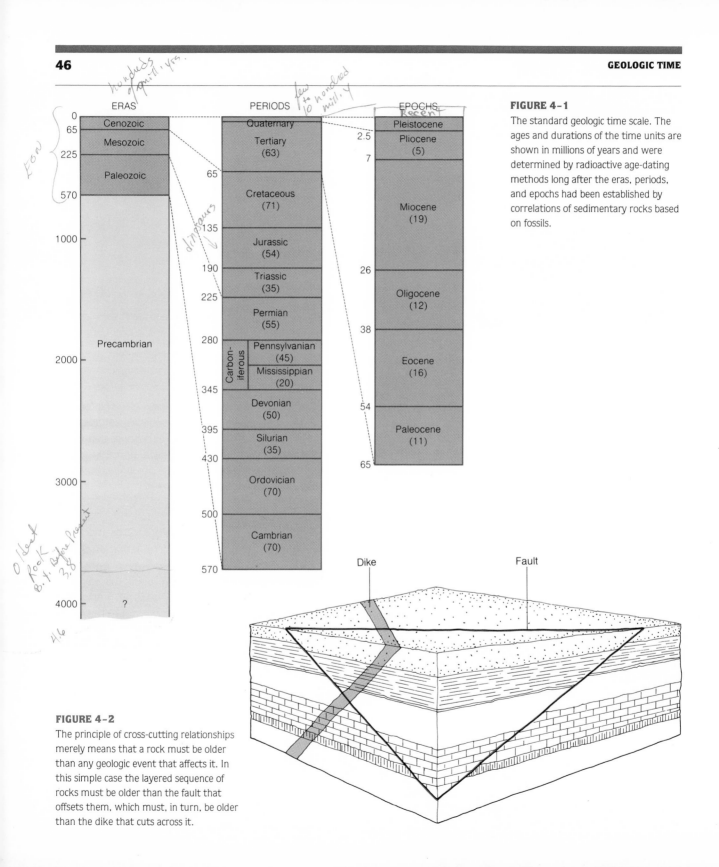

ERAS

0	
65	Cenozoic
225	Mesozoic
570	Paleozoic
1000	
2000	Precambrian
3000	
4000	?

PERIODS

	Quaternary
2.5	Tertiary (63)
65	Cretaceous (71)
135	Jurassic (54)
190	Triassic (35)
225	Permian (55)
280	Carboniferous — Pennsylvanian (45) / Mississippian (20)
345	Devonian (50)
395	Silurian (35)
430	Ordovician (70)
500	Cambrian (70)
570	

EPOCHS

	Recent
	Pleistocene
2.5	Pliocene (5)
7	Miocene (19)
26	Oligocene (12)
38	Eocene (16)
54	Paleocene (11)
65	

FIGURE 4–1

The standard geologic time scale. The ages and durations of the time units are shown in millions of years and were determined by radioactive age-dating methods long after the eras, periods, and epochs had been established by correlations of sedimentary rocks based on fossils.

Dike

Fault

FIGURE 4–2

The principle of cross-cutting relationships merely means that a rock must be older than any geologic event that affects it. In this simple case the layered sequence of rocks must be older than the fault that offsets them, which must, in turn, be older than the dike that cuts across it.

FIGURE 4-3
A dike of pale-gray granite cutting across dark gneiss at Caulfield Cove, British Columbia. The dike is full of angular fragments of gneiss that broke loose and floated into the liquid granite magma. The cross-cutting relationship and the floating fragments of gneiss both show that the granite is the younger rock. (D. W. Hyndman)

FIGURE 4-4
A cinder cone volcano perched on the rim of the Grand Canyon. The black basalt lava flow that poured down the canyon wall to the river shows that the eruption happened after the river had eroded the canyon almost to its present depth. (J. R. Balsley, U.S. Geological Survey)

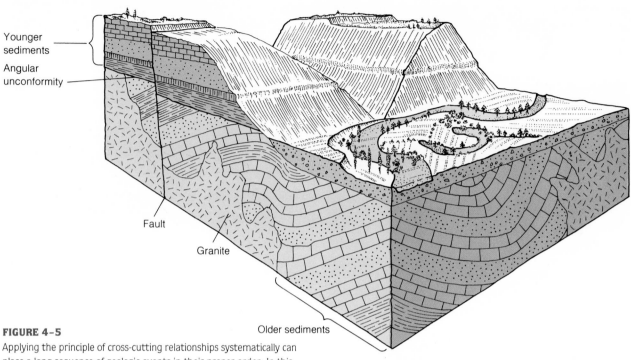

Younger sediments

Angular unconformity

Fault

Granite

Older sediments

FIGURE 4-5

Applying the principle of cross-cutting relationships systematically can place a long sequence of geologic events in their proper order. In this case the first event was deposition of the lower sequence of sediments, which were then folded and intruded by two large masses of granite. After a flat erosion surface had formed, a younger sequence of sediments accumulated, converting the erosion surface into an angular unconformity. Then the modern erosion surface formed as streams cut their valleys through the upper sequence of sediments and into the lower one with its granite intrusions. Finally, a fault offset both sedimentary sequences and the modern erosion surface.

quence, that one layer of sediments must exist before another can accumulate above it. This eminently useful principle seems almost absurdly elementary. Consider, however, that its formulation had to await the realization that sedimentary rocks are indeed lithified sediments, a relationship that did not seem at all obvious several centuries ago. The understanding that sedimentary rocks contain a record of geologic events that can be read in order from the bottom to the top of a sequence of layers as though they were pages in a book was one of the first major steps toward a rational view of the earth.

In utilizing the principle of superposition it is important to carefully stipulate that the sequence of strata be *undisturbed*. Movements of the earth's crust commonly tilt and in some cases actually overturn sedimentary layers. Therefore, geologists who work with severely deformed rocks must examine them carefully to determine which way is up; they thus become skilled observers of the subtle differences distinguishing the tops and bottoms of sedimentary layers.

The **principle of original horizontality,** another of Steno's contributions, is also an application of the principle of cross-cutting relationships. The original horizontality principle states that beds of sedimentary rock form from layers of sediment that accumulated in an essentially horizontal position. This insight also had to wait for the understanding that sedimentary rocks were originally common sediments, such as sand or mud. It was a major step toward rational analysis of rocks because it implies that any tilting of sedimentary strata must record movements of the earth's crust that occurred after their deposition (Figure 4–6). The principle of original horizontality must be applied with the realization that many sedimentary layers accumulate with an *initial dip,* which in some cases departs considerably from actual horizontality.

The principles of original horizontality and cross-cutting relationships make it possible to interpret situations such as those illustrated in Figures 4–7 and 4–8. Each of these figures show a younger sequence of sedimentary layers lying on an erosion surface planed across the tilted beds of an older sequence. This relationship is known as **angular unconformity.** The early geologists who first recognized angular unconformities were impressed by the phenomenon. The wreckage of an earlier world buried beneath that of a later period does convey a powerful sense of history. Years later, geologists became aware that many unconformities are not angular, that a buried erosion surface (Figure 4–9) representing an enormous passage of time may be sandwiched between two sequences of sedimentary layers with no sign of angular discordance. Indeed, the unconformities within most rock sequences account for more time than do the layers of rock.

James Hutton, a gentleman farmer of Scotland, arrived at one of the most profound insights of modern science almost two centuries ago. Hutton recognized that familiar processes of erosion could account for the shaping of the landscape if they were continued over a sufficiently long period of time. That is the principle of **uniformitarianism,** which holds that such natural elements as rivers, waves, wind, and ice operate now just as they did in the past. Uniformitarianism is often more gracefully expressed in the phrase, ''the present is the key to the past.'' The concept revolutionized scientific thought. It made the earth a much more orderly and understandable place than it had been when people attributed the environment to original creation modified by a few later catastrophes, such as the bibli-

FIGURE 4-6

These layers of sandstone and shale must have been nearly horizontal when they accumulated. Later crustal movements tilted them into their present orientation. The area shown is about 4 m across.

cal flood. The practical aspects of uniformitarianism are most vividly illustrated by those geologists who interpret ancient sedimentary rocks by comparing them to modern sediments. In this way analogies may be drawn between past and present sedimentary environments (Figure 4–10).

It must be true that inorganic chemical and physical processes operate now as they did in the past, on the earth as elsewhere in the universe. However, that does not necessarily imply that the earth's physical and chemical circumstances have always resembled those that prevail today. On the contrary, it is perfectly reasonable to suppose

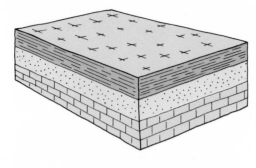

STEP 1:
Deposition of first
sedimentary sequence.

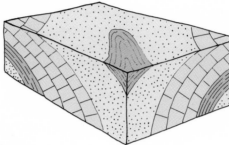

STEP 2:
Folding of first
sedimentary of sequence
and development of an
erosion surface.

STEP 3:
Deposition of second
sedimentary sequence
converting the old
erosion surface, now
buried, into an
angular unconformity.

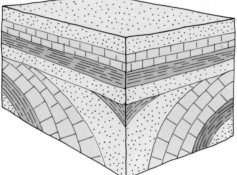

FIGURE 4–7
Stages in the development of an angular unconformity
recording deposition of an older sedimentary
sequence, a period of deformation, development of an
erosion surface, and deposition of a second
sedimentary sequence.

FIGURE 4–8
A thin bed of cobbles marks the erosion
surface that separates the tilted layers of
rock below from the horizontal layers
above.

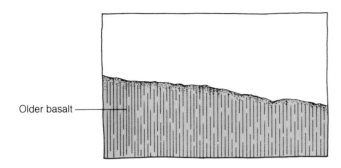

Older basalt

STEP 1: Eruption of
first basalt flow.

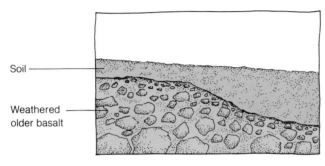

Soil

Weathered
older basalt

STEP 2: Weathering of
first basalt and
development of soil
horizons.

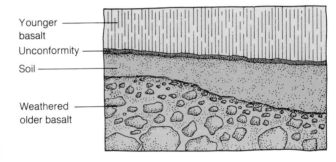

Younger
basalt

Unconformity

Soil

Weathered
older basalt

STEP 3: Eruption of
younger basalt flow

FIGURE 4-9

An unconformity. The basalt lava flow in the upper part of the photograph lies on an old soil that grades downward into weathered basalt. The soil is about 1 m thick.

(a)

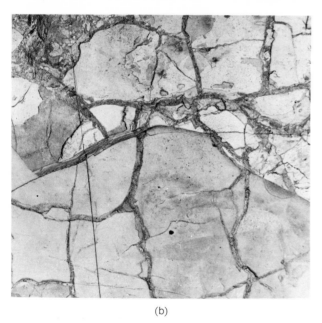

(b)

FIGURE 4-10

(a) A pattern of shrinkage fractures in sun-dried mud exposed on a modern tidal flat on the coast of Louisiana. (b) A pattern of shrinkage fractures in mudstone about 1 billion years old exposed in the northern Rocky Mountains.

that surface and internal temperatures, the composition of the atmosphere, climates, and many other factors were quite different during much of the geologic past. Furthermore, living plants and animals play a major role in many processes of erosion and deposition, and there is every reason to assume that species operated in different ways at different times as they evolved into different forms. The principle of uniformitarianism must be applied with discretion.

THE GEOLOGIC TIME SCALE

Intelligent application of the basic principles of superposition, original horizontality, uniformitarianism, and cross-cutting relationships makes it possible to decipher much of the historical record contained in rocks. Early geologists were quite able to read the geologic archives they found in rock outcrops. However, they had no way to relate the record they found in one exposure to that revealed in another. They were in the position of a person who discovers loose manuscript sheets without page numbers strewn about—it is virtually impossible to arrange these into a coherent story. During the 1790s, geologists in both England and France discovered how to number the pages of the geologic narrative by matching the fossils in the rocks.

They found that certain assemblages of fossils always occur together and that different assemblages always occur in the same vertical sequence, an observation known as the **law of faunal succession.** Evidently, the stack of sedimentary rock layers contains the fossil remains of a long succession of different populations of plants and animals that inhabited the earth during various periods of the past. Figure 4–11 illustrates how geologists use fossils to correlate layers of different kinds of rocks that formed in different places during the same period of time. Geologists specializing in the study of fossils are called **paleontologists.** These experts can generally assign an age to a sedimentary formation simply by examining a few fossils, such as those shown in Figure 4–12.

The consistent assemblages of fossils predictably stacked in the same vertical order was first interpreted as the record of a long series of special creations preceding the one described in Genesis. Because successive fossil assemblages typically differ considerably and because

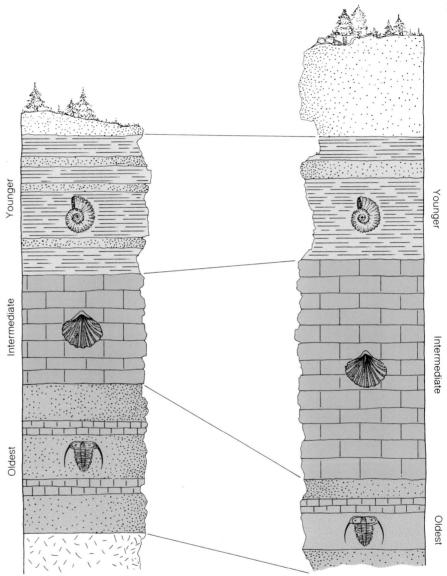

FIGURE 4–11
Matching fossils contained in layers of sedimentary rock exposed in widely separated outcrops makes it possible to identify those deposited during the same time period even though the rocks may be quite different.

they consist almost entirely of extinct species, early observers concluded that each population had been annihilated before the succeeding one was created. This view was widely applauded during the early nineteenth century. It was seen as evidence of close accord between science and religion, which incidentally furnished an object lesson in the fate of populations that failed to mend their sinful ways. The extinction-then-creation concept was also regarded as a major scientific advance that provided a key to the secrets of the earth's past. If God had indeed created and destroyed a series of distinct populations, then the story of the earth's history could be told by finding the fos-

FIGURE 4-12

These fossil clams identify the dark gray sandstone in which they are embedded as a Tertiary formation deposited in seawater. (D. W. Hyndman)

sil remains of each and arranging them all in their proper sequential order.

Early nineteenth-century geologists eagerly searched for sections of sedimentary rocks containing distinctive assemblages of fossils. They painstakingly described and named each one, convinced that they were identifying the wreckage of a series of past periods in which the Creator had populated the earth with living creatures, let them live for awhile, and then annihilated them in a universal catastrophe. So persuaded, the early geologists established the series of named periods that still survives as the geologic time scale shown in Figure 4-1.

Previously unknown assemblages of fossils were becoming increasingly difficult to find by the middle of the

nineteenth century. So it seemed that the great work of defining the periods of the earth's past was nearing completion. Unpleasant problems, however, had emerged. Certain sections of rock, for example, contained fossils that seemed to bridge the boundaries between geologic periods. That was theoretically intolerable—the geologic periods were then regarded as natural subdivisions of time that began in a general creation and ended in a universal catastrophe.

Evolution and the Geologic Time Scale

Charles Darwin published his theory of evolution in 1859, when the climate of scientific opinion was thoroughly ripe for his ideas. Geologists had helped prepare both Darwin and the scientific community for the concept of evolution by showing that geologic time was sufficiently long to permit slow processes to work great changes. Darwin's new theory explained differing fossil faunas as the result of slow evolutionary change rather than of a series of special creations and sweeping annihilations. Most geologists accepted the theory of evolution almost immediately, realizing as they did so that it deprived the geologic time scale of its original theoretical base. Evolution left no reason to suppose that the earth's history was naturally divisible into a series of periods bounded by real events.

However, old ideas persist even in the face of adverse evidence and superior reasoning. Most geologists, professed evolutionists though they were, could not abandon the concept that the geologic periods correspond to natural subdivisions of earth history. So they found another way to reinforce their former approach: They focused on a new kind of catastrophic event to mark the period boundaries— mountain building.

The diehard geologists argued that the earth is subject to periodic episodes of mountain building that raise the continents and drain the shallow seas away from their margins. These events would greatly restrict the shallow marine environment, presumably forcing rapid adaptive evolution, which would be reflected in widespread extinction of old faunas and appearance of new ones. Draining the low-lying areas of the continents would also expose them to erosion and interrupt deposition of marine sediments. Thus, the continuity of the sedimentary record would be broken with an unconformity. When the mountain-building "convulsion" ended and the continents subsided, shallow seas would again flood their lower regions. These floods would bring greatly evolved assemblages of animals and would deposit a new sequence of layered sedimentary rocks. That concept transformed the study of earth history into a matter of relating deposition of sedimentary rocks to episodes of mountain building. That approach continued with increasing futility until it finally collapsed during the 1950s under the accumulated weight of massive evidence to the contrary. By the 1950s study of sedimentary rocks had convincingly shown that period boundaries do *not* correspond to global unconformities. It had become clear that mountain building occurs constantly in one region or another and does not involve planetary convulsions.

Today we realize that the geologic periods are not, in general, natural time units but, rather, artificial subdivisions that must be arbitrarily separated. Period boundaries are now defined by the appearance or disappearance of certain marker fossils selected for that purpose by international commissions of geologists. The end of the Cretaceous period, for example, is defined as having occurred when the dinosaurs disappeared. Geologists working with sections of rock crossing that boundary often use fossils to locate it. They assign the layers containing dinosaur bones to the Cretaceous period and those above them, the ones that contain only mammal bones, to the succeeding Tertiary period. In many areas, the rocks are otherwise indistinguishable.

Therefore, the original geologic time scale survives as a set of arbitrarily and pragmatically defined periods tied to the evolution of living species rather than to physical events. Nevertheless, it remains a basic intellectual tool of geology: It provides a framework for quickly assigning an age to almost any fossil-bearing rock. Now that the ages of the periods are known in terms of years, fossils are more useful than ever for geologic dating. Nevertheless, a number of serious difficulties remain, and it would be a mistake to suppose that all the problems of deciphering the earth's past are now solved.

Fossils, and entire faunas do indeed appear and disappear abruptly, making it easy to understand why the early geologists concluded that they were studying the record of a series of creations and annihilations. A good example of a continuous evolutionary series of fossils is very difficult to find. That is due in part to local incompleteness of the geologic record, which almost invariably contains unconformities marking times when evolution was proceeding elsewhere. It also seems that new species tend to develop in small and isolated populations—where evolution may proceed rapidly—and then to spread by migration. Furthermore, many geologists now suspect that at least *some* of the abrupt changes in fossil faunas may actually record catastrophic annihilations.

Something extremely destructive to life in general evidently happened at the end of Cretaceous period. At that time, the dinosaurs and several other large, quite unrelated, and apparently prospering groups of animals vanished within a brief time, perhaps within a few years. Numerous theories have been proposed to explain what happened. These range from general loss of reproductive urges through drastic collapse of the ecosystem to a sudden flux of radiation coming from the sun or from a supernova explosion of a nearby star. Chemical analyses have recently shown that very thin zones of rock deposited at the time of the extinction contain abnormally large concentrations of platinum group metals. This observation is interpreted by some geologists to be evidence that the earth had been struck by a large meteorite. Conceivably, clouds of dust ejected from the explosion crater might have filled the sky for several years, obscuring the sun and halting photosynthesis. If so, then nothing would have survived except seeds and spores, and the plants and animals that could live on them or on decaying organic matter. The geologic record contains several other examples of apparent mass extinctions although none are quite so drastic as that which happened at the end of Cretaceous time.

The Precambrian

For reasons yet unexplained to the satisfaction of the scientific community, abundant and varied animal fossils first appear in rocks deposited at the beginning of Cambrian time, about 570 million years ago. Rocks a few tens of millions of years older locally contain sparse faunas of enigmatic animal imprints, and some geologists favor creating a new period—the Eocambrian—to accommodate them. The apparently complete absence of animal fossils from older rocks along with the virtual lack of evolutionary change in the primitive plant fossils they do contain makes it impossible to extend the geologic time scale into that remoter past. Therefore, the vast gulf of time that separates formation of the oldest known terrestrial rocks and the first appearance of abundant animal fossils is simply called the Precambrian. The Precambrian comprises some 85 percent of geologic time as recorded in the rocks. Despite the absence of animal fossils, it is possible to recognize certain distinctive sedimentary rock types formed during several long segments of Precambrian time. They may record stages in the evolution of the earth's atmosphere. We will examine that topic in Chapter 12.

CLOCK TIME

The geologic time scale arranges rocks and geologic events in their proper sequential order; it reveals nothing of their actual ages in terms of years. That lack combined with understanding of the principle of uniformitarianism—which immediately implied the passage of extremely long periods of time—led directly to efforts to measure the actual length of geologic time. Much of the history of geology from the 1790s to the 1950s could be written as an account of a science in search of a clock. The story is one of a long series of discouraging but helpful failures finally ending with the mid-twentieth-century success of radioactive age dating.

All clocks of whatever nature depend upon some process that operates at a known and constant rate and produces cumulative results that can be measured. Sand trickles through the orifice of an hourglass at a constant rate, producing a pile that reflects the passage of time as it grows (Figure 4–13). An electric clock consists essentially of a motor that turns the hands as it responds to the regular pulses of the alternating current. The fundamental problem in measuring geologic time concerns finding a natural process that provides the basis for a clock.

Various nineteenth-century geologists attempted to use the rates of erosion and sedimentary deposition, neither of which are constant, to measure geologic time. Although those attempts failed to produce consistent or useable re-

sults, they nevertheless convinced geologists that the earth must be hundreds of millions of years old. Lord Kelvin, a great physicist of the late nineteenth century, tried to measure the age of the earth by assuming that it had begun its career as a molten ball and had since cooled to its present temperature by radiating heat into space. Lengthy computation yielded a result close to 20 million years, a figure that could be stretched to as much as 40 million years by judiciously juggling the data. Even the longer figure did not impress the geologists of the time, who promptly rejected Lord Kelvin's results as being much too far below their own more intuitive notions of the earth's age. There ensued a lengthy, and at times unseemly, dispute between geologists and physicists, a completely futile argument in which neither side clearly understood what the other was saying. Neither geology nor physics yet knew of the existence of radioactivity, which provides the earth with its own internal source of heat thereby making it impossible to use the earth's temperature as a direct measure of its age. It now appears that the earth is close to thermal equilibrium, losing heat through its surface at about the same rate that radioactive decay releases heat inside.

Nothing seems to affect the rate of radioactive decay. Unstable atomic nuclei come apart in their own time regardless of extremes of temperature, pressure, and other physical circumstances. Each kind of radioactive atom has its own characteristic **half-life,** the period during which half of all the existing atoms of that kind decay (Figure 4–14). There is no reason to suspect that the half-lives of the radioactive isotopes have changed with time and every reason to believe that they have remained constant. Furthermore, radioactive decay transforms unstable atoms of the parent element into stable atoms of one or more daughter elements, causing their proportion to change progressively with time (Figure 4–15). Therefore, radioactivity has all the characteristics of a natural clock—it is a process that continues at the same rate and produces cumulative results.

The theoretical possibility of basing geologic clocks on radioactive decay was clearly understood as early as 1905. However, analytical instruments capable of fully converting the vision into reality did not appear until the 1930s. These instruments became widely available during

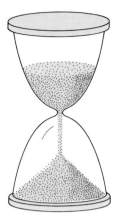

FIGURE 4–13

A clock consists basically of a process operating at a known rate, producing a cumulative result that can be measured. In the case of the hourglass, sand trickles from top to bottom at a constant rate, and the changing proportion of the amount of sand in the top of the glass to that in the bottom measures the passage of time.

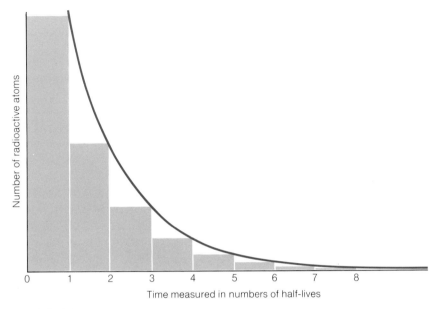

FIGURE 4–14

During each half-life, exactly one-half of the total number of a particular kind of radioactive atom decays. The length of a half-life depends upon the kind of atom. Length can range from a minute fraction of a second for some extremely unstable atoms manufactured in reactors, to hundreds of millions of years for most of those atoms used in radioactive age dating of rocks. Whatever the half-life, the total inventory of the radioactive atoms declines according to an exponential curve such as that shown here.

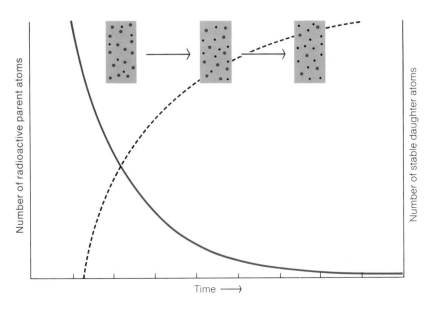

FIGURE 4–15

As time passes, the number of radioactive parent atoms diminishes and that of stable daughter atoms increases correspondingly. Therefore, the proportion of radioactive parent to stable daughter atoms changes progressively with time and can be used as a clock.

the 1950s when they enabled geologists to gain a firm grasp on the actual ages of rocks and therefore of geologic events.

Radiocarbon Dating

The first routinely successful radioactive age-dating method was developed during the mid-1930s. It depends upon the decay of carbon-14, which forms in the upper atmosphere through cosmic bombardment of nitrogen-14 and then reacts with oxygen to make carbon dioxide. If the rate of cosmic bombardment remains constant, so should the proportion of carbon-14 in atmospheric carbon dioxide.

Plants use carbon dioxide to generate their tissues and animals eat plants so both contain the same proportion of radioactive carbon-14 as the atmosphere. When a plant or animal dies, it no longer absorbs new carbon and the unstable carbon-14 decays back to gaseous nitrogen-14, which escapes. Therefore, it is possible to determine the ages of many specimens of wood, bone, or other organic matter by measuring their remaining content of carbon-14. Unfortunately, the method has limited geologic use because carbon-14 decays so rapidly that virtually none remains after about 40,000 years. Very few rocks are that young.

Uranium Dating

Uranium-235 and uranium-238 both decay radioactively to form lead, as does thorium-232. Fortunately, they form lead atoms with different atomic weights so it is possible to measure separately those created by decay of each kind of radioactive parent atom. It is also fortunate that uranium and thorium are chemically similar and tend to occur together so minerals that contain uranium also contain thorium. Each crystal of such a mineral contains three radioactive clocks running independently and simultaneously: two based on the decay of uranium to lead, and one based on the decay of thorium (Figure 4–16). The ratios of any pair of parent and daughter atoms tell the age of the mineral.

It is wildly improbable that three independent clocks could all be wrong and nevertheless show the same time. Therefore, if all three radioactive clocks in a uranium-thorium mineral give the same age, as they quite often do, then it seems that the age must be correct beyond any reasonable doubt. Such **concordant ages** obtained through agreement of independent radioactive clocks provide geologists with a firm basis for assigning ages in terms of years to rocks and therefore to the geologic periods in which they formed. Unfortunately, uranium minerals are rare and tend to be secondary and therefore younger than the rock that contains them. Radioactive age-dating techniques based on more abundant elements that occur in a wider variety of minerals and rocks offer more promise.

Potassium-Argon Dating

Potassium is among the most abundant elements and occurs in a wide variety of minerals and most kinds of rocks. A small percentage of all potassium consists of the radioactive isotope potassium-40, which decays very slowly to produce calcium-40 and argon-40. The minute amount of calcium-40 is lost in the vastly greater amount already present as an impurity in most minerals, so it is useless for radioactive dating. But the argon-40 is easy to identify because it is an inert gas that is not known to form chemical bonds and should therefore get into minerals only through decay of potassium-40. Atoms of argon gas leak out of most minerals but a few, most notably the micas, have crystal structures that trap argon atoms within them in the same way that a birdcage holds canaries. The proportion of argon-40 to potassium-40 increases steadily with time in such minerals and makes an excellent clock (Figure 4–17).

Rubidium-Strontium Dating

Rubidium is nowhere abundant but does occur in small quantities in many minerals in which it substitutes for potassium. A small percentage consists of the radioactive isotope rubidium-87, which decays very slowly to strontium-87. Almost all potassium minerals contain enough rubidium to make them useful for radioactive dat-

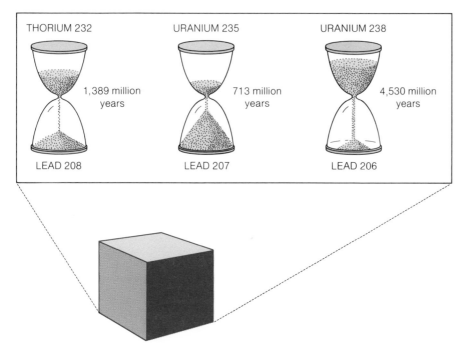

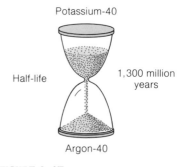

FIGURE 4–17

The decay of weakly radioactive potassium-40 into stable argon-40 makes an excellent geologic clock to measure the length of time elapsed since a potassium mineral cooled to a temperature below about 300°C.

FIGURE 4–16

The two kinds of radioactive uranium atoms usually occur together with radioactive thorium because uranium and thorium are chemically similar. Therefore, minerals that contain uranium and thorium have three radioactive clocks running within them quite independently and at different rates. Age agreement among all three clocks provides convincing evidence because it is difficult to imagine how each could give the same time unless all three are correct.

ing so the ratio of rubidium-87 to strontium-87 can be used to measure the ages of a wide variety of rocks. The method is especially useful because strontium, unlike argon, does not easily escape from minerals or rocks.

A Time Perspective

The late nineteenth-century geologists must have thought themselves dashing fellows indeed when they ventured to roundly speculate that the earth might be as much as several hundred million years old. Now, many dating laboratories routinely obtain ages greater than 3 billion years old on certain Precambrian rocks. Some of the samples that the astronauts collected on the moon are slightly more

than 4 billion years old and many meteorites yield ages between 4 and 5 billion years.

Such vast stretches of time are far beyond the grasp of the human imagination. We just manipulate the numbers without really comprehending their meaning (Figure 4–18). Consider, for example, that there are approximately 31.5 million seconds in a year. If each of those seconds were expanded to a year, that would account for enough time to take us back to the middle of the Tertiary period, but only half enough time to take us back to the end of the Cretaceous period when the dinosaurs vanished. Such span of time is incomprehensible by ordinary human reckoning but not that vast in geologic terms. Many Precambrian rocks are more than 3 billion years old—more years than there are seconds in a century.

RELATIVE AND ATOMIC GEOLOGIC TIME

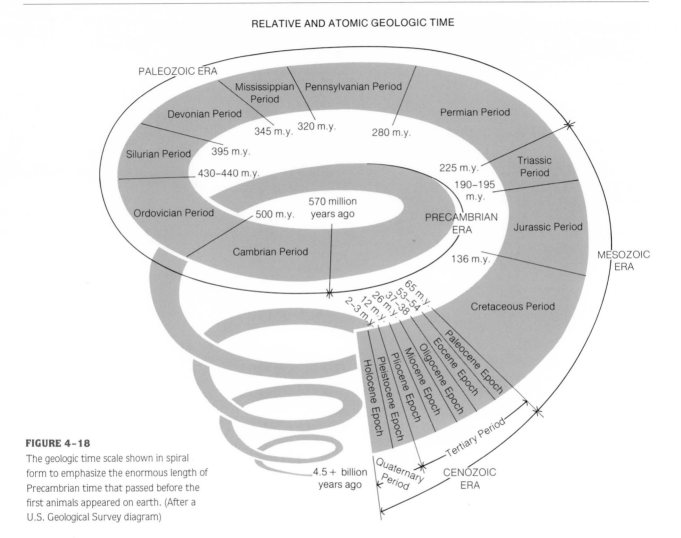

FIGURE 4–18

The geologic time scale shown in spiral form to emphasize the enormous length of Precambrian time that passed before the first animals appeared on earth. (After a U.S. Geological Survey diagram)

SUMMARY

The interpretation of geologic history relies primarily upon the principle that any event that changed some feature of a rock or landscape must have occurred after that feature already existed. Systematic application of this principle makes it possible to analyze the sequence of geologic events recorded in any particular exposure. Comparison of fossils makes it possible to relate the events recorded in one place to those recorded in another. Early geologists established a geologic time scale that divides the rocks formed during the last 570 million years into a series of named periods. These geologic periods are now arbitrarily defined on the basis of the appearance or disappearance of certain selected fossils.

Radioactive age dates, unlike the geologic time scale,

measure the ages of rocks and therefore of the geologic events that formed them in terms of years. Such dates are based on the decay of one radioactive isotope to another, a process that occurs at a constant rate and produces cumulative results and therefore serves as a natural clock.

Measurement of radioactive ages has shown that some Precambrian rocks are well over 3 billion years old and that most meteorites formed between 4 and 5 billion years ago, an age that presumably corresponds to that of the earth.

KEY TERMS

angular unconformity
concordant age
cross-cutting relationships
half-life
initial dip
law of faunal succession
paleontologists
Precambrian
principle of original horizontality
principle of superposition
uniformitarianism

REVIEW QUESTIONS

1. Why would you expect the concept of uniformitarianism to apply more generally to the study of igneous rocks than to that of sedimentary rocks?

2. If you were to separate the clastic, nonclastic, and diagenetic minerals from a sedimentary rock and then run separate age dates on each fraction, what might the results tell you?

3. Suppose you were to collect a sample from a thin layer of basalt sandwiched within layers of Cretaceous sedimentary rocks and send it to a laboratory for age dating. What kind of result would suggest that the basalt might be a lava flow? What kind would be evidence that it is a sill?

4. Suppose you were to find granite intruding sedimentary rocks containing early Cretaceous fossils and covered unconformably by other sedimentary rocks that contain early Tertiary fossils. What could you infer about the age of the granite? About the buried erosion surface in the unconformity?

5. Is it logical to expect to find an unconformity separating sedimentary rocks deposited during different geologic periods?

SUGGESTED READINGS

Albritton, C. C. Jr., ed. *Philosophy of Geohistory: 1785–1970.* Stroudsburg, Penn.: Dowden, Hutchinson & Ross, 1975.

A collection of original articles, each of which played an important role in the development of our concepts of geologic time.

Berry, W. B. N. *Growth of a Prehistoric Time Scale.* San Francisco: W. H. Freeman & Co., 1968.

A short book devoted to the development of the geologic time scale.

Dalrymple, G. B. and M. A. Lanphere. *Potassium-Argon Dating.* San Francisco: W. H. Freeman & Co., 1969.

A complete account of the method and its applications by two of the leading practitioners. Adequately technical but still fairly readable.

Eicher, D. L. *Geologic Time.* Englewood Cliffs, N.J.: Prentice-Hall, 1968.

An outstanding short book dealing with the measurement of geologic time.

Faul, H. *Ages of Rocks, Planets and Stars.* New York: McGraw-Hill Book Co., 1966.

A brief, fairly nontechnical authoritative review of absolute age dating. Easy to read.

Harbaugh, J. W. *Stratigraphy and the Geologic Time Scale.* Dubuque, Iowa: Wm. C. Brown Co., 1974.

A brief textbook on the geology of layered rocks. Excellent.

LaPorte, L. F. *Ancient Environments.* Englewood Cliffs, N.J.: Prentice-Hall, 1968.

A good discussion of the problems involved in interpreting sedimentary rocks.

Matthews, R. K. *Dynamic Stratigraphy.* Englewood Cliffs, N.J.: Prentice-Hall, 1974.

A superb textbook treatment of the geology of layered sedimentary rocks.

CHAPTER FIVE

Spheroidal weathering in diabase. Shells of
partially weathered rock peel away from
the fresher cores like the layers of an onion.

WEATHERING AND SOILS

Enduring though they seem, rocks break down into soil under the attack of the elements through a complex interplay of mechanical and chemical processes known collectively as **weathering.** Geologists customarily divide the processes of weathering into two large groups: (1) **mechanical weathering,** which breaks rocks without altering them chemically; and (2) **chemical weathering,** which destroys old minerals and creates new ones. Figure 5–1 shows a major consequence of mechanical weathering: the exposure of new rock surfaces to the processes of chemical weathering, some of which also break rocks mechanically.

MECHANICAL WEATHERING

The main activity of mechanical weathering is **frost wedging,** which occurs when water freezes in cracks, expands as ice, and breaks the rock (Figure 5–2). The same process splits frozen water pipes and cracks engine blocks in vehicles lacking adequate antifreeze. Frost wedging operates most actively in wet temperate climates. There, frequent cycles of freezing and thawing rapidly reduce rocks of all sizes to angular rubble (Figure 5–3) and shatter bedrock outcrops into sliding talus slopes. Such slopes are rare in warm regions where frost wedging is not active.

In places where abundant soluble salts exist they may diffuse into the pore spaces of rocks and crystallize there, wedging the mineral grains apart. This process is often called **salt cracking.** It is significant locally in deserts and in some coastal environments but does not generally operate in wet regions where abundant rainfall prevents soluble salts from accumulating. Nevertheless, many people who use salt to melt snow can watch their concrete sidewalks and roads crumble to rubble under the attack of salt cracking.

Intense forest fires commonly spall large flakes from rocks by heating their outer surfaces thereby expanding them and cracking off the cooler rock beneath. Although this process is not important, it has led geologists to speculate that the less intense daily heating of rock surfaces exposed to the sun might have a similar impact over a long period of time. However, numerous experimental attempts to break rocks by subjecting them to repeated cycles of heating and cooling have generally failed. Furthermore, field observation in hot desert regions reveals that the parts of boulders exposed to the sun are typically sounder than those buried in the soil. Evidently, heating and cooling is not an effective process of mechanical weathering despite its logical appeal.

FIGURE 5–1
Mechanical
weathering exposes
new surface to
chemical attack.

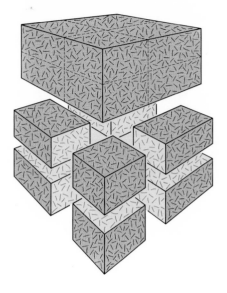

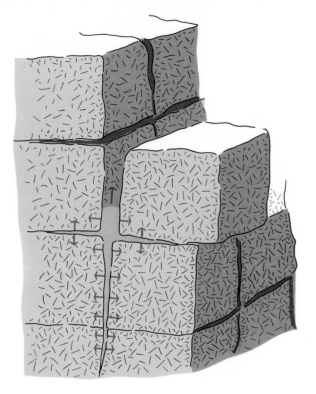

FIGURE 5–2 (right)
Water expands almost 11 percent as it freezes within fractures and
thus wedges blocks of rock apart. In theory, freezing ice can exert a
pressure of about 2,000 kilograms per square centimeter. It rarely
does so, however, because in most cases much of the expansion is into
free space instead of against solid rock.

FIGURE 5–3
A frost-shattered boulder of limestone at the mouth of Sun River Canyon, Montana. It must have
been in one piece when a glacier dropped it here at the end of the last ice age about 10,000 years
ago. The rock is 2 m high.

CHEMICAL WEATHERING

The processes of chemical weathering include a number of reactions that dissolve or otherwise transform the original minerals in a rock, removing some and changing others into soil minerals. All the chemical weathering processes involve reaction of minerals with water and therefore concentrate their attack where water can soak into exposed surfaces or seep into the rock along fractures. The simplest chemical weathering process is **solution,** which attacks all rocks to some extent and becomes an important factor in those composed of the minerals calcite and dolomite.

The carbonate minerals dissolve in acidic solutions. Ordinary rainwater is slightly acidic because it reacts with atmospheric carbon dioxide to form carbonic acid (Figure 5–4), which is familiar in more concentrated form as carbonated water. Rain that falls during thunderstorms also contains small amounts of nitric acid. This acid forms as the water absorbs nitrogen oxides created by lightning flashes that combine atmospheric oxygen and nitrogen. Moreover, rain that falls through polluted air commonly becomes rather strongly acidic. Such rain combines with abnormally high concentrations of carbon dioxide and nitrogen oxides as well as with sulfur dioxide, which forms sulfurous acid when it combines with water. Such acidic rains increasingly attack marble or limestone buildings and works of art, inhibit plant growth, and kill wildlife.

Even without the help of atmospheric pollution, the effects of solution weathering on exposed limestone and marble in a rainy climate become obvious within a few decades. Marble tombstones commonly lose their inscriptions within less than a century in many parts of eastern North America whereas similar monuments remain intact in the more arid climatic regions of the West. We see the same relationship between climate and solution weathering of carbonate rocks expressed on a larger and more vivid scale in the landscape (Figure 5–5). Formations of limestone, dolomite, or marble weather rapidly in wet climates—such as those that prevail in the Appalachian Mountains—where they typically form deep valleys. Similar formations weather very slowly, and typically form bold cliffs and prominent ridges in the much drier climates of the Rocky Mountain region.

FIGURE 5–4

Rainwater absorbs atmospheric carbon dioxide to form carbonic acid, which in turn dissociates to become a solution of bicarbonate ions and acidic hydrogen ions. The dissolved hydrogen ions attack calcite to produce a solution of calcium ions and bicarbonate ions. This sequence of chemical processes is chiefly responsible for the weathering of carbonate rocks. The other acids that form where rain falls through polluted air also liberate hydrogen ions, which attack calcite in precisely the same way.

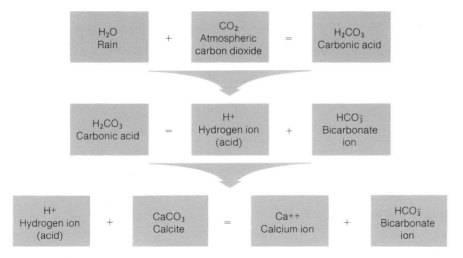

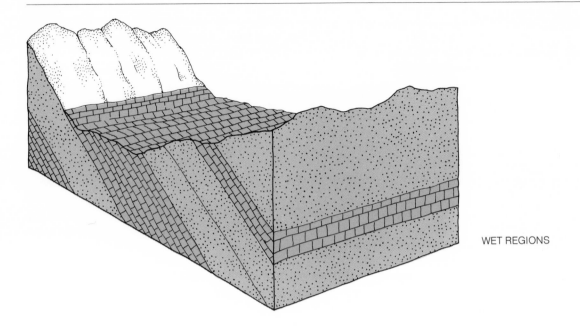

WET REGIONS

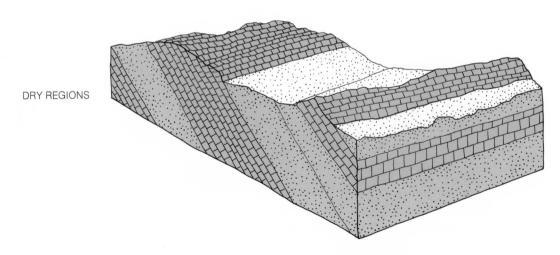

DRY REGIONS

FIGURE 5-5
Carbonate rocks weather rapidly in wet regions where they form valleys and slowly in dry regions where they form hills.

$$FeS_2 \text{ Pyrite} + O_2 \text{ Atmospheric oxygen} + H_2O \text{ Water} = SO_4^- \text{ Sulfate ion} + H^+ \text{ Hydrogen ion} + Fe_2O_3 \cdot nH_2O \text{ Goethite}$$

FIGURE 5-6

A typical oxidation reaction. Atmospheric oxygen dissolves in water and attacks pyrite to form dissolved sulfate, hydrogen ions, and the iron oxide mineral goethite. The hydrogen ion will attack calcite or other carbonate minerals if it encounters them, and the sulfate may react with dissolved calcium ions and water to form gypsum. The reaction shown here is not balanced because goethite contains variable amounts of water.

We tend to think of quartz as being among the most insoluble and chemically inert minerals, immune to almost every weathering process. However, quartz becomes significantly soluble in weak alkaline solutions and water does become slightly alkaline as it reacts with most kinds of rocks. Therefore, much of the water contained in rocks can dissolve quartz and it is not surprising to find this mineral leached out of many soils.

Oxidation is a chemical weathering process that commonly attacks minerals containing iron, manganese, or sulfur, all of which combine readily with oxygen. The colorful red and yellow iron oxide minerals hematite and goethite are conspicuous in many soils as is the black manganese oxide mineral pyrolusite. Oxidation weathering principally affects the sulfide minerals of which the iron sulfide pyrite is the most abundant.

Pyrite exists in small quantities in a wide variety of common rocks and typically weathers by oxidizing to form goethite and soluble sulfate anions (Figure 5-6). Goethite is extremely insoluble under most natural conditions and generally remains in the weathering rock or soil as a yellow stain. Many otherwise beautiful rocks cannot be used as ornamental stone because they contain crystals of pyrite, which weather into unsightly blotches of rusty goethite within a few years. However, the same process beautified the ancient buildings on the Acropolis in Athens. These edifices were constructed of a white marble that mellowed to a warm ivory tone as minute crystals of pyrite disseminated through it weathered into goethite. In wet regions the sulfate anions created during the oxidation of pyrite simply wash away in the rain. In dry regions they commonly react with calcium to form crystals of gypsum that may glitter like slivers of glass in the soil.

The light rains or dew that typically dampen exposed rock surfaces in arid regions do not furnish enough water to flush oxidation products off the rock. Instead, these products accumulate to form a surface crust of iron oxide minerals stained dark by the black manganese oxide pyrolusite. Such coatings are called *desert varnish* because they look like paint (Figure 5-7). In many arid regions nearly every exposed rock surface bears a coating of shiny desert varnish, which makes the entire countryside look as though it might have been sprayed with black paint. Similar coatings also form on some limestone cliff faces in humid regions. There, they suggest streaks of soot and have often been mistaken for smoke stains left by ancient campfires. They form because carbonate rocks turn water trickling across them into a slightly alkaline solution, which then precipitates any dissolved iron or manganese it may contain as the insoluble oxide minerals.

The most important chemical weathering processes are the **hydration reactions.** In these, water reacts with silicate minerals—except quartz—converting them into clays and releasing most of the positive ions that had held their silicate tetrahedra together as soluble salts (Figure 5-8). Hydration reactions most readily attack minerals—such as olivine and pyroxene—which have relatively little oxygen sharing between their silicate tetrahedra. As the degree of oxygen sharing in silicate minerals increases, they tend to become less vulnerable to hydration reactions. The order in which silicate minerals succumb to chemical weathering generally tends to reverse that in which they appear in a crystallizing magma.

Silicate mineral grains generally swell up as they react with water and turn into clay. Such expansion pries them apart (Figure 5-9) and destroys the fabric of the rock just

FIGURE 5-7

Indian carvings and later graffiti and bullet holes pecked into a dark desert varnish on the face of a sandstone cliff in Capitol Reef National Monument, Utah. (J. R. Stacy, U.S. Geological Survey)

as a wall would crumble if some of its bricks were to expand. Therefore, the hydration reactions mechanically dismantle rocks as they chemically alter them, and function as major agents of both chemical and mechanical weathering.

Water penetrates rocks from their exposed surfaces, so hydration affects the outer parts of a boulder more severely than it does the inner core. Therefore, the outer shell of a rock may expand as its silicate minerals weather into clays and spall off, or **exfoliate,** in thin shells, expos-

ing the less weathered rock beneath. Figure 5–10 shows blocks of coarsely crystalline silicate rocks exfoliating successive shells as though they were the layers of an onion.

Figure 5–11 shows that water penetrating an angular rock attacks the edges where two sides meet from two directions and the corners where three sides meet from three directions. Therefore, hydration affects the edges and corners more than the flat faces and preferentially removes them to convert angular blocks into round boulders, a process called **spheroidal weathering** (Figure 5–12).

FIGURE 5-8

An example of a simple hydration reaction. Orthoclase feldspar reacts with water to produce dissolved potassium ions, an essential fertilizer nutrient, dissolved hydroxyl ions, dissolved silica, and kaolinite clay, which occupies more volume than the original feldspar because it absorbs water as it forms. The hydroxyl ions make the water alkaline and therefore able to retain silica in solution so that it, along with the potassium ions, will leach out of the soil if the climate is very wet.

| $4KALSi_3O_8$ Orthoclase feldspar | + | $22H_2O$ Water | = | $4K^+$ Dissolved potassium ion | + | $4OH^-$ Dissolved hydroxyl ion | + | $8Si(OH)_4$ Dissolved silica | + | $Al_4Si_4O_{10}(OH)_8$ Kaolinite clay |

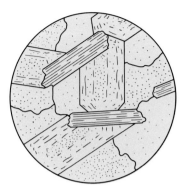

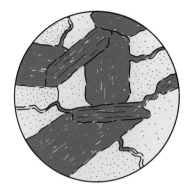

Fresh rock consisting of flaky grains of mica, blocky grains of feldspar and irregular grains of quartz.

Feldspar and mica grains expand as they begin to weather into clay fracturing crystals and wedging them apart along grain boundaries. The cores of feldspar grains often weather more rapidly than the rims.

Feldspar and mica almost completely weathered into clay leaving the quartz grains fresh but separated.

FIGURE 5-9
Crystals of silicate minerals swell as they weather into clay, prying the rock apart and causing it to disintegrate. The final result in this case will be soil composed of clay derived from hydration of the feldspar mixed with unweathered quartz sand grains remaining from the original granite.

FIGURE 5-10
Exfoliation of concentric shells of partially weathered rock.

FIGURE 5-11
Chemical weathering attacks rock surfaces, concentrating on edges and corners where two or three surfaces meet. It reduces angular chunks to rounded boulders thus minimizing the surface area.

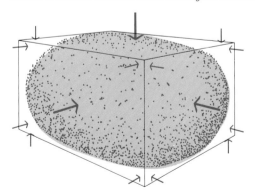

FIGURE 5–12
Spheroidal weathering of fractured granite creates rounded forms that may appear at the surface as boulders if erosion removes the soil formed along the fractures.

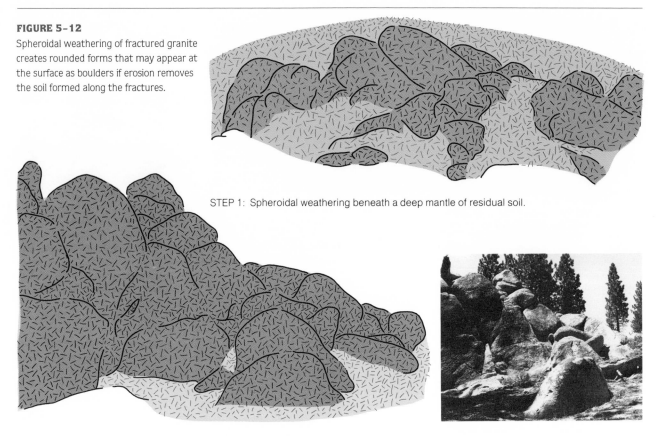

STEP 1: Spheroidal weathering beneath a deep mantle of residual soil.

STEP 2: Erosion has stripped off much of the original soil leaving the residual cores of unweathered rock as rounded boulders on the surface.

Exfoliation and spheroidal weathering operate together so that the block becomes rounder as it sheds each successive layer of partially weathered rock. Another way of viewing the rounding process is to consider that, since weathering tends to remove the rock surface, it should logically create spherical forms—the shape that has the minimum surface area in proportion to its volume.

The soluble salts released as silicate minerals weather into clays commonly include calcium, magnesium, sodium, potassium, and iron, along with numerous less abundant chemical elements. Most of the abundant elements released during weathering, as well as many of the less abundant ones, are essential fertilizer nutrients vital to both healthy plant growth and human nutrition. The fertility of

the soil that develops during weathering depends largely upon what happens to those soluble nutrients and that, in turn, depends mostly upon the wetness of the climate.

SOILS

If the products of weathering escape erosion and accumulate on the surface where they form, they become a mantle of *residual soil*. This type of soil may be opposed to *transported soils,* which consist of sedimentary deposits left by some agent of erosion. Residual soil mantles typically become thickest in humid regions, where a dense plant cover prevents rapid erosion. Arid regions generally lack enough plant cover to protect the ground from erosion. Thus, dry areas tend to have thin and discontinuous soil mantles concentrated in low places where eroded material accumulates.

The patterns of maps showing distributions of major soil types in large regions typically resemble those of climatic maps. They show little resemblance to patterns of geologic maps. Therefore, it seems that climate must play a much stronger role than parent rock type in determining the type of soil that develops. The influence of bedrock appears only on more detailed maps, showing the local distributions of more narrowly classified types of soil.

Most soils show a distinct vertical zonation, which makes them superficially resemble layered sedimentary deposits. However, the zonation in soils reflects the activity of water moving within them and the change in degree of weathering with depth. The uppermost zone, called the **A-horizon,** consists of thoroughly weathered material commonly full of roots and burrows and stained dark by **humus,** the partially decayed remains of plants. In most soils, the A-horizon grades downward within less than half a meter into an organically sterile **B-horizon.** The B-horizon consists entirely of minerals created during weathering. It is generally no more than a few meters thick although it may, in extreme cases, reach depths as great as 100 meters. At its base, the B-horizon grades downward into a **C-horizon.** This zone consists of partially weathered bedrock still retaining remnants of the original minerals and grades downward into fresh bedrock (Figure 5–13). Specialists in soil science generally recognize subhorizons within the major zones.

FIGURE 5–13
The C-horizon of the soil profile, top, grades downward into fresh bedrock; in this case, granite.

Pedocals

Dry regions typically receive light rains that dampen the upper levels of the soil but do not soak all the way through it. Soluble ions sink down into the soil with the rain during wet weather. These ions then return to the surface with the water as it evaporates during the sunny days that follow (Figure 5–14). The upper horizons of such soils accumulate soluble salts (Figure 5–15) and may contain concentrations sufficient to poison all but the most tolerant plants. In less extremely arid climates the rainfall is typically sufficient to wash the more soluble substances—such as sodium and potassium—out of the soil. The less easily soluble substances—such as calcium—remain behind.

The large and varied group of soils that forms in arid and semi-arid regions are collectively known as **pedocals,** a word coined to refer to their large calcium content. Such soils cover most of the western half of the United States and the prairie provinces of Canada. Their dominant color is pale gray although many have a very dark brown or black A-horizon full of humus.

Many pedocal soils contain unleached calcium ions which react with atmospheric carbon dioxide to form calcite, $CaCO_3$, which appears as crusty white deposits called

FIGURE 5–14

Heavy rains rinse soluble constituents out of wet climate soils whereas light rains concentrate them in desert soils.

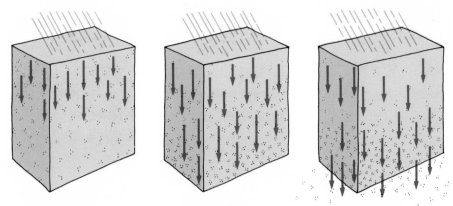

In a wet climate heavy rains soak down through the soil leaching its soluble constituents.

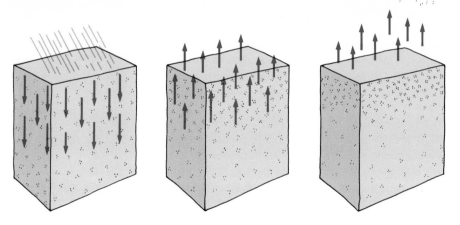

In a dry climate light rains wet the upper part of the soil and then the water evaporates concentrating the soluble constituents high in the profile.

caliche (Figure 5–16). In semi-arid regions where relatively little unleached calcium remains in the soil, the caliche deposits consist of little more than white crusts on the undersides of pebbles, which make them look as though they might have been whitewashed. In extremely arid regions where large amounts of calcium carbonate form in the soil, caliche deposits develop into massive stony ledges that defy all attempts short of dynamite to dig post holes. Those deposits force many western ranchers to string their barbed wire between triangles of posts set on the surface.

The old land and irrigation project promoters probably spoke more honestly than they realized when they ex-pansively promised to ''make the desert bloom like a rose.'' Pedocal soils generally produce abundant and highly nutritious crops if provided with water. Their natural fertility is obvious even without irrigation in herds of grazing animals flourishing on sparse but highly nutritious forage.

Cultivation of arid pedocal soils, especially irrigated cultivation, must be managed carefully. Otherwise, an increase in moisture content may mobilize the soluble salts and bring them to the surface. In extreme cases, those salts may not only poison the soil but also contaminate irrigation runoff water, making it unfit for further use downstream. For example, runoff from fields in southern Califor-

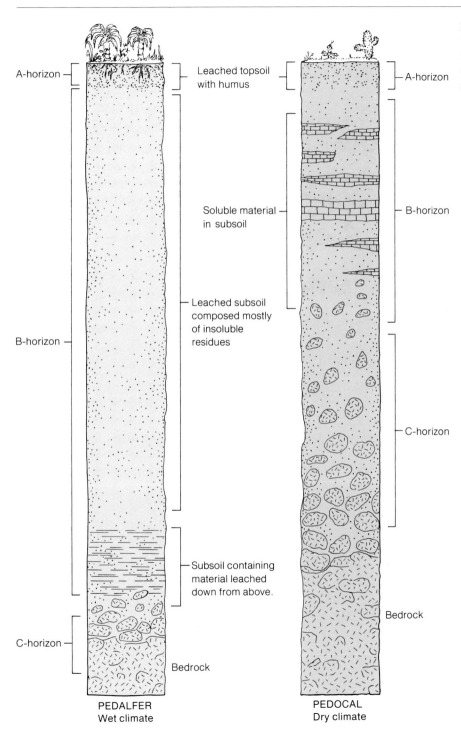

A-horizon

Leached topsoil
with humus

Soluble material
in subsoil

Leached subsoil
composed mostly
of insoluble
residues

B-horizon

Subsoil containing
material leached
down from above.

C-horizon

Bedrock

PEDALFER
Wet climate

A-horizon

B-horizon

C-horizon

Bedrock

PEDOCAL
Dry climate

FIGURE 5–15
Deep soils composed mostly of insoluble
residues left by leaching develop in wet
climates. Dry climate soils tend to be much
shallower and to contain concentrations of
unleached soluble material.

FIGURE 5–16

A thick crust of white caliche partly covers a black boulder of porphyritic basalt on Abert Rim in the desert of southeastern Oregon.

nia and Arizona has made the lower course of the Colorado River so salty that Mexican farmers are unable to use the water. Dry land farming techniques that depend upon summer fallowing instead of irrigation can create similar problems: They also increase the amount of water in the soil. Approximately 50 years of dry land wheat farming in the northern high plains has greatly increased the amount of water in the soil. As a result in some places water is seeping down, picking up soluble salts along the way, and emerging elsewhere as saline springs, which annually poison large areas of soil and contaminate surface water supplies (Figure 5–17). The obvious way to eliminate excessive amounts of salt in the soil is to flush them out with water. However, this technique is hardly feasible if the problem arises in a dry region.

Pedalfers

In wet regions copious flows of rainwater percolate through the soil, leaching the soluble fertilizer nutrients re-

leased by the hydration reactions. In many cases clay minerals also wash down out of the upper horizons and into the lower levels of the soil. There, they may accumulate to form dense hardpans almost impervious to water. The compositions of soils in wet regions change progressively with depth in ways that reflect the chemical activity of the water soaking down through them. The degree of leaching and of hydration decreases with increasing depth.

Wet temperate regions typically have rusty yellowish soils consisting largely of clays stained by goethite. Most of the many types of soil that fit within that rather broad general description are called **pedalfers,** a word coined to refer to their large content of aluminum and iron. The concentrations of both elements in pedalfer soils typically exceed those in the parent bedrock. Most pedalfers are considerably depleted in quartz, evidently because quartz escapes from the weathering horizon in solution. Pedalfers typically develop under a forest cover. They occur throughout a broad region of eastern North America as well as in the wetter and more heavily forested areas of the west. The pedalfer soils of temperate regions vary greatly in their natural fertility; some retain reasonable concentrations of fertilizer nutrients whereas others are severely leached and require considerable fertilization.

Laterites

In very warm wet regions of the tropics the weathering reactions that produce pedalfer soils in more moderate climates proceed to their logical end. They create profoundly leached soils called **laterites.** Such soils have lost most or all of their quartz. They consist essentially of the extremely insoluble iron oxides goethite and hematite mixed in varying proportions with an equally insoluble aluminum oxide mineral called gibbsite and the clay mineral kaolinite. Gibbsite and kaolinite are both white but the goethite and hematite generally give the soil a bright color somewhere in the range between rusty yellow and barn red.

True laterites cover vast regions of the wet tropics, and red and yellow soils of distinctly lateritic affinities cover large areas of the southeastern United States. Laterites and closely related soils also exist in large regions along the Pacific Coast, ranging as far north as southern British

Columbia. Most of the western laterites appear to have formed between about 15 and 25 million years ago, when the climate of that region was much warmer and perhaps wetter than it has been since.

Rural poverty and lateritic soils are consistently associated in the combination of red dirt with unpainted buildings, sagging fences, wretched livestock, and ragged people. All owe their sad condition to the impoverishing rain that leaches mineral nutrients out of the soil, leaving it incapable of growing nutritious food crops. A common sight in wet tropical countries is that of bone-lean cattle grazing in lush fields of deep grass. The grass provides barely enough nutrition to keep the herds alive—the leathery quality of the beef is notorious. The only plants that flourish on such sterile soils are those adapted to subsist on an extremely meager supply of mineral nutrients. Such plants do not include food crops.

Extensive clearing of tropical jungles, now proceeding at an alarming rate, cannot significantly increase the world supply of agricultural land: Most jungles grow on lateritic soils. The lack of agricultural potential in such soils is strikingly evident in the scarcity of animal life in tropical jungles. Despite their spectacular lushness, these jungles contain little nutritious vegetation. It seems obvious that soils unable to support much animal life in their natural condition can hardly be expected to do so under cultivation. Forestry is the only kind of agriculture that consistently produces good crops on laterite soils.

Fertilization of lateritic soils provides little help. The reason is that kaolinite, unlike the smectite clays of pedocal soils, does not store fertilizer nutrients—such as potassium, calcium, and magnesium—between its silicate layers. Therefore, fertilizers wash through the soil quickly and do not provide lasting benefits. In any case, people who live on laterites rarely have money to buy fertilizer. Volcanoes that regularly spread blankets of fresh ash across the countryside do constantly replenish the soil and thereby support many of the agriculturally productive regions of the tropics.

Some laterites have the curious property of hardening permanently when they dry. This phenomenon would never happen in their natural forest-covered state but may well occur if they are cleared and cultivated. Evidently, the goethite loses water and changes into hematite, which

FIGURE 5-17

This fence post with its heavy encrustation of salt stands in a formerly fertile field now ruined by excessive water.

cannot absorb water to revert back into goethite. In some regions people make bricks by cutting rectangular slices of laterite and letting them dry in the sun. Buildings made of such brick have lasted for hundreds of years in the rain forests of Southeast Asia. More commonly, laterites harden in farm fields converting them into expanses of rocky rubble that look like broken brick and serve approximately as well for agricultural purposes.

Laterite Ores

If lateritic soils develop on bedrock that contains little aluminum, they consist almost entirely of iron oxide minerals. Most of the world's major deposits of high-grade iron ore consist of such soils, which are mined in large open pits. If, on the other hand, lateritic soils develop on bedrock that contains little iron, they consist mostly of the aluminum oxide mineral gibbsite. Such soils are called **bauxites.** They provide the world's only source of aluminum ore, which is likewise stripped off the ground surface in large open pits.

SUMMARY

Rocks exposed to the elements decompose into residual soils through a complex of weathering processes. The mechanical processes break rocks without altering them chemically, whereas the chemical processes destroy old minerals and create new ones. The two classes of processes depend upon each other. Those that operate mechanically expose new surfaces to chemical attack, and those that operate chemically may also cause mechanical breakage.

Soils generally display a conspicuous sequence of horizontal zones varying in composition in ways that reflect the patterns of water movement. In wet climates where water percolates down through the soil, the zones are most heavily leached and weathered near the surface. They become progressively less so with increasing depth. In dry climates where water tends to evaporate from the surface of the soil and only rarely percolates through it, soluble substances accumulate near the surface.

The chemical weathering processes release soluble salts, the mineral nutrients required by plants. Therefore, the natural fertility of soils depends largely upon how well they retain mineral nutrients. The abundant rains of wet regions impoverish their soils by leaching out the mineral nutrients. Those same soluble salts not only remain in the soils of dry regions but may actually accumulate to poisonous concentrations. The most useful agricultural soils form in regions that receive enough rain to nurture crops without excessively leaching the soil.

KEY TERMS

A-horizon	laterite
B-horizon	mechanical weathering
bauxite	oxidation
C-horizon	pedalfer
caliche	pedocal
chemical weathering	salt cracking
exfoliate	solution
front wedging	spheroidal weathering
hydration reaction	weathering
humus	

REVIEW QUESTIONS

1. How do mechanical and chemical weathering contribute to each other?

2. Why do carbonate rocks stand topographically high in dry regions and low in those with abundant rainfall?

3. Suggest an explanation for the common observation that weathered surfaces on rocks exposed in sea cliffs often look quite different from weathered surfaces on similar rocks exposed inland.

4. Although most of central and western Australia is a desert, much of the region is covered by lateritic soils. What does that situation imply?

5. Granitic rocks and sandstones that contain large amounts of feldspar both tend to weather into rounded boulders whereas pure quartz sandstones do not. Can you suggest an explanation?

6. Why is runoff water from irrigated fields often salty?

SUGGESTED READINGS

Carroll, D. *Rock Weathering.* New York: Plenum Press, 1970.

A comprehensive treatment of weathering in all its complexity.

Degens, E. T. *Geochemistry of Sediments.* Englewood Cliffs, N.J.: Prentice-Hall, 1965.

Includes an excellent discussion of the chemistry of weathering within its much broader scope.

Garrels, R. M. and F. T. Mackenzie. *Evolution of Sedimentary Rocks.* New York: W. W. Norton & Co., 1971.

The discussion of the chemistry of weathering and soils is outstanding.

Hunt, C. B. *Geology of Soils.* San Francisco: W. H. Freeman & Co., 1972.

An excellent short text dealing with the geological aspects of soil development and occurrence.

Ollier, C. *Weathering.* New York: American Elsevier Publishing Co., 1969.

A comprehensive and rather technical text on weathering processes and soils.

CHAPTER SIX

Accordant summits and uniform slope
angles in a densely forested ridge and
ravine landscape in the Appalachians.
(J. K. Hillers, U.S. Geological Survey)

HILLSLOPE EROSION

Hills bear the marks of the erosion that carved them just as surely as markings on a stump reveal the tool that brought the tree down. Each process of hillslope erosion leaves its distinctive mark on the landscape as it carries soil and weathered rock down to the streams, which haul the eroded material away.

The following sections discuss several of the more important processes of hillslope erosion, relating the mechanics of each to the landforms it creates. For convenience the processes will be considered in two large categories: (1) **rainsplash and surface runoff** erosion, which depend upon surface water flowing across a slope; and (2) **mass wasting** of soil and rock moving under the influence of gravity without the aid of a fluid medium, such as water or wind.

RAINSPLASH AND SURFACE RUNOFF EROSION

Wherever the ground is so sparsely sheltered by plants that raindrops splash directly onto bare soil, rainsplash and surface runoff erosion may become important. In the modern world these processes dominate regionally in deserts, and locally where soil has been bared by fire, overgrazing, clearcutting, or any other of the numerous agents that can devastate plant cover.

Rainsplash and surface runoff erosion must have dominated landscape development everywhere until the first land plants appeared during Silurian time. We can imagine the domain of such processes shrinking steadily since then as plants evolved to adapt to a wider range of habitats. Except for Greenland and Antarctica, which are covered by ice, the earth's deserts are the only regions still not sheltered beneath a more or less continuous plant cover. The deserts are the only regions where we can see landscapes resembling those that must have existed nearly everywhere during most of the geologic past.

Rainsplash

Raindrops strike hard! They make a drumming sound on the roof and they splatter particles of soil into the air wherever they hit bare ground. Everyone has surely seen a hard rain splatter soil from a new garden onto the side of a house or from a dirt road onto a car. That process alone can move considerable quantities of soil but it generally acts in concert with surface runoff, which greatly accelerates erosion.

Surface Runoff

Watch a hard rain fall on a bare surface—the drops splattering as they first come down and then plunking as a film of water forms. That surface film is too thin to protect the soil from the impact of splashing raindrops but thick enough to carry particles of soil with it as it flows downslope, eroding a pattern of little rills.

Surface Sealing

Farmers and gardeners are familiar with the hard crust that forms on cultivated soil after a few heavy summer rains. It develops as raindrops "puddle" the soil into a muddy slurry and then hammer that into a tightly compacted crust, a process called **surface sealing.** The sealed surface is not compacted tightly enough to prevent raindrops from detaching particles of soil but it does effectively prevent water from soaking into the ground. Therefore, surface sealing increases both the volume of surface runoff and the rate of soil erosion. Midsummer cultivation to break the sealed surface helps rainwater to soak into the ground and thus minimizes both runoff and erosion.

Rapid Erosion

Rainsplash and surface runoff erosion can, on occasion, strip as much as several centimeters of soil from a vulnerable surface during a single storm. Evidence of such catastrophic erosion often appears in the form of little pedestals of soil capped by pebbles that protected the soil beneath them from splashing raindrops. Figure 6–1 shows some that formed during a single afternoon. Figure 6–2 shows much larger pedestals that illustrate truly catastrophic soil erosion. When that happens, the eroded soil runs into the streams, making them very muddy.

The devastating efficiency of rainsplash and surface runoff erosion explains why the rate of erosion is typically greatest in dry regions (Figure 6–3). It also accounts for why streams draining from barren watersheds are typically muddy whereas those flowing from well-vegetated watersheds tend to be clear. Consider, for example, the striking contrast between the muddy waters of the Missouri River, which flows out of the semi-arid northern high plains, and

FIGURE 6–1

Pebbles capping pedestals of soil they protected from rainsplash erosion. The pedestals, about 5 cm high, formed during a single afternoon of heavy rain.

FIGURE 6-2

A tree stump in southeastern Missouri standing on a pedestal of soil that it sheltered from rainsplash erosion. Torrential surface runoff from the bare ground eroded the closely spaced gullies. (F. E. Mathes, U.S. Geological Survey)

FIGURE 6-3

Mount Garfield, Utah, an outpost of the Book Cliffs. The intricate network of small drainages is typical of desert landscapes carved by rainsplash and surface runoff erosion. (S. W. Lohman, U.S. Geological Survey)

the relatively clear water of the Ohio River, which drains green countryside of the eastern United States.

Flash Floods

The surfaces of most hillslopes in sparsely vegetated regions consist primarily of bare rock and sealed soil, both of which shed water. Only a small proportion of the rain in those regions soaks into the ground except locally where it consists of water-receptive material, such as gravel. A large proportion of the rain runs off the surface and into the streams, filling them and causing sudden flash floods that flow while the rain falls and subside as the sky clears. Knowing that makes it easy to understand why inhabitants

of dry regions believe their occasional rains are cloudbursts. In fact, dry regions only rarely receive heavy downpours; even a moderate rain can cause flash flooding if most of it runs off.

Man-Made Deserts

If the plant cover is destroyed and does not soon regenerate, rain may seal the ground surface, making it shed water that would otherwise soak in. In many cases the ground no longer absorbs enough water to support the native vegetation and therefore remains permanently barren. Semi-arid regions commonly contain local areas of badlands, such as those shown in Figure 6–4—miniature

FIGURE 6–4

Badlands are miniature deserts often set in otherwise green countryside. They form in places where the plant cover is destroyed and does not regenerate itself because the bare slopes shed most of the rain that falls on them. The foreground of this view in the Makoshika Badlands of eastern Montana includes several tall pedestals capped by a layer of sandstone that protected the soft shale beneath from rainsplash erosion. Compare them to the miniature pedestals shown in Figure 6–1.

deserts that bleed torrents of muddy water into the streams with every rain. Perhaps such badlands mark areas that once lost their plant cover to intense fires and then developed a sealed ground surface before new plants could sprout.

People have inadvertently converted large regions of formerly productive semi-arid countryside into deserts through wanton destruction of the plant cover. Such man-made deserts exist all around the Mediterranean Sea, in central China, and in other regions where people have denuded the land through overgrazing or by cutting the forests for lumber and fuel. Now the ground in those regions sheds water that it would otherwise have absorbed. The streams in such places now behave as desert streams, suddenly filling with muddy water when it rains and drying up as the rain stops. The muddiness of their flash floods testifies to the rate at which soil erodes off the barren slopes.

Most man-made deserts can be restored to much of their original productivity by planting trees and shrubs and then watering them until they establish themselves. As the new plants begin to shelter the ground, it again absorbs enough water to support them. The plants arrest surface runoff as well as the flash flooding and rapid soil erosion it causes. Such restoration projects have been especially successful in Israel and in parts of China.

Barren Landscapes

Sparsely vegetated landscapes shaped dominantly by rainsplash and surface runoff erosion consist of gullied hills with smoothly graded plains sloping gently away from their flanks (Figure 6–5). Landscapes of that type form on a large scale in deserts and in miniature in badlands, eroding roadcuts, piles of fill dirt, neglected vacant lots, and nearly any other place where bare ground lies exposed to the rain. Erosion strips most of the soil off the barren slopes (Figure 6–6), making the hills of the desert as variable in form and color as the exposed bedrock that composes them.

As rainsplash and surface runoff erosion proceed, the hills waste away and the smooth plains that flank them grow steadily larger. Ultimately, the plains consume virtually the entire landscape, leaving only isolated remnants of the original hills, such as those shown in Figure 6–7. Im-

FIGURE 6–5

A miniature desert landscape developing in an eroding roadcut. An intricately dissected slope rises steeply above a smooth and gently sloping desert plain in the foreground. The sagebrush near the top of the picture is about 50 cm high.

mense regions of the Australian desert have reached that advanced stage of erosion as have much smaller areas of the deserts in the southwestern United States and northern Mexico.

Desert Plains

The upper parts of the plains created by rainsplash and surface runoff erosion exist where hills once stood and must therefore be erosional surfaces. They are often called

FIGURE 6-6
Rainsplash erosion stripped the soil off these jagged crags of dark volcanic rock in the Sutter Buttes, California. Natural gas field installations appear in the foreground.

FIGURE 6-7
Isolated small hills of granite standing as remnants of erosion on a desert plain surface in central Wyoming.

pediments if they expose bedrock. In many cases the lower parts of the same plains are deeply underlain by deposits of sediment that bury the bedrock and must therefore be depositional surfaces. Many geologists call such surfaces **bajadas.** Streams typically dump large quantities of sediment where they emerge from the hills onto the plain, building **alluvial fans** such as that shown in Figure 6–8. Alluvial fans are shaped like segments of cones laid against the flank of the hill. However, pediments and bajadas typically form a continuous topographic surface, and it is often difficult to determine exactly where alluvial fans blend into bajadas. Therefore, it is convenient to use the term **desert plain** (Figure 6–9) to refer to the entire surface and eliminate the need to distinguish its parts. Desert plains are truly remarkable topographic surfaces.

Although all parts of the desert plain are typically gullied and therefore irregular on a small scale, the size of the gullies is small compared to the extent of the surface. Thus, desert plains are actually extremely smooth, as smooth in comparison to their size as a finely finished machine part. Furthermore, most desert plains have a simple form in which the steepness of the slope increases steadily with

elevation. How can running surface water create such perfectly smooth and mathematically simple land surfaces?

Photographs of the same desert plain taken over periods of decades generally show that the intricate pattern of closely spaced gullies covering the surface changes almost from year to year. Evidently, the gullies work back and forth across the surface as they erode their banks and as old ones fill and new ones form. The gullies shift position frequently because the water that flows through them is so heavily laden with sediment that it chokes its own channel, fills it with sediment, and then spills over the banks and carves a new course. That same process also explains why desert plain gullies tend to branch downstream into distributaries (Figure 6–10) instead of picking up tributaries as most streams do. The effect of the constantly shifting pattern of gullies is to make the entire desert plain a sort of extended streambed because every part of the surface is a gully at one time or another. Therefore, the problem of understanding the origin of desert plains can be pursued in terms of how running water carries sediment.

Because the long profile of most desert plain surfaces,

FIGURE 6–8

An alluvial fan banked against a mountainside at the mouth of a canyon in the desert near the south end of Death Valley, California. (D. W. Hyndman)

FIGURE 6–9
View across a desert plain rising gently onto the flank of the Furnace Mountains east of Death Valley, California. (D. W. Hyndman)

FIGURE 6–10
Large alluvial fans spread from the mouths of canyons in deeply eroded mountains of the Mojave Desert in southern California. An intricate network of gullies that branch downslope into distributaries covers their surfaces. (J. R. Balseley, U.S. Geological Survey)

FIGURE 6-11

View across a smoothly graded desert plain in southern Nevada. The craggy mountains in the background are composed of soilless bedrock. The upward concavity of the desert plain profile is clearly shown. (D. W. Hyndman)

like that of most stream channels, is concave to the sky (Figure 6–11), deposition of sediment anywhere along the profile will build it up to a steeper gradient (Figure 6–12). The steepened gradient speeds the flow of water thus improving its ability to carry sediment. Continued deposition will eventually steepen the slope enough to enable the flow of water to carry its load of sediment. Conversely, if the runoff water erodes the surface, that process reduces the slope to a locally flatter gradient that will eventually slow the water enough so that it will no longer be able to erode. Therefore, deposition and erosion on desert plains have opposite effects that lead to the same result: A surface so adjusted that the water flowing across it after a rain can carry its load of sediment without net deposition or erosion. All streams, whether they flow across a desert plain or not, adjust the gradients of their channels in the same way to achieve a similar equilibrium between water flow and sediment load, a condition geologists refer to as *graded*. Therefore, desert plains and streams are self regulating transport mechanisms that automatically adjust their

FIGURE 6-12

Deposition of sediment on the concave surface of a desert plain steepens its slope and therefore increases the velocity of water flowing across it. Erosion of the surface has the opposite effect. All streams, not just those that flow across desert plain surfaces, adjust their gradients in the same way.

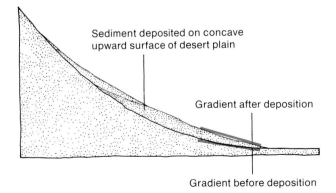

Sediment deposited on concave upward surface of desert plain

Gradient after deposition

Gradient before deposition

gradients to respond to changes in the proportion of sediment to water in the runoff they carry.

Desert plains also respond to the grain size of the sediment that washes across them. Fine-grained sediments, such as silt or clay, tend to travel in suspension; water flowing slowly down a gentle gradient can carry them. Coarser sediments, such as sand and gravel, travel by bouncing or rolling and require a much faster flow to keep them in motion. Desert plains with fine-grained sediments on their surfaces slope more gently than those covered with coarser material.

TRUE DESERTS—
THE EFFECT OF ARIDITY

Suppose we were to strip the plant cover from the state of Georgia and somehow prevent its regeneration. The landscape that would then develop as rainsplash and surface runoff erosion ran its course in that wet region would resemble that of true deserts in most but not in all respects. Desert landscapes—that of Nevada, for example—owe some of their characteristics directly to their aridity rather than merely to their sparse plant cover.

Desert Sedimentation

Arid regions are unique in having a high rate of erosion combined with too little water to maintain enough stream flow to carry the eroded sediment away. Therefore, the quantity of sediment overwhelms the limited carrying capacity of the streams, causing them to deposit sediment in the desert instead of carrying it to the ocean as they would if the climate were wetter. Deserts are typically regions of widespread terrestrial deposition in which a number of distinctive kinds of sedimentary rocks form.

With a few exceptions such as the Rio Grande and the Colorado that start in humid watersheds, desert streams typically dry up and vanish as they lose water to evaporation and seepage into the ground. Therefore, the valleys in mountainous deserts generally become closed basins of internal drainage, sometimes called **bolsons,** where sediment eroded from the surrounding hills accumulates. Many desert valleys in the western United States contain more than 1,000 meters (m) of such basin fill sediments. The

eroding mountains that surround them are slowly burying themselves in their own debris (Figure 6–13).

Playas

Each undrained desert valley contains in its lowest part an area called a **playa,** which generally consists of a broad expanse of salt-glazed mud as level as a billiard table. Most playas flood temporarily during wet weather with a sheet of water that may be as much as several kilometers across and as little as a few centimeters deep. Some playas also contain permanent lakes, such as the Great Salt Lake of Utah. Playas have no outlets except through evaporation and therefore function as the chemical sumps of a landscape without drainage. The soluble salts released during weathering concentrate in them, making their waters alkaline brines unfit for use except by a few specially adapted plants and animals.

Some playas accumulate deposits of limestone and others beds of evaporite sediments that typically include gypsum among other minerals. Exactly which other minerals depends primarily upon the kinds of soluble salts the bedrock in the surrounding mountains releases as it weathers. A few playa evaporite deposits contain valuable chemical raw materials, such as borax, trona, or epsom salt, all of which support large mining operations.

The High Plains Surface

Deserts without mountains develop into vast expanses of smooth and nearly level desert plain floored by layers of water-washed sand, silt, and gravel. Occasional heavy rains flood large regions of such deserts, starting sudden streams on a headlong race to nowhere and filling vast playas with shallow lakes that may extend to the horizon until they dry up a few days later. Sediment moves from place to place during those wet periods but little actually leaves the desert. Much of central and western Australia and large regions of North Africa typify such deserts.

The high plains region of western North America had a desert landscape of that type until the end of Pliocene time about 3 million years ago. Vast remnants of old desert plain still survive in that region as flat upland surfaces now largely planted in grain fields, such as those shown in Figure 6–14. Those level expanses are floored by layers of old

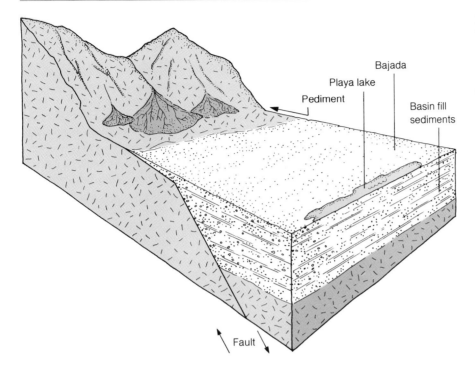

FIGURE 6-13
The basic anatomy of a mountainous desert. The pediment surface encroaches into the mountains as they erode and their debris accumulates in the valley, progressively burying the lower part of the pediment.

FIGURE 6-14
The high plains surface—an old desert plain floored by a veneer of stream-rounded gravel. Although it is more than 3 million years old, much of the surface remains undissected by streams because water soaks into its porous gravel cover instead of running off. (J. R. Balsley, U.S. Geological Survey)

desert plain sediments, mostly gravel, which are as much as 100 m thick. They are called the Oglalla formation throughout much of the region. The gravel deposits protect the surface from erosion by absorbing water and thus preventing surface runoff.

MASS WASTING PROCESSES

Many mass wasting processes move rock and soil down hillslopes in a variety of ways that defy simple classification. Such processes involve slow or rapid movement of all possible combinations of wet or dry soil and rock down steep or gentle slopes. Each style of movement contributes distinctively to the landscape. The following sections discuss several mass wasting processes. Some are important because they play a major role in shaping erosional landscapes; others interest us because they may cause property damage.

Soil Creep

In most heavily vegetated regions the dominant erosional process is **soil creep,** which is actually a complex of sev-

eral processes working together. They move soil down the slope almost as though it were an extremely viscous liquid with the surface levels moving fastest and the rate of movement diminishing rapidly with depth. Soil creep is typically an extremely slow process that can dominate development of a landscape only if a dense plant cover or other conditions preclude rainsplash and surface runoff erosion.

Expansion and contraction of the soil mantle probably contributes most to creep. Soil expands as it gets wet, as it gets warm, and when ice forms in it. It contracts as it dries, cools, or thaws. As Figure 6–15 shows, soil moves directly away from the slope as it expands and then drops vertically under the pull of gravity as it contracts. It cannot return to its starting point upon contraction because nothing constrains it to move laterally. Therefore, every cycle of expansion and contraction hitches the soil mantle a short distance downslope. Repeated cycles walk it off the slope one short step after another.

A number of other processes also contribute to soil creep although probably in a minor way. Burrowing animals, for example, dump their excavated soil downslope and then more soil moves downslope when the burrow finally collapses from the top down (Figure 6–16). Roots also burrow through the soil and must likewise contribute

FIGURE 6–15

Expansion of the soil mantle heaves it perpendicularly outward from the slope; contraction moves it vertically downward. Since all soil particles move the same way, it follows that every cycle of expansion and contraction must move the entire soil mantle a short step downslope.

Soil expands, heaving particle up perpendicularly to the slope.

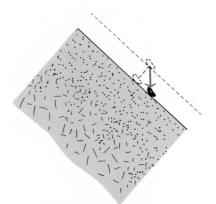

Soil contracts, permitting particle to sink vertically downward.

to its downslope movement. All of the processes that make the soil creep are most active near the surface. Thus, the upper levels of the soil move fastest and the rate of movement slows with increasing depth. That style of movement, shown in Figure 6–17, distinguishes soil creep from other mass wasting processes.

Although creeping soil moves so slowly that we normally regard it as stable, the process does betray its activity in familiar ways, some of which are shown in Figure 6–17. Posts set vertically in a creeping slope lean downhill as the years pass because the more rapidly moving surface soil pushes them over. Retaining or basement walls set flush against a cut in a creeping slope eventually collapse from the top down for the same reason.

The burrowing animal dumps its excavated soil downslope.

The abandoned burrow collapses from the top down.

FIGURE 6–16
Every animal burrow moves twice its volume of soil a short distance downslope.

FIGURE 6–17
The upper levels of the soil are the more active so the rate of creep is fastest near the surface and rapidly diminishes with increasing depth. The more rapid creep of the upper levels makes retaining walls fail from the top down, pushes fence posts over so they tilt downslope, and causes growing trees to develop a downhill slouch.

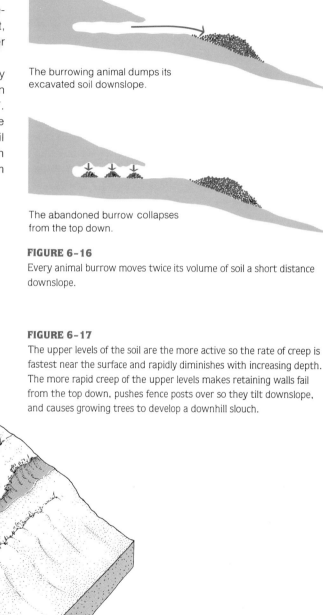

FIGURE 6-18

Ridge and ravine landscape in the Shenandoah Valley, Virginia. (I. C. Russell, U.S. Geological Survey)

FIGURE 6-19

Soil creep erosion develops hillslopes that consist of a concave toe, straight midsection, and convex summit.

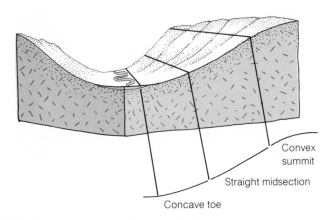

Ridge and Ravine Landscapes

Soil creep erosion produces hilly topography (Figure 6–18), which is perfectly described by the expression **ridge and ravine landscape.** Beautifully developed examples exist throughout the eastern half of the United States south of the limit of ice age glaciation, which approximately follows the lines of the Ohio and Missouri Rivers. Similar but more rugged ridge and ravine landscapes exist in the forested mountain ranges of the western United States and Canada that escaped glacial erosion during the ice ages.

Ridge and ravine landscapes typically contain smoothly rounded hills having few gullies or rocky outcrops on their slopes because they shed little surface runoff and possess a thick soil mantle. Such hills typically have a cross-sectional profile (see Figure 6–19), which is concave to the sky at the base, straight in the midsection, and convex at the crest. Their convex summits lend such hills a swollen look, which contrasts sharply to the hollowed appearance typical of hills shaped by rainsplash and surface runoff erosion.

Ridge and ravine landscapes soak up water as though they were sponges, store it underground, and then let it seep slowly into the streams. That water storage capacity greatly stabilizes stream flow, minimizing flooding during wet weather and maintaining flow during dry seasons. And those streams are typically clear except during floods because soil creep is a slow process that strips relatively little sediment off the hills.

Systematic measurement of slope angles in ridge and ravine landscapes shows that they typically remain quite uniform throughout large areas and change with the character of the underlying bedrock. Many geologists interpret that observation as evidence that the slopes are adjusted so that the processes of weathering and erosion—which create and remove soil—proceed at exactly equal rates. Unfortunately, we do not now have enough information about the many variables that control the rates of weathering and erosion to demonstrate whether that is actually the case.

Even Skylines

People who climb a hill to view a ridge and ravine landscape often see ridge after ridge cresting at about the same elevation. Those ridges make a rather flat scene (Figure 6–20), a landscape that suggests a floor with valleys gouged into it. For many years geologists interpreted such even skylines as evidence that the modern streams had carved their valleys into an old erosion surface that was nearly flat. In many cases that interpretation is demonstrably correct: The ridge crests do contain independent evidence that they are indeed remnants of an old erosion surface. Vast regions of the southeastern and south central United States have accordant ridge crests capped with layers of stream-rounded sand and gravel similar to the Oglalla gravels beneath the high plains. However, the southeastern gravels, older than the Oglalla formation, contain Miocene fossils. The hills on the low western slopes of the Sierra Nevada in California likewise rise to an even skyline underlain by deposits of stream-rounded sediments that must have covered an old erosion surface. However, that is not the only way ridge and ravine landscapes can acquire an even skyline.

FIGURE 6–20

Accordant summits in a ridge and ravine landscape in the western Sierra Nevada, California. In this case the hills are capped with layers of stream-rounded gravel, which suggests that the even skyline is indeed an old erosion surface. (D. W. Hyndman)

Maps show that streams of similar size are typically rather evenly spaced in ridge and ravine landscapes. Recall that hillslopes in such landscapes stand at consistent angles and then imagine building hills between the evenly spaced streams. Since their base width is fairly uniform and their sides have fairly uniform slopes, the hills must crest at fairly uniform elevations. Figure 6–21 illustrates the point.

Therefore, the even skyline of a ridge and ravine landscape may be the remains of an old erosion surface or the result of an elementary exercise in the geometry of making hills. Close field examination of the landscape to determine whether the ridge crests contain evidence of having once been an old erosion surface will generally resolve the issue.

Landslides

Figure 6–22 shows a landslide that consists of a mass of soil and weak or weathered bedrock moving on one or more slip surfaces while the material below remains essentially in place. Although many landslides move slowly or re-main stationary for long periods, most are capable of moving rapidly and many cause massive devastation. In most regions landslides are relatively uncommon and occur only locally on abnormally steep or wet slopes, in many cases on slopes steepened or saturated through development. However, there are regions of deep soil or weak bedrock in which landsliding becomes the dominant process of hillslope erosion.

Figure 6–23 shows that there is a limit, called the **angle of repose,** to how steeply loose objects will stack before they begin to roll over each other and slide. Piles of dry sand stand at their angles of repose, which vary according to the angularity of the grains. Well-rounded sand grains roll more easily than angular ones so their angle of repose is correspondingly flatter. Talus slopes also stand at their angles of repose, which vary in the neighborhood of 30° depending upon how easily the pieces of sliderock move. Pyramidal stacks of oranges in grocery store displays likewise stand at their angle of repose. They illustrate the limitations that slope imposes by sliding if anyone plucks an orange from the bottom of the pile or adds one to its side.

FIGURE 6–21

In a ridge and ravine landscape the hills commonly rise at consistent slope angles from more or less evenly spaced streams to nearly uniform summit elevations that give the landscape an even skyline.

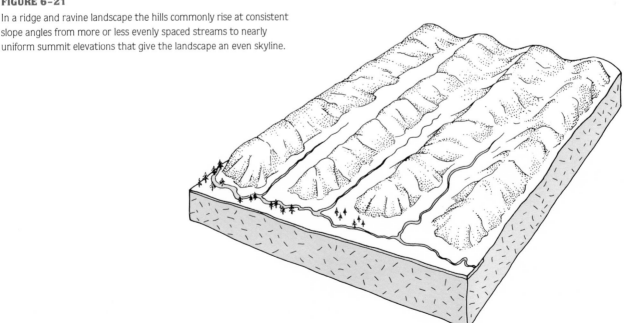

FIGURE 6-22
A landslide covered a highway with a thick deposit of wet mud. (Montana Highway Department)

FIGURE 6-23
The mechanical behavior of sand—or soil—depends largely on its water content. (a) Dry sand grains roll freely over each other until they reach their angle of repose, usually near 30°. (b) Films of water that wet the grains in moist sand stick where they touch, binding the pile together. (c) Water completely submerges the grains in saturated sand, buoying them up and thus reducing the weight they exert on each other.

DRY SAND   MOIST SAND   SATURATED SAND

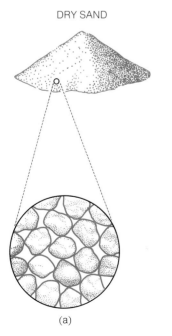
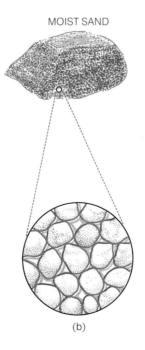
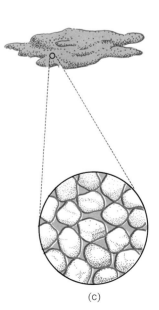

(a)   (b)   (c)

Most grocers pile their oranges at slopes much less steep than their angle of repose so that addition or removal of an orange will not cause a catastrophe. Most hills, except talus slopes, likewise stand at slopes considerably flatter than the angle of repose of the soil that covers them. Thus, it is generally possible to dig or fill on a hillside—the equivalent of removing or adding an orange to the pile—without causing serious problems. However, any slope will slide if steepened to an angle greater than the angle of repose of its soil cover or, as more commonly happens, if the soil on the slope is weakened so that its angle of repose becomes less than the angle of the slope. Recall the lessons of the sandbox, which everyone learns at an early age.

Children quickly discover that it is much easier to build sand castles if the sand is moist. That is true because the films of water that coat the grains also stick them together. Being human, most children then conclude that if a little bit of water improves the sand, a great deal more water would make it even better. So they turn on the hose and saturate the sand only to discover that it then loses most of its strength and slides. The explanation of that experience also explains most landslides.

FIGURE 6–24

A cut-and-fill terrace weakens a hillslope by steepening it in both of the two possible ways: by undercutting the slope above the cut and loading that below the fill.

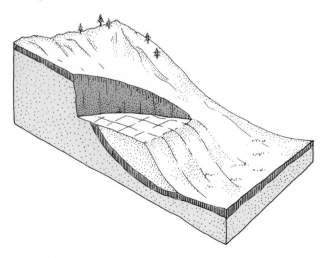

Water-saturated sand or soil loses virtually all its strength. This loss of strength happens partly because the water exerts a buoyant effect on the mineral grains and partly because its pore pressure tends to separate them. Like all submerged objects, mineral grains are buoyed up by the mass of the volume of water they displace. Every grain in saturated soil is submerged. It therefore weighs less than it would if surrounded by air, and it bears down less heavily on the grains beneath. That reduces internal friction and thus permits sliding to occur. If the water in the soil is under significant pressure, as it will be if the slope is long, that further diminishes contact between grains and thus further weakens the soil. A slip surface forms and sliding begins when water weakens the soil enough to reduce its angle of repose to a value less than the actual gradient of the slope. That happens as an unintended consequence of development in many cases.

The usual first step in building on a slope is to level the site by digging into the hillside and piling the excavated dirt below the hole to make a cut-and-fill terrace. As Figure 6–24 shows, that simultaneously steepens the slope in both of the two possible ways: by undercutting the part above the terrace and loading the part below. Although cutting and filling tends to threaten slopes, it rarely causes immediate sliding because most hills stand at angles considerably less steep than the angle of repose of their cover of soil and weathered rock.

Generally, the greater danger lies in slow saturation of the slope after development. Water soaks into the ground from such varied sources as lawn and garden irrigation, drain fields, leaking sewer pipes, and disruption of the normal surface drainage. Over a period of years enough may accumulate to saturate a large section of the slope, making its soil cover too weak to stand on the hill.

Landslides begin with development of a slip surface, which is typically—although not invariably—shaped like the bowl of a spoon. Figure 6–25 shows how movement on such a curving slip surface rotates the slide mass, making formerly vertical posts, walls, and trees lean back into the slope instead of pitching forward as they do if soil creep moves them. In some cases small *sag ponds* form at the heads of such rotating slides thus contributing to further saturation of the slope below. Most large landslides move as a series of slices on a nested set of concentric slip surfaces instead of as a single mass (Figure 6–26).

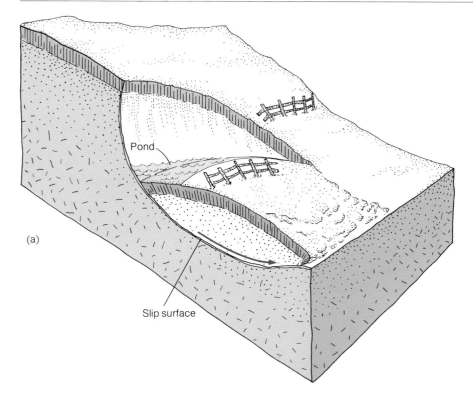

(a)

Slip surface

Pond

FIGURE 6–25
(a) A simple rotational slump. The material below the curving slip surface remains in place while that above moves, tilting the ground surface back to a flatter gradient. Vertical walls, posts, and trees tilt back toward the slope instead of pitching forward as they do if soil creep moves them. In some cases the head of the slump tilts back far enough to impound a pond on the side of the hill. (b) A small rotational slump in a roadcut that was graded to an angle too steep for the soil on the slope.

(b)

FIGURE 6-26
Many slides move on a nested set of more or less concentric slip surfaces.

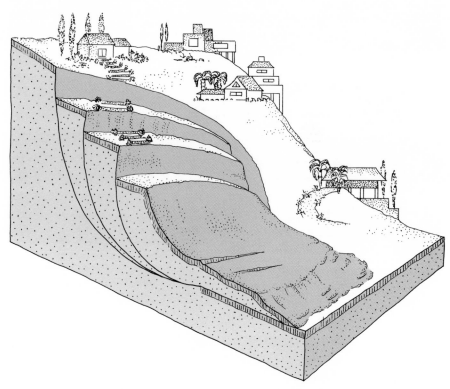

Most landslides give ample warning of their existence long before they move catastrophically. Small movements open cracks and rotate structures and trees on the upper part of the slope while raising bulges in the lower part. Generally, such landslides may be stabilized before they cause major property damage. Stabilization may be accomplished either by draining the excess water out of the slope or by loading large masses of soil or rock onto the toe of the slide mass to increase friction along the slip surface. On the other hand continually adding water to the slide mass, or excavating its toe, are likely to trigger catastrophic movement, which may cause enormous property damage and even loss of life. The best precaution against landsliding is to minimize cutting and filling on potentially unstable slopes and to avoid saturating them with water.

Rockfalls

Technically, rockfalls differ from landslides in that they move on surfaces of weakness within the bedrock, such as fractures or bedding planes. However, the distinction is more easily drawn in principle than in practice. Many slides involve both soil and bedrock, and weak or closely fractured bedrock behaves mechanically almost as though it were soil.

The simplest kind of rockfall is simply a large block that moves slowly downslope, opening a gap such as that shown in Figure 6–27 behind it. Such gliding blocks may present a serious hazard. That happened, for example, in 1978 when a large rock started moving toward a major highway and several expensive homes in Malibu Beach, California. In that case, as in many similar ones, the rock was broken and removed before it could fulfill its threat.

Rockslides are simply masses of rock that slide down a slope and land in a heap at its base. They tend to occur where the orientation of fractures or bedding surfaces lies nearly parallel to the slope. Rockslides become a problem where excavation in such a slope leaves large masses of rock poised to slide into the cut. Geologic examination of the site can usually predict such problems (Figure 6–28) and make it possible to mitigate them by adapting the design to the site.

FIGURE 6-27

A small stream flowing along a perfectly straight course through the narrow gap opened behind a block glide. Sunrift Gorge in Glacier National Park, Montana. (D. W. Hyndman)

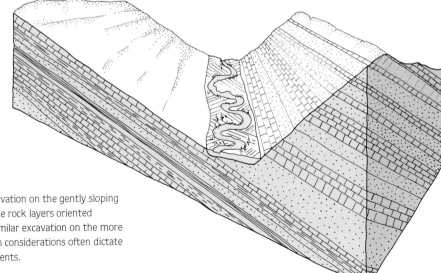

FIGURE 6-28

Rock slide hazard. A roadcut or other excavation on the gently sloping left bank of the stream would undercut the rock layers oriented parallel to the slope and invite a slide. A similar excavation on the more steeply sloping right bank would not. Such considerations often dictate the locations of roads and other developments.

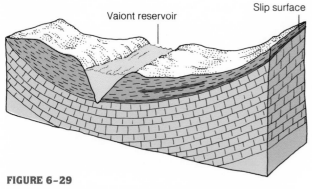

FIGURE 6-29

Diagram illustrating the conditions that led to the Vaiont Reservoir catastrophe. The Vaiont River had undercut tilted sedimentary beds oriented nearly parallel to the topographic surface. Then the water in the reservoir provided buoyant support for the lower part of the slide mass, which triggered rapid movement by reducing friction along the slip surface.

The Vaiont Catastrophe

In 1963 an enormous rockslide destroyed the Vaiont Reservoir in northern Italy after having demonstrated ample warning signs for more than three years. Its history illustrates several interesting points.

The Vaiont Reservoir flooded a valley eroded along the axis of a fold that bent layers of sedimentary rocks, mostly limestones, down into a trough so they sloped toward the valley from both sides (Figure 6-29). That fold placed the sedimentary layers nearly parallel to the topographic surface on both sides of the valley. It therefore made them liable to slide if they were undercut as they had been by the Vaiont River. Nevertheless, the dam was built and the reservoir filled with the expectation that any possible rockslides would be small.

After the water began to rise behind the new dam in 1960, a few small rockslides did indeed occur and a large segment of slope on the south side of the reservoir began to move several centimeters per day. That large slope movement had not been expected so the supervising engineers watched it closely and observed that it accelerated whenever the water level rose to a new height. They devised a plan to fill the reservoir slowly, hoping to thus avoid destabilizing the slide mass. Movement did slow down and the situation appeared to be under control until the slide

accelerated again in early October, 1963. At that time grazing animals fled from their pastures on the slide mass, something that often happens when large slides begin to move.

Then a heavy rain fell on October 8 adding more water to the slide, which accelerated to a rate of about 20 cm per day. On the night of October 9, the last restraining rocks broke to the accompaniment of loud cracking noises. The entire slide mass splashed into the reservoir with effects parallel to what might happen if an elephant were to jump into a small swimming pool. A wave approximately 219 m high washed over the dam and down the valley, destroying several villages. The dam survived but more than 2,000 people did not and the reservoir was half-filled with broken rock.

Evidently, the rising water destabilized the slide mass by providing buoyant support for its toe. That effectively decreased the weight of the lower part of the slide and therefore reduced friction along the slip surface. In other words filling the reservoir had a mechanical effect comparable to that of cutting the toe of a slide, an effect exactly opposite to that of loading rock onto the toe of a slide to stabilize it. In hindsight it seems clear that the strategy of filling the reservoir slowly to avoid destabilizing the slide mass could not have succeeded because it did not address the basic problem.

Rock Avalanches

In March, 1903 a mass of shattered rock weighing more than 30 million tons poured down the side of Turtle Mountain and obliterated part of the little town of Frank in southwestern Alberta. The slide spread more than 3 km across the valley and splashed more than 100 m up the mountain on the other side in less than two minutes. In this particular case a coal mine that had undercut the lower slope of the mountain caused the disaster. In another case a large earthquake in August, 1959 jolted a mass of rock off the south wall of the narrow canyon of the Madison River a few kilometers west of Yellowstone Park (Figure 6-30). It poured down the mountain and across the river so rapidly that its momentum carried it well up onto the opposite valley wall before it came to rest a few seconds later. Part of that slide mass dammed the Madison River, creating Quake Lake, Montana.

Large rockfalls occasionally develop into such **rock avalanches,** which pour down slopes as great sheets of shattered rock that move almost as though they were fluid. Rock avalanches may travel several kilometers at speeds that reliable investigators soberly estimate to be well in excess of 100 kilometers per hour (km/hr)—in some cases over twice that fast! They finally come to rest as an undulating plain of broken rubble, such as that shown in Figure 6–31.

One theory concerning the great mobility of rock avalanches holds that they ride on a cushion of compressed air. Trees near the edges of large rock avalanches blow down with their tops pointing away from the swath just as they would had they been felled by a blast of compressed air escaping from beneath the moving slide mass. Furthermore, passing rock avalanches may skip over patches of hillslope leaving them quite untouched, which also suggests that the mass may have moved on a cushion of air.

FIGURE 6–30

The Madison Canyon slide, Montana. The large area of slope on the far side of the canyon failed during the severe earthquake of August, 1959. It poured across the river as a very mobile rock avalanche, suddenly burying a U.S. Forest Service campground. The slide also dammed the river, creating Quake Lake. (J. R. Stacy, U.S. Geological Survey)

FIGURE 6-31
The undulating field of angular rubble in the foreground is a rockfall that tumbled from the volcanic hills on the skyline. Chaos Jumbles in Lassen National Park, California.

However, orbital photographs show a few landforms on the moon that look very much like terrestrial rock avalanches. If that is indeed what they are, then the air cushion theory does not provide a general explanation of how rock avalanches move because the moon has no atmosphere. An alternative theory holds that rock avalanches keep themselves in suspension for the few seconds of their great mobility simply through the internal impacts of their rock fragments clattering against each other. That could happen without the help of an air cushion but does not necessarily imply that none exists.

Mudflows

The name defines itself. Mudflows are thick slurries of soil and water that pour down slopes, posing great danger to human life and causing enormous property damage in developed areas. The commonest sources of mudflows are desert canyons and active volcanoes but they occur anywhere that large volumes of sloping and water-soaked ground exist. Many landslides turn into mudflows as they move downslope.

Mud is much denser than water—commonly more than twice as dense—and it exerts a correspondingly greater buoyant effect on rocks submerged in it. Therefore, large rocks are much more easily transported by mudflows than by clear water flowing at the same rate. Many mudflows carry boulders weighing several tons without significantly rounding them, which suggests that the mud must raft the boulder along instead of rolling it. Mudflow deposits typically consist of an unsorted mixture of boulders, pebbles, sand, and mud stirred indiscriminately together. It is difficult to distinguish such material from that deposited by glacial ice.

The torrential runoff from desert rains often gurgles out of the mouths of canyons and across alluvial fans as a thick mudflow. Many alluvial fans consist largely of mudflow deposits accumulated to great thickness. Desert mudflows pour through many communities laid out on alluvial fans in the southwestern United States nearly every spring, causing great property damage. Those desert mudflows most commonly occur when heavy rains follow large brush fires that left the mountain slopes barren and even more exposed than they usually are to rainsplash and surface runoff erosion.

A thick blanket of freshly erupted ash on the flanks of a volcano makes excellent material for a mudflow, which needs only water to complete the mixture. Snow and ice melted by the heat of the eruption usually provide ample water. If they do not, rain probably will because erupting volcanoes vent large volumes of steam, which condenses to form clouds. The most dangerous aspects of many volcanic eruptions are large mudflows that pour down stream valleys at speeds as great as 50 km/hr.

Tundra Erosion

The term **soil flowage** refers to downslope movement of extremely fluid masses of water-soaked soil, a process that most commonly occurs when the subsoil is frozen.

Frozen ground is absolutely watertight because any water that may start to seep through it freezes and plugs its own channel with ice. When the upper levels of frozen ground begin to thaw in the spring, they become extremely

wet because the still-frozen subsoil beneath prevents downward percolation of surface water. The sloppy and weak surface soil tends to flow downslope sandwiched between the matted roots of the sod above and the hard and watertight frozen subsoil beneath (Figure 6–32).

In most heavily populated regions where the appropriate conditions last only a week or two during the spring, soil flowage merely creates a few lumpy slopes and may occasionally tilt a small tree. In cold regions where the subsoil remains permanently frozen—a condition known as **permafrost**—soil flowage may continue for nearly half the year and become the dominant process of hillslope erosion. In permafrost regions the so-called **active layer** of thawed surface soil drains off the high places and fills the low ones until virtually no relief remains. The result is a nearly featureless plain, the typical tundra landscape of large regions in arctic latitudes. Tundra landscapes also exist on a much smaller scale on the tops of high mountains where they generally appear as large areas of low relief. Many geologists suspect that soil flowage may have dominated development of landscapes in large regions around the great continental glaciers of the ice ages.

Large areas of the northern part of the midwestern United States and central Europe have flat landscapes underlain by glacial deposits, which date from earlier ice ages but not the latest one. Glaciers do not leave their deposits as flat plains but instead create intricate landscapes

FIGURE 6–32

The water-saturated, weak soil in the active layer flows downslope between the frozen subsoil below and the sod above, creating shallow hollows and corresponding low bulges on the slope.

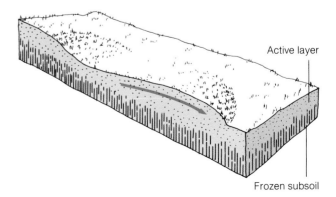

Active layer

Frozen subsoil

containing a variety of hilly landforms. Many geologists interpret the flat plains underlain by older glacial deposits as fossil tundra landscapes leveled by soil flowage during later ice ages.

Permafrost makes a treacherous foundation. Paved roads and airstrips destroy themselves by absorbing enough heat to thaw the ground beneath and then sinking into the active layer. That is why so much heavy transport in the Arctic moves over winter roads and why so many permanent roads are left unpaved even though they carry heavy traffic. Most buildings are set on piers so they will not thaw the ground beneath them and then sink unevenly into the muck. All construction on permafrost requires designs that will preserve the heat balance in the soil.

Permafrost generally develops a distinctive network pattern of cracks or stone stripes, which outline four, five, or six-sided figures that vary in size according to the severity of the climate. The smaller polygons, which could fit into an average room, form near the southern boundary of the permafrost regions. The larger polygons, such as those shown in Figure 6–33, form in the colder regions of the Arctic and could easily surround a large house.

Soil polygons appear to begin their formation when shrinkage cracks develop in frozen ground as it contracts during extremely cold weather. The cracks fill with ice during the winter and rocks topple into them as the active layer thaws during the spring. The rocks establish the cracks as persistent zones of weakness in the soil, zones that open repeatedly every winter and then accumulate more rocks the next spring. Soil polygons become conspicuous features outlined by rock-filled cracks a meter or more deep and nearly half that wide at the top.

That network of rock-filled cracks makes an excellent system of natural gutters. The soil polygons carry runoff water during the warmer months thus performing the same function that small stream tributaries serve in other types of landscapes. Therefore, streams draining tundra landscapes tend to lack small tributaries and to have a map pattern that resembles a severely pruned tree.

EROSION AND COMPENSATING UPLIFT

The streams that drain most landscapes haul their burdens of eroded sediment out of the region and dump them into the ocean. Since the continents float on the earth's mantle, they rise to compensate for the material removed by erosion. The interaction between erosion and compensating

FIGURE 6–33
Patterned ground near Point Barrow, Alaska. (T. L. Pewe, U.S. Geological Survey)

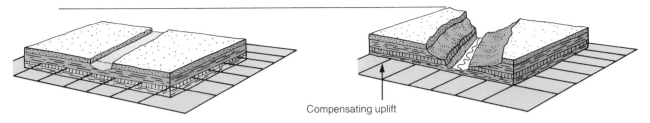

FIGURE 6-34

The crust of the earth floats higher to compensate for erosional denudation. If most of the eroded material comes from the valleys, the compensating uplift will increase both local relief and the absolute elevations of the hilltops. On the other hand, if the processes of erosion remove material from the divides at the same rate as from the valleys, they will reduce both the relief and elevation of the landscape regardless of compensating uplift.

uplift affects the evolution of all landscapes except those that develop in undrained deserts where eroded sediment does not leave the region.

Continental rocks, which have an average density approximately 2.7 times that of water, float on the heavier rocks of the mantle, which have density approximately 3.3 times that of water. The heavier mantle rock flows in beneath the continent to replace the lighter rock eroded from its surface so the amount of compensating uplift must be in the ratio of those densities. Therefore, every 100 m of eroded rock will be compensated by approximately 80 m of continental uplift—if we assume that the continent is not rising or sinking for other reasons that have nothing to do with erosion.

If erosion and compensating uplift could continue indefinitely, the continents would have vanished long ago. Continents survive despite billions of years of erosion because the ordinary erosional processes cut no deeper than sea level. Erosion and compensating uplift of a landmass must therefore continue at a decreasing rate until both processes nearly cease as the elevation approaches sea level. Vast areas of the more stable continental surfaces do indeed stand at elevations less than about 200 m above sea level.

It is tempting to imagine the hills melting away before the processes of erosion as though they were so many sugar cubes in a rainstorm. Unfortunately, the metaphor is misleading. Recall that compensating uplift will maintain much of the original elevation in the eroding landscape. Also re-call that a balance between the rates of weathering and erosion seems to dictate the slopes of hills in ridge and ravine landscapes. Therefore, simply determining the rate of erosional denudation cannot possibly lead to a direct estimate of how long the hills will last before they finally disappear. Neither does such an estimate provide any useful insight into how the hills may change in shape as erosion proceeds. Consider a few possibilities.

If the processes of erosion denude all parts of the landscape at the same rate, then it seems that the hills may remain essentially unchanging as they slowly approach sea level. Under those conditions, every 100 m of erosion will simply reduce the general level of the landscape by about 20 m without affecting its topographic relief. However, if the rate of erosion is not uniform on all surfaces, the landscape will evolve quite differently.

For example, if most of the soil creeping down the slopes of a ridge and ravine landscape comes from the ridge crests, then erosion will reduce topographic relief more rapidly than it reduces the general level of the landscape. On the other hand, erosion may remove more material from the valley floors than from the hilltops. Then compensating uplift will increase both topographic relief and the elevations of the hill crests (Figure 6–34) so that erosion and compensating uplift will make the hills grow higher in all respects. Unfortunately, geologists still know too little about the relative rates of erosion on different parts of a landscape to make general statements about the way in which they change with time. However, there are a few specific cases in which it is possible to determine what is happening.

Consider, for example, Hell's Canyon of the Snake River along the border between Oregon and Idaho. This canyon is the deepest in North America. Hell's Canyon is eroded through a thick section of lava flows that comprise the volcanic Columbia Plateau. It is possible to show that the plateau surface on both rims of the canyon is almost untouched by erosion. Clearly, then, erosion has been removing material from the canyon while leaving the adjacent drainage divides intact. The surface of the Columbia Plateau comes to its highest elevation at the rim of Hell's Canyon, evidently because it has been raised there by uplift compensating for erosion of the adjacent canyon. An exactly similar situation exists in the case of the Grand Canyon of the Colorado River where a plateau surface also reaches its highest elevations along the canyon rim.

SUMMARY

Each process of hillslope erosion leaves its characteristic imprint on the landscape.

Rainsplash and surface runoff erosion operate where the soil is so sparsely covered by plants that raindrops can strike directly onto bare soil. Those processes operate primarily in deserts but also become effective in other regions if the plant cover is damaged. Rainsplash and surface runoff erosion create intricately dissected hills flanked by smoothly graded desert plains, which grow as erosion reduces the hills. Such landscapes are drained by a network of closely spaced streams that depend upon surface runoff for their flow and therefore remain dry except during wet weather.

Soil creep erosion moves the entire soil mantle almost as though it were an extremely viscous liquid—the upper levels move fastest and the rate of motion diminishes with depth. Landscapes shaped by soil creep erosion typically consist of hills that have convex crests, straight midsections, and concave toes. Streams draining such ridge and ravine landscapes depend upon stored ground water instead of surface runoff for most of their water supply and therefore flow during all seasons. The clarity of the streams reflects the slow rate of soil creep erosion.

Landsliding moves large masses of soil and weathered rock on slip surfaces that develop where the steepness of the slope exceeds the angle of repose of the soil. Landslides most commonly occur where the soil mantle has been weakened by water saturation. They may also develop on slopes that have been oversteepened either by undercutting at the toe or loading at the top.

Rockfalls move on surfaces of weakness within the bedrock. They may involve slow movement of large blocks or more rapid movement of masses of broken rock. Some develop into rock avalanches that move as though they were fluid.

Mudflows are dense slurries that exert a much greater buoyant effect than clear water and therefore may carry large boulders. Mudflows are commonest in deserts and around active volcanoes but may occur anywhere large areas of sloping ground become watersoaked.

Soil flowage generally involves movement of thawed and mobile surface soil on a base of frozen subsoil. It levels the landscape by draining soil off the high places and into lower ones. Tundra plains leveled by soil flowage exist on a large scale in the Arctic and on a much smaller scale in high mountains.

Continents float so erosional denudation causes compensating uplift to maintain floating equilibrium. The relationship between the style of erosion and compensating uplift determines how the landscape will change with time.

KEY TERMS

active layer	desert plain	rainsplash erosion	soil flowage
alluvial fan	mass wasting	ridge and ravine landscape	surface runoff erosion
angle of repose	pediment	rock avalanche	surface sealing
bajada	permafrost	rockslide	
bolson	playa	soil creep	

REVIEW QUESTIONS

1. In the Rocky Mountains it is commonplace to see alluvial fans dissected by streams. What sequence of changes in climatic and watershed conditions does that situation suggest?

2. Why do vehicle tracks and unpaved roads tend to increase the volume of surface runoff and to become the sites of new gullies?

3. What steps would you suggest taking to arrest the movement of a landslide?

4. How would you determine—by looking at a hillslope—whether soil creep, rainsplash and surface runoff, or landsliding is the dominant process of hillslope erosion?

5. What properties make certain surfaces relatively immune to rainsplash and surface runoff erosion regardless of whether or not they have a dense plant cover?

6. Why is large-scale terrestrial sedimentation typical of desert regions?

7. What are the distinctive features of a landscape shaped primarily by permafrost processes?

SUGGESTED READINGS

Carson, M. A. and M. J. Kirkby. *Hillslope Form and Process.* London: Cambridge University Press, 1972.
  A sophisticated treatment of hillslope erosion.

Davies, J. L. *Landforms of Cold Climates.* Cambridge, Mass.: The M.I.T. Press, 1969.
  An easily readable text devoted entirely to landforms that develop on permanently frozen ground.

Garner, H. F. *The Origin of Landscapes: A Synthesis of Geomorphology.* New York: Oxford Press, 1974.
  Just what the title offers—a truly heroic synthesis of the entire subject.

Leopold, L. B., M. G. Wolman, and J. P. Miller. *Fluvial Processes in Geomorphology.* San Francisco: W. H. Freeman & Co., 1974.
  Deals mostly with streams but also includes an excellent section on hillslope erosion.

Mabbutt, J. A. *Desert Landforms.* Cambridge, Mass.: The M.I.T. Press, 1977.
  An excellent treatise on desert landforms.

Sharpe, C. F. S. *Landslides and Related Phenomena.* New York: Columbia University Press, 1938.
  A small classic. Still perfectly useable.

Twidale, C. R. *Analysis of Landforms.* Sidney, Australia: John Wiley, Australasia Pty. Ltd., 1976.
  A comprehensive treatment of the entire subject of landscapes. Excellent.

Young, A. *Slopes.* Edinburgh, Scotland: Oliver & Boyd, 1972.
  A technically detailed work on hillslopes.

CHAPTER SEVEN

Rounded boulders and sculptured bedrock
in the floor of a narrow canyon. (J. K.
Hillers, U.S. Geological Survey)

STREAMS

We tend to identify with our land in terms of the stream that gives it a unique character in our minds. It would be difficult, for example, to find a stream that passes through more varied landscapes than those along the Snake River. It starts in the cold high country of Yellowstone Park, loops through the bleak deserts of the Snake River Plain, and then passes through the green depths of Hell's Canyon before finally joining the Columbia River. Yet the people who live within its basin think of themselves as residents of the "Snake River Country."

Streams contribute to development of the landscape by eroding their beds and by functioning as conveyor belts to haul away the debris dumped into them by hillslope erosion operating above their floors. Figure 7–1 illustrates the relative importance of the two functions. Streams are difficult to study because they tend to perform most of their geologic work during occasional great floods when direct observation is difficult. Our knowledge is still insufficient to enable people who attempt to manage streams to correctly anticipate all the consequences of their actions.

Streams flow from drainage basins that supply them with a flow of water and a burden of sediment that vary seasonally and even from day to day. The steepness of the stream's gradient and the shape and course of its channel change in response to variations in the supply of water and sediment. Geologists generally assume that streams tend to exist in a *graded* condition, a state of equilibrium in which the supply of water suffices to carry the load of sedi- ment without net deposition or erosion in the channel. The mechanisms that maintain that condition are the same as those that adjust desert plains—which are also graded surfaces—to their supply of water and sediment.

In the following sections, we will examine various aspects of stream behavior, illustrating many of them with examples of stream response to human interference. Those are the cases in which the cause and effect relationship is most readily apparent. The picture that will emerge portrays a complex geologic agent predictable in many ways but not in all.

STREAM FLOW

The history of the Blackwater River in Missouri illustrates several interesting aspects of stream flow. In 1910 the channel was straightened and deepened, an operation often called "channelization," in an effort to speed its flow and thus control flooding. By 1970 the stream had eroded its bed as much as 8 m deeper, a process called **entrenchment,** and broadened it by as much as 60 m in some places. In so doing it destroyed many of the bridges along its length and forced extensive rebuilding of those that survived. Tributary streams were also affected even though they had not been part of the original project because they had to entrench their own beds to keep pace with the river. Channelization projects commonly cause such responses.

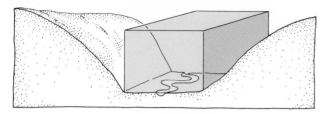

FIGURE 7-1
Because it operates on its valley floor, the stream can be directly responsible for eroding only the material enclosed within the colored volume. Hillslope processes must have eroded the rest and dumped it into the stream to become part of its sediment load.

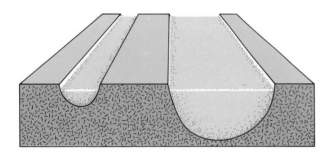

FIGURE 7-2
Friction with the channel and to a much lesser extent with the air directly affects proportionately more water in small than in large streams. In general, streams flow fastest in the parts of their channels most remote from frictional restraint.

FIGURE 7-3
Even though their cross-sectional areas are comparable, the broad shallow channel exerts more frictional restraint than the more nearly semicircular one. The reason for this is that a much larger proportion of the water flows through the broad channel in close contact with the streambed. The semicircular form offers the least frictional resistance because it encloses the maximum volume of water within the minimum area of surface.

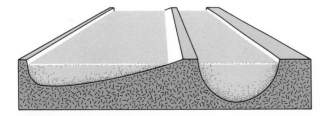

They increase the stream's flow speed and therefore its capacity to erode in two different ways: (1) by steepening the slope or **gradient** of the streambed; and (2) by reducing the amount of friction between the water and its channel.

Stream Gradient

Straightening bends steepens the gradient of a stream because it shortens the travel path between two points at different elevations. A hill likewise seems steeper if you walk straight down instead of following a zigzag course.

Streams tend to straighten their channels when an increase in their supply of water or sediment makes it necessary for them to flow faster. During floods, streams generally take shortcuts across bends, eroding channels that carry little water during periods of normal flow. Streams also straighten their courses when an excessive burden of sediment causes deposition in the channel—both the straightening and the deposition steepen the stream gradient and speed the flow of water.

Channel Friction

Water flowing through a stream channel, like a toboggan sliding down a hill, responds to the accelerating pull of gravity and the restraining drag of friction. Both the water and the sled travel at speeds determined by the steepness of the slope they descend and the amount of frictional resistance it offers.

Stream channels are typically steepest near their headwaters so the pull of gravity on the water and its tendency to accelerate both diminish downstream. However, friction between the water and its channel is also greatest in the headwaters. Friction diminishes downstream partly because the rocks in the bed, which determine the roughness of the channel, decrease in size downstream and partly because the size of the channel, which determines the amount of contact between the water and its bed, generally increases downstream (Figure 7–2). Furthermore, most streams tend to develop channels with a more nearly semicircular cross section in their downstream reaches and, as Figure 7–3 shows, they offer less frictional resistance to flow than does a broad shallow channel. The general tendency for channel friction to decrease downstream explains why most streams reach their maximum velocities near their mouths. Whitewater rapids in the upper reaches of a stream reflect high frictional resistance to flow rather than high velocity.

Channel friction explains the distribution of flow velocities within streams. The water generally flows most slowly where it is closest to the streambed and faster as the distance from the bed increases. The air also offers slight frictional resistance so we generally find the highest flow velocities just below the stream surface in the part of the channel most remote from the bed.

Volume of Flow

Geologists use the term **discharge** to refer to the total amount of water a stream carries, a quantity they determine by measuring the cross-sectional area of the channel and multiplying that by the average flow rate (Figure 7–4). When the watershed furnishes an increased supply of water, the stream responds by rising, thus increasing the cross-sectional area of its channel. That also increases flow velocity by reducing the proportion of the water in contact with the streambed, thus diminishing channel friction. When the amount of water furnished to the stream be-

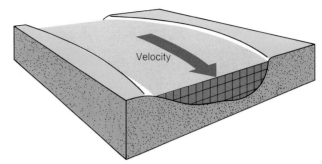

Discharge = Flow velocity × Channel cross-section area

FIGURE 7–4
The amount of water flowing through a stream can be calculated by multiplying the flow velocity times the cross-sectional area of the channel.

comes very great, it further enlarges its channel by picking up the material in its bed and carrying that along with the water to the accompaniment of distinctly audible grinding noises. It is then that streams carry most of their load of sediment. Figure 7–5 shows a typical sequence of changes in channel cross section during a large flood that

FIGURE 7–5
How the channel of the San Juan River near Bluff, Utah responded to a flood in 1941. The outline of the low water channel on September 9 is shown superimposed in color on the flood channels. On September 15 the river surface had risen and the old channel was nearly full of sediment. On October 15 the river had risen further and deeply scoured its bed. By October 26 it had begun to fall and to fill the scoured channel.

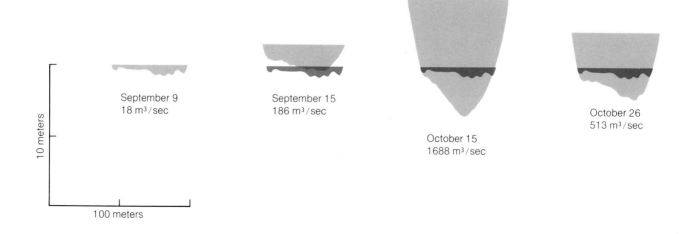

September 9
18 m³/sec

September 15
186 m³/sec

October 15
1688 m³/sec

October 26
513 m³/sec

10 meters

100 meters

FIGURE 7-6

An aftermath of the 1948 flood on the Columbia River. (F. O. Jones, U.S. Geological Survey)

scoured the streambed. Figure 7–6 shows a bridge wrecked when the streambed beneath its piers moved during such a flood.

Floods and Floodplains

People who live on a floodplain are in fact choosing to live in the river rather than beside it as they generally assume!

Most streams flood from time to time when their drainage basins deliver more water to them than their channels can accommodate. The area of valley floor inundated during such periods of high water is called the **floodplain.** Figure 7–7 shows an aerial view of a typical floodplain covered with an intricate scrollwork of abandoned stream channels now in various stages of filling with sediment. It vividly illustrates why many geologists argue that the floodplain is an integral part of the river no less vital to its natural functioning than the stream channel itself. Extensive residential or industrial development of floodplain property converts perfectly natural and otherwise harmless floods into costly disasters, such as that shown in Figure 7–8. That in turn creates demand for expensive flood control projects, which may cause at least as many problems as they solve.

Considered from an operational point of view, the floodplain is a storage facility. It holds water on its surface during floods and within the pore spaces in the deposits of sand and gravel beneath it during all seasons. Water seeps into the pore spaces within the floodplain sediments as the stream begins to rise and finally pours out over the floodplain surface as it overflows its banks. When the stream begins to fall, it drains water first off the surface of the floodplain and then out of the sediments beneath it. Therefore, water stored on and within the floodplain helps stabilize stream flow during wet and dry seasons alike. Efforts to control flooding by taking parts of the floodplain out of use with dikes or landfills merely displace water from one area to another and relocate the problem without solving it. The better solution—which is now being adopted in many areas—is to restrict floodplain development through zoning.

Geologists classify floods according to their **recurrence interval,** the probability that a flood involving a certain amount of water will occur within any given year. There is 1 chance in 10 that a 10-year flood will occur in any year, and 1 chance in 50 that the same year will see a much larger 50-year flood. It is reasonable to expect one 10-year flood every decade and 10 every century. There is, however, a chance that a decade may pass without bringing a 10-year flood just as there is a chance that several successive tosses of a coin may all come up heads. The recurrence interval method of describing floods predicts nothing: Last year's flood tells nothing about what may happen this year just as the last flip of a coin does not foretell the result of the next. It is perfectly possible for 50-year floods

FIGURE 7-7

A vertical air view of the floodplain of the Big Hole River in the Beaverhead Valley of Montana taken from an elevation of about 1,800 m. The present channel follows the southern edge of the floodplain in the lower part of the picture. Numerous sloughs and old meander scars mark former locations of the channel. The floodplain narrows abruptly near the right hand edge of the picture where the river has cut a narrow gorge through a ridge of resistant quartzite it encountered as it deepened its valley. Irrigation ditches show as wave-white lines in the northern part of the floodplain. (U.S. Bureau of Land Management)

to occur in two successive years and equally possible for a century to pass without bringing one such flood.

Flood classification still rests on an uncertain footing in many regions where stream records cover less than a century and may not include many long interval floods. That makes it necessary to estimate recurrence intervals from a short data base supplemented by such circumstantial evi-

dence as the location of old flood debris, the ages of flood damaged trees, and the recollections of elderly people. Therefore, attempts to estimate recurrence intervals, especially the longer ones, commonly lead to legitimate differences of scientific opinion, which may be resolved in court now that floodplain zoning and land use planning are becoming widespread realities.

FIGURE 7–8

A major bridge washed out by a flood on the Feather River in California. The dike in the foreground
has also been breached. (W. Hofmann, U.S. Geological Survey)

Flood Control

Local flood control projects generally shift the water from one part of the floodplain to another, making one person's flood protection another's flood hazard. In such cases a decision to protect people living in one part of the floodplain inevitably includes an implicit decision to inflict greater flood hazard on others living elsewhere.

The most commonly used methods of local flood control involve constructing a dike to keep the water off part of the floodplain or simply raising the ground level with fill. Neither affects the amount of water in the stream; they merely displace it from one part of the floodplain to another. If the river is denied part of its storage area, it must back the water up deeper in the part still available. The only alternative,

which is rarely pursued, is to provide alternative storage areas to replace those removed by the dike or fill.

Years may elapse between construction of a dike and the season that brings a flood large enough to demonstrate the cause and effect relationship between flood control in one area and increased flooding in others. Then the people who suffered from the displaced flood generally demand a new dike for their own protection. Our cultural tradition, which demands that we control natural events, and our legal system, which regards natural disasters as acts of God, compound the problem.

As dikes and landfills proliferate, they force floodwaters to rise ever higher in the diminishing area of floodplain still available for their storage. Therefore, many record flood crests do not mean that the stream is carrying more water than ever before but rather that less floodplain storage area is available than formerly. Furthermore, dikes confine the floodwaters to a channel, which offers relatively little frictional resistance to flow and therefore greatly increases the rate at which flood crests move downstream.

STREAM TRANSPORT AND EROSION

Although there are exceptions, streams typically remain relatively clear during periods of low water and move large quantities of sediment only during floods, especially during large floods. The available data do not resolve the question of whether the enormous amounts of sediment that move as streams scour their beds during great floods outweigh the smaller amounts that move more frequently during lesser floods.

Stream Load

The least conspicuous part of the burden streams carry, and in most cases the least important part, is the **dissolved load,** which consists of material traveling in solution. We tend to overlook the dissolved load because it is invisible. It probably, though, amounts to about 20 percent of all the material streams carry although the actual proportion varies greatly from one stream to another. In some cases, dissolved load constitutes the entire load and is far from small. For example, the perfectly transparent waters of Silver Springs in Florida carry about 1 ton of calcium carbonate dissolved in about 40,000 tons of water every three minutes. That is approximately the equivalent of rolling a spherical boulder of limestone 90 cm in diameter down the spring run every three minutes—quite a respectable load.

The **clastic load,** which consists of eroded particles of soil or rock, accounts for most of the sediment streams carry. Silt and clay, the smaller size fractions of the clastic load, generally travel in suspension because the turbulently flowing water contains currents that move upward at rates fast enough to prevent their settling. Although sand may also travel in suspension during floods, it more commonly bounces in short hops along the stream bed, a mode of transport called **saltation.** Sand is part of the **bed load.** The rest of the bed load consists of larger clastic particles—pebbles, cobbles, and boulders—that move when the stream scours its bottom during floods. The larger bed load particles generally remain in place between floods and accumulate scummy coatings of algae that could not grow on them if they were moving.

We often see sand grains bounding along the bed of a clear stream that promptly becomes muddy if we stir the bottom. That raises the question of why the stream moves sand grains but not the much smaller particles of silt and clay even though it is obviously capable of carrying them in suspension. The answer lies mostly in the great cohesiveness of silt and clay that makes them difficult to erode and also explains why you can make a mudball more easily than a sandball. The rest of the explanation lies in the existence of a paper-thin layer of bottom water that flows in straight lines instead of turbulently because it is kept nearly still by friction. Water in *laminar flow*, as that straight line motion is called, neither erodes the bottom nor carries sediment. The grains of sand are large enough to project through the layer of bottom water into the turbulent flow above whereas the smaller particles of silt and clay do not. That bottom layer of laminarly flowing water disappears during floods and then the stream becomes muddy as it picks up silt and clay.

Geologists use the term **competence** to refer to the largest particles of sediment a stream can move. Stream competence increases sharply with flow rate but has relatively little relationship to discharge. The term **capacity** refers to the total clastic load a stream can carry. Stream capacity increases rapidly with discharge and depends to a much lesser extent upon the flow rate.

Stream Erosion

John Wesley Powell, the pioneer geologist who led the first boat exploration of the Grand Canyon during the 1870s, ignored the gratuitous advice of local people who warned of dangerous waterfalls ahead. Powell was among the first to clearly understand that it is the abrasive sediment streams carry, not the flowing water itself, that carves bedrock. Figure 7–9 shows bedrock and large boulders sculptured by muddy water. Therefore, Powell was confident that the muddy waters of the Colorado would quickly carve away the lip of any rock ledge they might cross thus ensur-

ing an absence of waterfalls. Have you ever wondered why waterfalls always seem to occur in clear streams?

When low water exposes large areas of bedrock stream-bed, **potholes** such as that shown in Figure 7–10 often become visible. They form beneath persistent eddies that turn pebbles on the bottom to grind cylindrical holes ranging in size from that of a tin can to occasional giants several meters in diameter. It is sometimes possible to watch pebbles turning in the bottom of a pothole.

Quartz, the hardest abundant rock-forming mineral, comprises a major part of the load of most streams, providing them with an abrasive capable of cutting through any kind of rock. We commonly see streams, such as that shown in Figure 7–7, forming wide valleys where they pass through weaker rocks and narrow gorges in the more resistant formations. No rock can actually stop a stream. However, the hardness of quartz does not fully explain why streams do not always follow the weaker rock units as common sense suggests they should.

Many streams first began to flow across rocks well above their present levels. Then, as hillslope and stream erosion reduce the general level of the landscape, the stream cuts down into rocks that had been buried when it started to flow (Figure 7–11). If some of those rocks are more resistant to hillslope erosion than others, they will form hills as erosion preferentially removes the less resistant rocks. However, the stream must cut its valley through the more resistant rocks as it encounters them. It cannot flow uphill to escape from its valley and seek an easier route across less resistant rocks. Such **superimposed streams,** as they are called, exist in many parts of North America. Their tendency to cut through ridges of resistant rocks instead of going around them is especially striking in the Appalachian and Rocky Mountains.

Although they are less common, it is also possible to find examples of **antecedent streams** that cut through mountains because the stream was there before the range rose. Such streams simply maintain themselves at grade as the land rises at a slower rate than the stream can erode its bed. Several streams pass right through the northern part of the Sierra Nevada Range in California and must have been flowing in essentially their present courses before the

FIGURE 7–10

Potholes in the bed of a small stream. Pebbles turning beneath persistent eddies drill such holes into the bedrock.

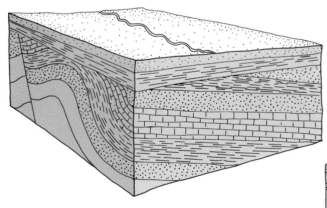

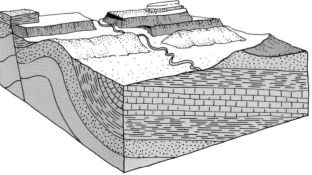

FIGURE 7–11

An example of how streams develop courses that cut across rocks of varying hardness instead of following the path of least resistance along the more erodible rocks. In this case the stream started flowing on a flat-lying sequence of sedimentary strata then cut down into tilted layers lying beneath an unconformity. Remnants of the upper sequence of sediments remain on a few hilltops to tell the story.

range began to rise several million years ago. Numerous deep canyons in the Colorado plateau, including the Grand Canyon, were carved by streams that maintained themselves in their original courses as the land rose during the past several million years.

CHANNEL BEHAVIOR

Meanders

Many streams loop back and forth down their floodplains in an intricate series of sweeping curves called **meanders** (Figure 7–12), which have an almost constant radius pro-

portional to the width of the stream. That habit is intriguing because it contrasts so sharply to the behavior of solid objects that move in a straight line unless some force deflects them. Why streams meander is a question geologists approach from different directions, which leads to strongly contrasting interpretations.

We can watch meanders form in the laboratory by observing what happens as water flows through an initially straight channel moulded in erodible material. Sooner or later one of the banks will cave, deflecting the flow against the opposite bank—the flow then continues to rebound from one bank to the other (Figure 7–13), eroding alternate areas on opposite banks and thus developing a sinuous

FIGURE 7–12
The Laramie River in Wyoming meandering across a floodplain covered with a complex pattern of old meander scars and abandoned channels. (J. R. Balsley, U.S. Geological Survey)

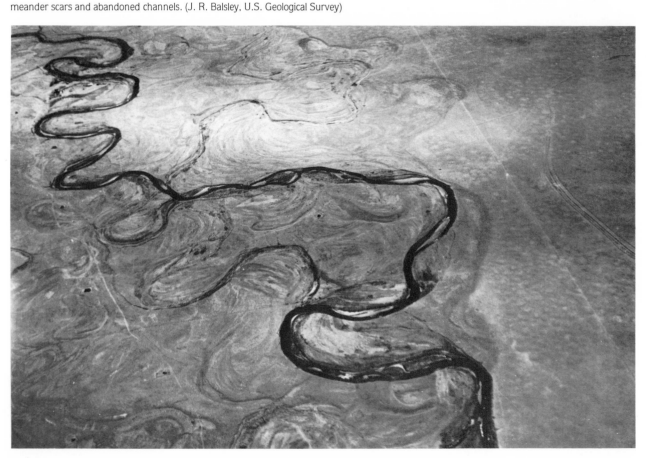

FIGURE 7–13

The flooding Walla Walla River resumes a sinuous course as it breaches dikes that had confined it to a straight channel near Milton-Freewater, Oregon. (A. O. Waananen, U.S. Geological Survey)

course. Therefore, it appears experimentally that any irregularity in an initially straight channel will tend to propagate itself downstream in a series of bends. However, many observations show that the meandering habit is not confined to ordinary streams flowing within banks.

Infra-red imagery received from weather satellites shows ocean currents meandering in their courses, and it seems that the jet stream meanders in its wanderings through the upper atmosphere. The explanation for such behavior cannot depend upon a mechanism involving interaction between the moving fluid and its banks. Perhaps streams meander for some other reason.

Mathematicians have shown that the most probable random path between two points on a surface is a meandering curve identical in form to that followed by many streams. Of course, mathematical descriptions of natural phenomena do not necessarily constitute explanations. They can, however, and at least a few geologists believe that in this case the description is indeed an explanation. If so, then it would seem that streams meander simply because they are following random paths down their floodplains, that the tendency to meander is inherent in the nature of streams and requires no further explanation. Instead of asking why so many streams meander, we should perhaps ask why some do not.

Whatever the causes for meandering, the habit is best developed in streams that flow on a very gentle gradient. Geologists have interpreted that observation in several different ways. Some argue that meanders form as the stream lengthens its course to develop a flat gradient. Those who regard meandering as random behavior contend that streams flowing on a very gentle gradient meander simply

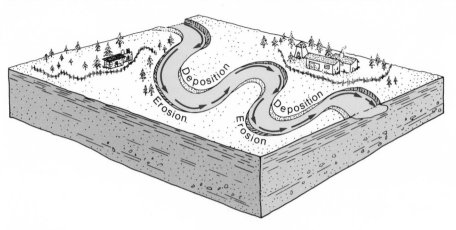

FIGURE 7–14

Stream meanders tend to acquire exaggerated forms because the main thread of the current always clings to their outside banks. Eventually, the stream cuts across their narrowed necks and leaves them abandoned on the floodplain as oxbow lakes.

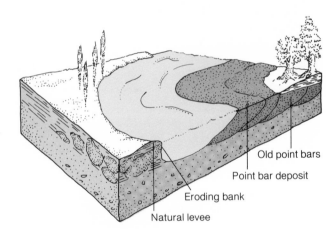

FIGURE 7–15

Streams erode the outside banks of their bends, making them steep; they deposit sediment on the inside banks.

because they are more free to wander than those descending a steeper slope. Still other geologists argue that streams develop very flat gradients and strongly meandering courses where circumstances force them to direct their erosive energies laterally rather than vertically. Such circumstances may happen as a stream approaches a **base level** that limits its ability to erode its course deeper. Resistant rock formations that slow the rate at which streams erode their channels can be regarded as temporary and local base levels. The ultimate base level is sea level.

Bank Erosion

The momentum of the water makes the main current of a stream cling to the outsides of meander bends (Figure 7–14) just as a car tends to drift to the outside of a curve. Therefore, streams erode the outsides of their bends, carving their deepest channels there and undercutting the bank (Figure 7–15). The undercut outside banks eventually cave into the stream and most of the material moves to the **point bar** along the inside bank of the next bend downstream. It is possible to demonstrate that the sediment goes from the outside bank of one bend to the inside bank of the next by using marked gravel to trace its course.

Erosion on the outside bank combined with deposition on the inside bank enlarges meander bends into exaggerated forms. The stream finally ends that growth by cutting across the neck of the bend, leaving the old meander loop abandoned on the floodplain as an **oxbow lake,** which eventually fills with sediment. Therefore, the stream lengthens its channel as it erodes the outside banks of meanders and shortens its channel as it cuts across their necks to prune them off. The two processes generally cancel each other so the total length of the stream channel, the stream gradient, and the regular relationship between meander radius and stream width all remain essentially constant.

People resort to a variety of techniques in their efforts to stabilize streams in their channels by arresting bank erosion. Lining the outside banks of meanders with rip rap consisting of large blocks of rock is generally effective if the blocks are larger than anything the stream is competent to carry during a flood. Old car bodies, such as those shown in Figure 7–16, rarely prevent bank erosion except in small streams. However, they do greatly increase channel roughness and may therefore increase flood hazard upstream.

FIGURE 7–16

Old car bodies line the banks of the Bear River near Brigham City, Utah. The pair of faint benches low on the distant mountainside are old lake shorelines formed during the Ice Age when abundant rainfall filled the valley with water.

Natural Levees

As muddy floodwaters rise out of the stream channel and pour onto the floodplain, they encounter great frictional resistance to flow and become almost stationary. Air photos of large floods typically show the stream channel as a winding line of muddy water threading its way through broad expanses of clear water stored on the floodplain. If the muddy water that overflows the channel is clear by the time it moves well out onto the floodplain, then sediment must be accumulating along the stream banks. Those accumulations become an important factor in the behavior of streams that are building up their floodplains through deposition—a process called **aggradation.**

Continued deposition along the stream banks builds **natural levees** (Figure 7–17), which are typically the highest and therefore the driest parts of the floodplain. As aggradation continues, the floodplain develops a broadly convex cross-sectional profile (Figure 7–18) with the stream following the crest of the convexity. That is an unstable situation because it places the stream at an elevation higher than much of its floodplain.

Eventually, the stream erodes through one of its natural levees. It then flows downslope onto the lower parts of its floodplain where it must cut a new channel all the way to its mouth. It cannot return to any part of its old channel because that would require water to flow uphill. High natural levees most commonly fail during floods. Simple ditches, however, have on occasion diverted large streams into entirely new channels and greatly shifted the locations of their mouths.

As the stream cuts its new channel, it leaves the old one

FIGURE 7–17
High natural levees stand above floodwaters on the lower Mississippi delta in southern Louisiana.

abandoned on the floodplain as a winding lake called a **slough** in most regions of North America, or a **bayou** in some parts of the southern United States. The Mississippi River, for example, has changed its course many times, leaving its old channels behind in the form of numerous bayous that thread its delta region. It now flows above the level of its floodplain for a distance of more than 1,000 km above its mouth. The Mississippi now threatens with every major flood to abandon its present channel in favor of an old course called the Atchafalaya River, actually a bayou. If that were to happen, the mouth of the river would abruptly shift almost 100 km west. Its lower course—with its enormous investment in navigation improvements and wharfage facilities—would then be left as a stagnant bayou rapidly filling with sediment. Major economic incentive inspires efforts to maintain large rivers in their present channels. However, that causes other problems because the floodplain cannot drain into the river and therefore becomes increasingly marshy.

Braided Streams

The channels of streams that carry extremely heavy loads of sediment become so unstable that they develop a **braided pattern** consisting of a restlessly shifting network of small channels separated by gravel bars and islands (Figure 7–19). Most such streams are actively aggrading their beds although some are merely attempting to cope with a local oversupply of sediment furnished from unstable and rapidly eroding banks. Braided channel patterns most commonly develop in streams that carry a heavy burden of sediment as bed load. In such cases the braided pattern is efficient because it puts a large proportion of the water in direct contact with the streambed where it can roll pebbles along.

Braided streams are common in dry regions where the scanty plant cover permits rapid soil erosion and the scarcity of rainfall limits the volume of stream flow. They are rare in ridge and ravine landscapes where the rate of soil erosion is typically low and the volume of stream flow ample. The few braided streams that do exist in ridge and ravine landscapes drain watersheds that supply excessive quantities of sediment, typically from melting glaciers or large areas of heavily logged or overgrazed land.

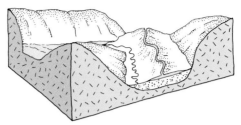

Muddy floodwaters deposit much of their sediment load in natural levees along their banks. Repeated floods build the floodplain to a slightly convex profile with the stream at its highest part.

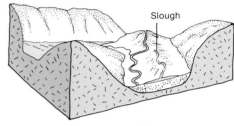

Eventually, the stream breaches its natural levees to find a new course on a lower part of the floodplain, leaving the old channel behind as a slough.

FIGURE 7–18
Many streams shift their courses as deposits of sediment raise the old channel above the general elevation of the floodplain. Large sediment loads accelerate the process.

FIGURE 7–19
The braided channel of Poison Creek, Wyoming. (E. R. Colby, U.S. Geological Survey)

Stream Terraces

Streams generally leave remnants of old floodplains along their valley walls as they erode their channels deeper. Those remnants form smoothly graded benches called stream terraces (Figure 7–20) that flank the modern floodplain and slope in the same direction although not at the same gradient. Many streams have several terraces in their valley floors that rise like a flight of broad steps from the modern floodplain to the valley wall (Figure 7–21). In such cases the highest terrace is the oldest and they become progressively younger as they step down in elevation.

Figures 7–22 and 7–23 illustrate the distinction between paired stream terraces that match in elevation on opposite sides of the valley and unpaired terraces that do not. Paired terraces are clearly remnants of a formerly continuous surface that extended all the way across the valley. In contrast unpaired terraces evidently form as the river channel sweeps back and forth across the valley, eroding first one side and then the other. Some terraces record major geologic events, others local events significant only

to the specific watershed, and others merely reflect the internal dynamics of the stream system.

Some streams along the south Atlantic and Gulf Coasts of the United States have terraces along their floodplains that merge downstream into old shorelines now above sea level. Because the emergent shorelines match the elevations of others elsewhere in the world, it seems likely that they, and the stream terraces associated with them, record ancient higher stands of sea level. Similar stream terraces along the West Coast of the United States merge downstream into emergent shorelines that do not match the elevations of others elsewhere. In that region the old shorelines and their associated stream terraces evidently record local uplift of the land rather than fluctuations of sea level.

In many cases sets of terraces in nearby streams do not correspond to each other, indicating that they formed independently and record events that affected only their own watersheds. In one especially well documented case, that of Douglas Creek in northwestern Colorado, researchers have shown that a complex assemblage of small unpaired terraces formed during the last century. The stream was

FIGURE 7–20

The flat-topped benches in the foreground and middleground of this picture are stream terraces along the South Fork of the Shoshone River near Cody, Wyoming. The lower terrace must be the younger. (W. G. Pierce, U.S. Geological Survey)

FIGURE 7-21

A spectacular flight of terraces rising above the floodplain of the Madison River in southwestern Montana. The highest terraces are the oldest; they become progressively younger as they step down toward the river. (J. R. Stacy, U.S. Geological Survey)

entrenching its channel in response to a single event—overgrazing in part of its watershed. Evidently, the large number of terraces does not record a long sequence of events. The number simply reflects internal instability in the stream system as it returned to a graded condition after having been disturbed.

Placer Deposits

Heavy minerals lag behind during stream transport and concentrate locally to form deposits called **placers.** Gold is the most notable of those minerals but they also include platinum, sapphires, diamonds, certain rare earth minerals,

FIGURE 7-22

Paired stream terraces match in elevation on opposite sides of the valley and are obviously remnants of an old floodplain surface. They record a time when the stream flowed at a higher elevation; in many cases because it was carrying a heavier sediment load.

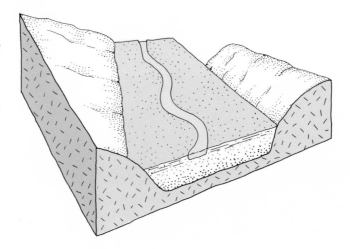

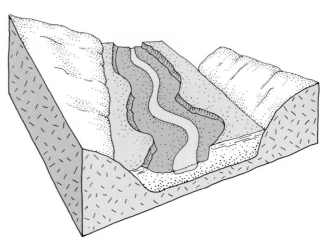

FIGURE 7-23

Unpaired terraces do not match on opposite sides of the valley. They probably form as the stream migrates from one side of the valley floor to the other while it erodes its bed to a flatter gradient.

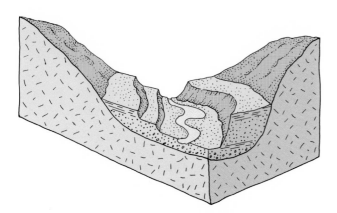

and many others, including some such as garnet and magnetite that have little or no commercial value. The heavy minerals sink to the bottom of the bed load as it moves during large floods so streams that contain rich deposits may show little or no surface indication of the wealth they contain. Although miners have worked placer deposits with simple hand tools for thousands of years, efficient operation generally requires enormous dredges (Figure 7–24). They strip the gravel down to its bedrock base and commonly leave the floodplain furrowed with bouldery ridges, such as those shown in Figure 7–25. Such spoil heaps typically contain rounded boulders far larger than any on the

FIGURE 7–24
A derelict gold mining dredge slowly sinking into a pond it dug in the North Boulder River of Montana. It worked until the 1950s when the high cost of operation put it out of business.

FIGURE 7–25
An old gold mining dredge pulled these boulders from the gravel bed beneath a small stream in the Sierra Nevada and stacked them in ridges. Some of the larger boulders are about the size of watermelons and vividly show the competence the stream achieves during occasional great floods. They would not be rounded if the stream had not moved them. (D. W. Hyndman)

undisturbed surface of the streambed, eloquent testimony to the great competence the stream achieves during occasional great floods.

MAINTAINING EQUILIBRIUM

The Case of the Sierra Nevada

The gold miners who eagerly rushed to California in 1849 to work stream placers in the Sierra Nevada promptly started to prospect their way upstream with pan and shovel, looking for the bedrock source of the nuggets. Although they did eventually find the bedrock "mother lode," they first discovered that much of the gold in the modern streams had come from old placer deposits in ancient stream channels exposed high on the hillsides. The miners then diverted water from the modern streams into long systems of ditches. These ditches carried the water to the high gravels, which they worked by washing them down with streams of water directed through huge nozzles at high pressures (Figure 7–26). This technique is called hydraulic mining. After flushing the gravels through sluice boxes to recover their gold, the miners simply washed them on into the nearest stream.

The water used to work the high gravels was the same that would have flowed through the stream naturally. Thus the hydraulic mining greatly increased the sediment load dumped into the streams without providing a corresponding increase in their water supply. Therefore, the streams responded by assuming a braided channel pattern and building their beds up to a steeper gradient by laying down deposits of sediment. By the 1870s the steepened stream gradients had so greatly increased the flow rates of the rivers that snowmelt water was passing from the high mountains to the Sacramento Valley in a fraction of its former time, causing sudden floods after the first warm days of spring. The excessive sediment loads also caused rapid

FIGURE 7–26

A hydraulic gold mining operation in the Rogue River basin of Oregon. The enormous jets of water flush the banks of old stream gravels through sluices to recover gold nuggets. (K. N. Phillips, U.S. Geological Survey)

development of high natural levees in the Sacramento Valley. These levees impaired floodplain drainage and caused outward seepage from the streams to convert large areas of formerly productive farmland into marshes.

Although their detailed understanding of the problem must have been somewhat imperfect, the farmers in the Sacramento Valley correctly guessed the source of their trouble. They filed suit, asking for a court order forbidding the miners to dump their spoils into the streams. After a long legal battle fought through a succession of courts, the California Supreme Court finally did issue such an injunction in its Sawyer Decision of 1884, which effectively ended hydraulic mining in California. The gold placers are still there but they are not rich enough to repay the investment needed to build dams to retain the gravel. Within a few years after the mining stopped, the unusual flooding began to abate and the streams of the northern Sierra Nevada have since returned almost to the level they were flowing on before gold mining began.

Remnants of the old gravel deposits washed down from the hydraulic mines remain as stream terraces flanking the restored modern floodplains. They are lasting mementos of the exuberant days of large scale hydraulic mining and they represent a capsule social history of early California as it affected the streams. Similar terraces have formed in other parts of North America during the period of recorded history as the results of changing agricultural and forestry practices.

The Impact of Dams

People build dams to impound water but dams also trap sediment if the stream is muddy. The sediment will eventually fill the reservoir and convert the dam into an artificial waterfall. Meanwhile, clear water that has left its sediment load in the reservoir pours through the spillway into a channel that had been graded to carry both water and sediment. The clarified water then entrenches the stream channel below the dam (Figure 7–27) as it erodes it down to a flatter gradient appropriate to the reduced sediment load of the stream.

In an especially dramatic recent example in Egypt, the clear water pouring over the Aswan Dam rapidly entrenched the bed of the Nile River, leaving extensive systems of new irrigation ditches too high above the river to receive the water that had been stored for them. Similar

(a) Initial stream profile graded to carry muddy water.

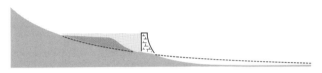

(b) Deposition of mud in the reservoir creates clear water that regrades the stream profile below the dam. Remnants of the old floodplain will remain as stream terraces.

FIGURE 7–27
The effect of building a dam across a muddy stream.

problems have inspired construction of diversion dams downstream from irrigation storage reservoirs in many parts of the world. In the case of the Red River below Lake Texoma on the Texas-Oklahoma border, the amount of sediment eroded from the riverbed below the dam almost exactly matches that deposited in the reservoir. However, it is not reasonable to assume that a similar relationship exists in all cases because it obviously depends upon the erodibility of the streambed.

Streams draining watersheds that contain large residential developments commonly entrench their beds for basically similar reasons. Roofs and roads shed water that would otherwise infiltrate the ground thus increasing runoff without increasing the amount of sediment. The increased runoff makes storm drains necessary and generally causes the stream that receives it to entrench its channel, undercutting bridge abutments and other bankside installations.

STREAMS AND THE OCEAN

Deltas

Many streams build deposits of sediment called **deltas** where they dump much of their load as they enter a lake or the ocean. Streams that carry heavy sediment loads and dump them in relatively sheltered locations build large deltas, such as those of the Mississippi and Mackenzie Rivers. Other streams do not build deltas. The St. Lawrence River is an example of a stream that does not build a delta because it carries very little sediment. The Columbia River does not build a delta because it empties into the ocean at a place where strong currents sweep the sediment away.

Figure 7–28 illustrates an idealized delta of the kind that may form where a small stream dumps its load of sediment into a quiet lake. The stream spreads thin layers of fine sediment called **topset beds** across the upper surface of such a delta while thicker layers of coarser sediment called **foreset beds** accumulate on its steep underwater surface. Meanwhile, thin layers of fine sediment called **bottomset beds** spread across the fan surface on the lake floor. As the delta grows new foreset beds cover old bottomset beds as they are themselves covered by new topset beds. Few deltas are actually that simple.

Large streams emptying into large lakes or the ocean build elaborate deltas complicated by shifts of the river mouth and the effects of wave action on the stream-deposited sediments. Everytime the stream breaches its natural levees and shifts the position of its mouth, it forms a new subdelta that adds another piece to the mosaic of deposits that compose the entire structure. Meanwhile, the waves erode and rework the older stream deposits. A long sequence of such events creates a complex pile of wave- and stream-deposited sediments that geologists distinguish by studying their primary sedimentary structures (Figure 7–29) and their fossils.

Many large deltas contain important reserves of oil and natural gas stored in the pore spaces between grains in sandstones. Geologists try to locate those sandstones by using data gathered from wells and natural exposures to prepare maps that reconstruct the delta as it existed during various stages in its development. Since sand accumulates mostly in specific sedimentary environments, such as stream channels and beaches, geologic maps help predict the locations of sandstone bodies and thus minimize the amount of drilling needed to explore the delta.

Submarine Canyons

Detailed mapping of the ocean floor became possible with the development of echo-sounding techniques during the 1930s. Some of the first maps to emerge clearly revealed the existence of large valleys closely resembling those on land that cross the continental shelf and descend the continental slope to the deep ocean floor. That discovery astonished geologists who had always assumed that the story of rivers ended at the shore where they deliver their water to the ocean and their sediment to the beach.

The discovery of submarine canyons started a vehement debate. Geologists did not know the age of the canyons and could not immediately determine whether they had been eroded underwater, or on land as normal stream valleys that were later submerged. The range for theoretical speculation narrowed as dredge samplers brought up chunks of hard rock broken off ledges exposed in submarine canyon walls. Some of those rocks contained extremely young fossils, which proved that the canyons are even younger and must therefore be forming under modern conditions. Many geologists suspected that the canyons are eroded by **turbidity currents,** great clouds of muddy water that occasionally pour down the continental

FIGURE 7–28

An idealized diagram of a simple delta. The set of beds shown in color are all the same age.

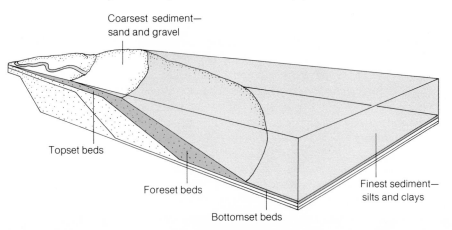

Coarsest sediment— sand and gravel

Topset beds

Foreset beds

Bottomset beds

Finest sediment— silts and clays

Slope. But there were doubts about this type of current's ability to erode solid rock.

Those doubts faded after an ingenious group of researchers ransacked old telephone company records to show that a series of submarine cables on the Grand Banks off Newfoundland had snapped in orderly downslope sequence after an earthquake shook the area in 1929. They argued that the earthquake had started a turbidity current that raced downslope at an average speed of about 50 km/hr, breaking the cables as it reached them. Many geologists now question that interpretation and suggest that the cables may have succumbed instead to a series of underwater slumps. Nevertheless, turbidity currents are now generally regarded as the erosional agents

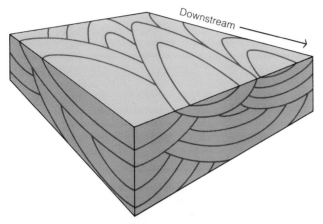

FIGURE 7–29

Streams scoop channels in their beds and then fill them to create a complex festoon pattern of cross beds, one of the distinctive primary sedimentary structures that help to identify stream-deposited sediment. These uncommonly large examples are in a deposit of dark basaltic sand laid down by a catastrophic flood that swept the Snake River during the last Ice Age. (H. E. Malde, U.S. Geological Survey)

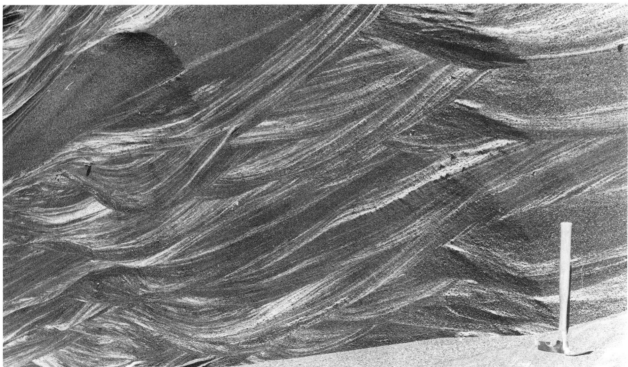

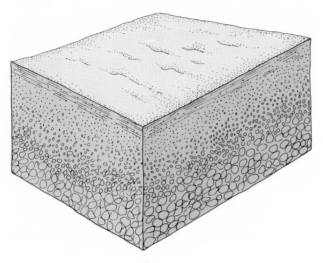

FIGURE 7–30

As clouds of muddy water pour out of the mouths of submarine canyons and across the deep ocean floor, the coarser grains settle first and the finer grains last. This process makes a layer of sediment that grades upward from sand or gravel at the base to clay at the top. Such layers may be less than a centimeter or more than a meter thick.

responsible for submarine canyons. A current of sediment-laden water flowing downslope underwater is closely comparable to a river flowing on land. Therefore, it seems reasonable to regard the system of submarine canyons as underwater river valleys that continue stream transport of sediment all the way to the deep ocean floor.

Turbidites

Turbidity currents deposit much of their sediment load in enormous **abyssal fans** that spread outward from the mouths of the submarine canyons. The forms of abyssal fans resemble those of alluvial fans on land except in their vastly greater size and much gentler slopes. The coalescing abyssal fans make a gently sloping underwater surface often called the *continental rise* that grades seaward into vast **abyssal plains,** which are nearly level and virtually featureless. The abyssal plains are deeply underlain by clastic sediments derived from the continents and carried to the ocean floor in turbidity currents.

As each successive turbidity current billows out over the ocean floor, the coarse material settles first and the fine last to make a **graded bed,** which is coarsest at the base and finest at the top (Figure 7–30). Deposits formed in that way are called **turbidites.** Similar deposits can be made experimentally by shaking a jar full of muddy water and then letting the mud settle. They typically contain the fossil remains of animals that live in shallow water mixed with those indigenous to the ocean floor. Turbidites can be broadly described as muddy sandstones even though individual layers may grade from clean sand at the base to fine clay at the top. Oceanic turbidite sequences occur on land wherever deepsea sediments have been incorporated into the continents. Their interpretation was a perplexing problem until modern coring techniques enabled geologists to study their modern counterparts on the ocean floor.

SUMMARY

Streams are complex systems that constantly adjust themselves to enable the available water to carry its burden of sediment without net deposition or erosion. Their velocities depend upon a balance between the pull of gravity and the drag of friction and generally tend to increase downstream as channel friction diminishes. An increase in the amount of water flowing through a stream enlarges the cross-sectional area of the channel, which diminishes friction and increases velocity.

Increasing flow velocity greatly increases the maximum size of the sediment particles a stream can carry; increasing discharge greatly increases the amount of sediment in transit. Therefore, sediment transport occurs mostly during large floods and nearly ceases during periods of low water.

Floodplains function as storage reservoirs that hold excess water until the stream channel can drain it away. Dikes and landfills displace water from one part of a floodplain to another, protecting some landowners at the expense of others. Streams wander freely on their floodplains as they abandon old channels to create new ones and as they erode the outside banks of bends while depositing sediment on the inside banks. Channel instability is especially marked in streams that are depositing sediment on their floodplains because they rapidly raise their banks above the level of the nearby valley floor and then spill out of them into a new course.

Many streams build deltas at their mouths. Such deltas are typically complex structures consisting of a mosaic of subdeltas built in different areas as the mouth of the stream shifts from one location to another. They also contain large amounts of stream sediment reworked by waves.

Stream erosion and transportation of sediment do not stop at the water's edge. A system of streams starts a few kilometers offshore, gathers its tributaries, and plunges down the continental slope to the floor of the ocean in submarine canyons. Much of the deep ocean floor is covered by layers of sediment deposited from the currents of muddy water that occasionally pour down those canyons. Stream transport of sediment continues from the mountaintops to the deepest parts of the oceans.

KEY TERMS

abyssal fan	clastic load	natural levee
abyssal plain	competence	oxbow lake
aggradation	delta	placer
antecedent stream	discharge	point bar
base level	dissolved load	potholes
bayou	entrenchment	saltation
bed load	floodplain	slough
bottomset bed	graded bed	stream terrace
braided pattern	gradient	superimposed stream
capacity	meanders	turbidite

REVIEW QUESTIONS

1. What changes could you make in a stream channel that would tend to increase the flow speed of the water?

2. What conditions must exist for a delta to form?

3. Why are braided streams much more common in arid than in humid regions?

4. Sediment accumulates all over the floors of most reservoirs even though the surface water remains clear. How is that possible?

5. Why does an increased sediment load tend to make streams more likely to shift their channels?

6. Explain why you would expect a dam to have greater effect on the downstream behavior of a muddy stream than on that of a clear stream.

7. Could a stream draining a watershed in which only pure limestone is exposed continue to entrench its channel after it had eroded downward onto a mass of granite?

SUGGESTED READINGS

Bloom, A. L. *Geomorphology*. Englewood Cliffs, N.J.: Prentice-Hall, 1978.

An excellent general textbook on landscapes. Includes a good treatment of streams.

Cooke, R. U. and R. W. Reeves. *Arroyos and Environmental Change in the American South-West*. Oxford, England: Clarendon Press, 1976.

Details specific examples of streams responding to historic changes in their watersheds.

Dunne, Thomas and L. B. Leopold. *Water in Environmental Planning*. San Francisco: W. H. Freeman & Co., 1978.

A comprehensive review of both surface and groundwater with emphasis on relationships to human activities.

Leopold, L. B., M. G. Wolman, and J. P. Miller. *Fluvial Processes in Geomorphology*. San Francisco: W. H. Freeman & Co., 1964.

A modern treatment of streams; rather technically presented.

Morisawa, M. *Streams: Their Dynamics and Morphology*. New York: McGraw-Hill Book Co., 1968.

An authoritative and readable account of streams.

Schumm, S. A. *The Fluvial System*. New York: John Wiley, 1977.

An advanced treatment of nearly all aspects of stream behavior. Outstanding.

CHAPTER EIGHT

This pair of fancy stalagmites stands on the floor at Lewis and Clark Cavern, Montana. The taller one is almost 3 m high. Small stalactites hang from the ceiling.

GROUND WATER

The ground-water reservoir contains a major portion of the earth's inventory of freshwater (Figure 8–1). It is a vital natural resource absolutely essential to the routine daily lives of untold millions of people as well as to the future economic development of large regions of North America and other parts of the world. If we wish to continue to have adequate supplies of clean water, it will be necessary to manage the ground-water supply intelligently.

Energy from the sun evaporates water from the oceans into the atmosphere where it condenses and falls on the continents as rain or snow (Figure 8–2). Part of that water runs back to the ocean through streams, part of it returns directly to the atmosphere through evaporation and the leaves of plants, and part of it soaks into the ground and goes into temporary storage in the ground-water reservoir.

THE GROUND-WATER SUPPLY

The Water Table

Nearly all ground water exists in the pore spaces between mineral grains in soil and in rocks. It moves by seeping from one pore space to another just as water poured into the soil at the top of a flower pot soon seeps out the bottom. There is a depth called the **water table** (Figure 8–3)

where the ground becomes saturated. Water can seep out of the completely filled pore spaces below the water table into wells, springs, rivers and lakes (Figure 8–4). Upward seepage into small pores keeps the ground constantly moist in a zone called the **capillary fringe** which extends a meter or so above the water table. However, the wetted surfaces of mineral grains hold that water so tightly that none is free to move.

Ground-water Recharge

In humid regions where rainwater recharges the ground-water reservoir by soaking down through the soil, the water table typically follows the shape of the landscape in a subdued way. It rises gently beneath the hills and intersects the ground surface wherever there is a lake, spring, or flowing stream. In such landscapes streams get most of their water supply through seepage from the ground-water reservoir and commonly increase in volume downstream even in reaches between tributaries.

In dry regions where rain rarely soaks all the way through the soil, the ground-water reservoir recharges mostly by seepage through streambeds and the water table normally stands highest beneath the largest streams (Figure 8–5). Streams in such regions generally flow above

FIGURE 8-1

Approximately 97.3 percent of the earth's inventory of surface water is in the oceans and most of the balance of 2.7 percent is on land. Glaciers of all kinds account for approximately 77 percent of the water on land, ground water for about 22 percent, and all the water in lakes and rivers for the balance of about 1 percent. (Data from "The Control of the Water Cycle" by J. P. Peixoto and M. A. Kettani, Scientific American, 1973)

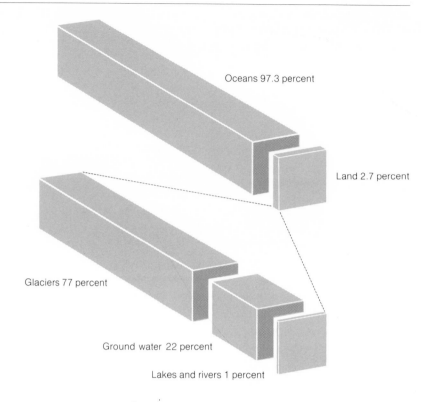

Oceans 97.3 percent

Land 2.7 percent

Glaciers 77 percent

Ground water 22 percent

Lakes and rivers 1 percent

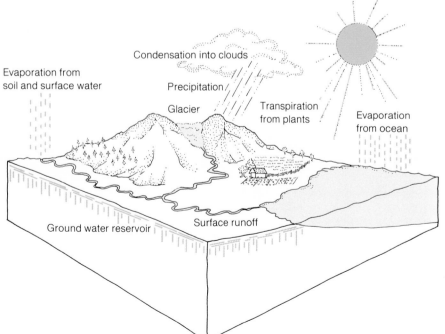

Evaporation from soil and surface water

Condensation into clouds

Precipitation

Glacier

Transpiration from plants

Evaporation from ocean

Ground water reservoir

Surface runoff

FIGURE 8-2

The hydrologic cycle. Heat from the sun evaporates water from the ocean into the atmosphere. There, it condenses to form clouds that drop rain and snow on the land, from where it eventually returns either to the atmosphere or to the ocean. Some precipitation returns directly to the ocean through streams and some is stored temporarily in glaciers, lakes, and the ground-water reservoir. Part of the stored water finally returns to the ocean as runoff. The rest returns to the atmosphere through evaporation from the soil, from bodies of surface water, and through transpiration by plants.

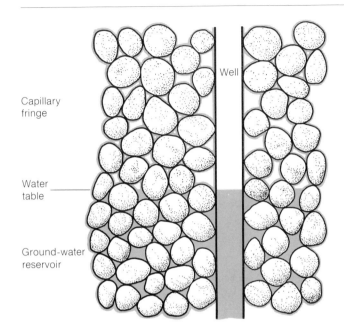

Capillary fringe

Well

Water table

Ground-water reservoir

FIGURE 8-3

All the water in the moist ground above the water table is bound up in wetting mineral grain surfaces and none is free to flow into a well. Part of the water in the saturated ground below the water table moves freely through completely filled pore spaces and can percolate into a well.

FIGURE 8-4

The water table is a continuous surface. Where it lies above ground level, it is visible as the water surface in lakes, swamps, streams, and springs. Where it lies below ground level, it is visible as the standing water level in wells. Highly porous and permeable materials, such as the sediments in the floodplain and stream terrace, are continuously saturated below the water table. A well drilled into them anywhere will succeed. In solid bedrock the ground-water reservoir may exist only as water-filled fractures.

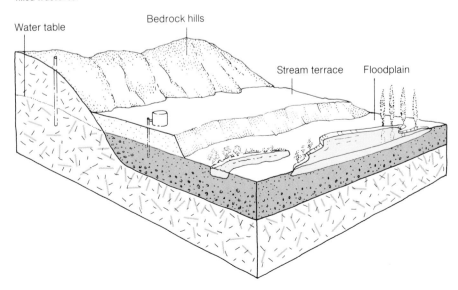

Water table

Bedrock hills

Stream terrace

Floodplain

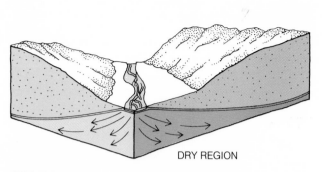

DRY REGION

WET REGION

FIGURE 8–5

In wet regions the ground-water reservoir discharges into the streams and the water table is a subdued replica of the topography. In dry regions streams recharge the ground-water reservoir and the water table rises to its highest elevations beneath their beds. The arrows indicate directions of ground-water flow.

FIGURE 8–6

A ledge of basalt in the Modoc Plateau of California. Although the numerous gas bubbles give this rock high porosity, their isolation from one another makes its permeability quite low except along the fractures. (D. W. Hyndman)

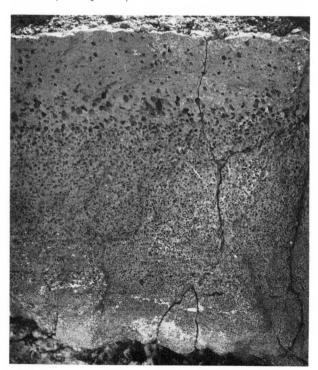

the level of the water table and lose water by seepage into the ground-water reservoir. Such streams typically dwindle downstream and in some cases they dry up completely.

In all regions the water table fluctuates seasonally. It rises during wet periods as the ground-water reservoir fills, and drops during dry weather as water seeps out of ground-water storage to maintain springs, streams, and lakes.

Aquifers

An **aquifer** is a body of rock capable of delivering significant quantities of water to a well. It must contain enough void space, or **porosity,** to hold significant quantities of water. The void spaces must be well enough connected to permit the water to flow freely through the rock, a quality called **permeability.** As Figure 8–6 shows, it is possible for a rock to contain a large volume of open pore space but to have very low permeability because the pores are poorly connected. On the other hand, a fractured rock might have little porosity—just the volume of the fractures—but high permeability.

Many fine-grained sedimentary rocks combine high porosity with low permeability because the void spaces between mineral grains are very small (Figure 8–7). Water contained in such small spaces is largely bound up in wetting the surfaces of mineral grains and is therefore not free to move. In coarser-grained rocks with correspondingly larger voids between the grains relatively little water is bound up in wetting grain surfaces and much more is free to move. Deposits of gravel make ideal aquifers.

Figure 8–8 illustrates the relationships between the degree of roundness and sorting in sedimentary deposits and their qualities as aquifers. Sediments that are sorted well—that is, those in which all the grains are nearly the same size—contain far more porosity than poorly sorted deposits, in which the smaller grains fill the spaces between the larger ones. Well-rounded sedimentary particles pack together as though they were spheres with regularly shaped

and connected voids between them. Angular particles, on the other hand, tend to fill void spaces with their projecting corners thus reducing both the porosity and permeability of the deposit.

Certain kinds of rocks—notably mudstones, shales, and most kinds of igneous and metamorphic rocks—that permit very little water to move through them are called **aquicludes.** In many regions an aquiclude forms the base of the ground-water reservoir. In regions underlain by thick sections of sedimentary rocks, deeper aquifers—called

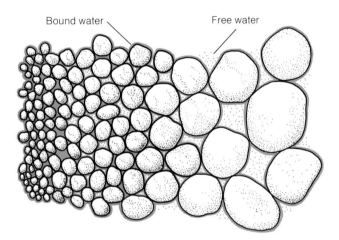

Bound water · · · · · · · Free water

FIGURE 8–7
The film of water that wets the surfaces of mineral grains is too tightly bound to move. Free water exists only in the larger pore spaces. Therefore, the permeability of fine-grained sediments is much less than that of those composed of larger grains even though their porosities may be about the same.

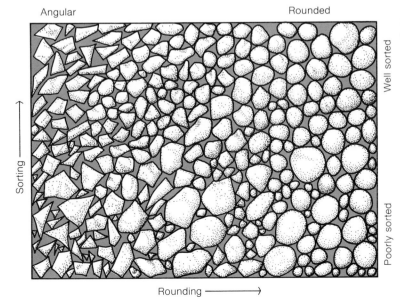

Angular   Rounded

Sorting ⟶

Well sorted

Poorly sorted

Rounding ⟶

FIGURE 8–8
Porosity and permeability both improve with increase in the degree of rounding and sorting of a sediment.

confined aquifers—may exist beneath aquicludes or be sandwiched between them. Confined aquifers recharge as water percolates into them from their outcrop areas instead of by infiltration from above.

Figure 8–9 shows a rather common situation in which an aquifer rests on top of an aquiclude and contains **perched ground water.** Finding such perched reservoirs is commonly quite profitable because they minimize drilling and pumping costs. Such reservoirs are commonest in situations in which sedimentary rocks still remain in their original horizontal position. Perched ground water commonly reveals its existence and location by seeping through lines of springs high on a hillside along the upper surface of the layer of impermeable rock.

In **artesian aquifers** the water is under pressure that makes it rise in the well bore above the level of the confined aquifer from which it comes. In some cases (Figure 8–10) artesian wells flow to the surface and all require less pumping than they would if the water were not under pressure. Figure 8–11 shows that artesian aquifers are precisely analogous to a city water supply that furnishes water under pressure from a high standpipe. Artesian aquifers recharge in a topographically high area, which corresponds to the standpipe, and then slope underground beneath a capping layer of impermeable rock, which corresponds in its function to the pipes in the city water system.

The water rises to a level called the **piezometric surface,** which would correspond to the elevation of the recharge area if there were no flow within the aquifer simply because water tends to seek its own level. However, the actual level of the piezometric surface is always lower than the elevation of the recharge area because water flowing through an aquifer loses pressure to friction. The effect is the same as that which causes water pressure to drop in the kitchen when the lawn sprinklers are working except in that case it is due to friction with the pipes and hoses. The pressure within an artesian aquifer progressively drops as more wells increase the draft of water and many aquifers that once flowed freely now require pumping.

One of the more spectacular of the many artesian systems in North America is the Floridan aquifer, which lies beneath large regions of the coastal plain in South Carolina, Georgia, Florida, and Alabama. The aquifer is a thick section of cavernous Eocene limestone exposed at the surface in a wide outcrop area running through the coastal plain and dipping gently seaward beneath layers of shale. Wells drilled into the limestone in low-lying coastal areas produce flowing water as do a number of large artesian springs that mark places where the aquifer is leaking naturally. Several large cities, including Jacksonville and Tampa in Florida, obtain their water supplies from the Floridan aquifer as do numerous smaller communities, industrial plants, and private households.

Wells

Pumping water from an unconfined aquifer creates a funnel-shaped dimple in the water table called a **cone of depression** (Figure 8–12). It expands until the rates of ground-water flow and pumping come into equilibrium and production from the well stabilizes at a constant rate, which can be sustained indefinitely. If the pumping rate is increased, the cone of depression will deepen thus steepening the slope of the water table and therefore the rate at which water seeps into the well. When pumping stops, the cone of depression fills thus restoring the water table to its static level. The difference between the depth to the static water table and the base of the cone of depression is called the **drawdown.** For any given rate of production, the drawdown will be less for wells producing from more permeable aquifers.

The productivity of a well begins to decline after a drop in the water table brings the base of the cone of depression to the bottom of the hole. If the water table continues to drop, the cone of depression must become progressively shallower and production will continue to decline. The well finally goes completely dry when the level of the water table reaches the bottom of the hole. The usual cure is to drill the well deeper to restore an adequate cone of depression or perhaps reach a deeper aquifer. Unfortunately, that solution cannot be extended indefinitely.

In many regions the ground-water reservoir is underlain by impermeable rocks that extend to great depth and offer no hope for deep wells. Prospects for deep water are much better in regions underlain by thick sections of sedimentary rocks that may contain a number of confined aquifers. However, the deeper sedimentary rock aquifers, those below a depth of several hundred meters in most regions,

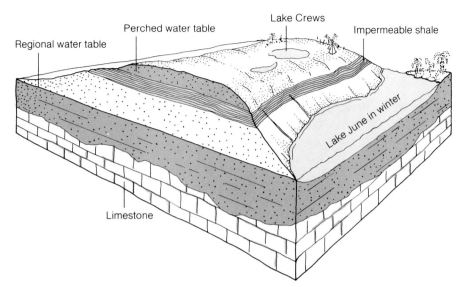

FIGURE 8-9

Ground water perched on a layer of impermeable shale maintains Lake Crews and nearby swamps at an elevation about 15 m above that of Lake June in winter, which is at the level of the regional water table. Ground water leaking out of the perched reservoir supplies the small streams that flow down the slope into Lake June in winter. This area is near Sebring in southern peninsular Florida. (Map from Lake June in Winter Quadrangle, U.S. Geological Survey)

FIGURE 8-10

A flowing artesian well. (Siebenthal, U.S. Geological Survey)

FIGURE 8-11

An artesian aquifer is analogous to a city water main. Both deliver water under pressure because they rise to an elevation greater than the well or tap. The actual pressure surface is lower than the highest elevation in the system only because water loses pressure to friction as it flows through the pipes or aquifer. A well drilled into the unconfined aquifer above the impermeable layer would produce ordinary ground water, which would have to be pumped to the surface.

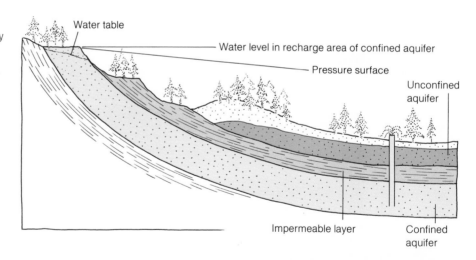

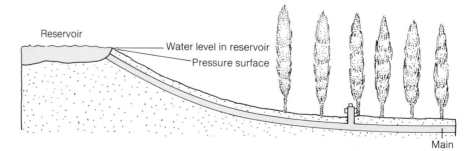

generally contain old seawater, called **connate water,** that was trapped in the sediments as they accumulated and has never been flushed out. In many dry regions and in most coastal areas, the reservoir of fresh ground water floats on a deeper reservoir of saltwater, and wells fail by going brackish instead of by going dry.

MANAGING GROUND WATER

The ground-water reservoir can be managed either as an annually renewable resource that will be perpetually available or as a mineable resource not renewable within an economically meaningful period of time. So long as ground-water production does not exceed the annual recharge, the ground-water reservoir functions as though it were a storage tank that fills during wet seasons and provides a continuing supply of water during dry periods. Under those conditions the water table fluctuates seasonally but does not drop progressively from one year to the next. However, if production of ground water exceeds annual recharge, the water table will drop progressively until the reservoir of freshwater is exhausted having been, in effect, mined out. Of course, rainfall may eventually refill the emptied aquifer but that may require centuries. In many parts of North America, especially in the dry western regions, extensive agricultural, industrial, and residential development depends upon mined water and can last only as long as the freshwater.

The panhandle region of north Texas and nearby areas of western Oklahoma, southeastern Colorado, and northeastern New Mexico provides a classic example of large scale ground-water mining. The principal aquifer there is the Ogallala Formation, an old desert plain deposit of silt, sand, and gravel laid down during Pliocene time that mantles the surface to depths as great as 100 m. Production of ground water for irrigation started in 1911 and continued on a modest scale until the late 1930s when it began to increase rapidly as vast areas of formerly dry land were brought into irrigated cultivation. By about 1960 the water table had declined alarmingly throughout the region; estimates showed that some counties in northern Texas had already exhausted more than one-fourth of their total reserves and that all were clearly headed for a disaster. Conservation measures to increase recharge and limit the overdraft of water from the aquifer have slowed the rate of

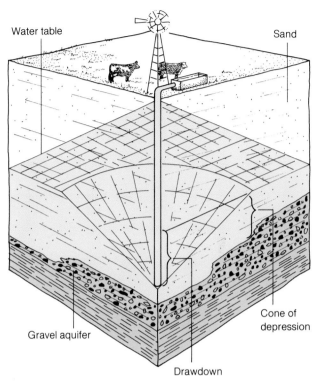

FIGURE 8-12

The depth of the cone of depression that develops in the water table around a producing well is called the drawdown. This well would have been more productive and its drawdown less had it been drilled a bit deeper into the underlying gravel aquifer.

depletion. However, the long-term downward trend continues and the threat of economic disaster remains.

Excessive withdrawal of ground water from unconsolidated sediments commonly causes compaction of the aquifer and corresponding subsidence of the ground surface. One of the most spectacular examples is in Mexico City. There, production of water from a thick deposit of lake sediments has caused buildings to tilt at odd angles as the ground surface subsides unevenly. In the Houston, Texas area, heavy production of ground water caused the water table to drop some 106 m and the ground surface to subside as much as 1 m between 1900 and 1960. In that case a program of injecting water into the aquifer to replace part of that withdrawn reversed the decline in the water table and greatly reduced the rate of surface subsidence.

Saltwater Intrusion

Unconsolidated sedimentary deposits along coastlines and in oceanic islands generally contain a reservoir of fresh ground water floating on slightly denser seawater. The interface between the two is typically quite sharp because the slow movement of water within the aquifer inhibits mixing.

The density difference between freshwater and seawater is so slight that every 1 m rise of the freshwater table above sea level pushes the saltwater interface about 40 m below sea level. For the same reason, a log so watersoaked that it barely floats will project much farther below than above the water surface. As Figure 8–13 shows, a cone of depression in the freshwater table relieves pressure on the saltwater beneath, permitting a corresponding cone of salt-

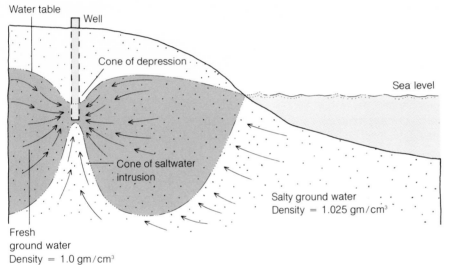

FIGURE 8–13

In coastal areas a reservoir of fresh ground water floats on slightly denser, salty ground water that moves in as the freshwater is withdrawn. Artificial recharge of the fresh ground-water reservoir creates a mound in the water table that helps prevent saltwater encroachment. Sewage effluent that would otherwise be run out to sea is, in many cases, a good source of recharge water.

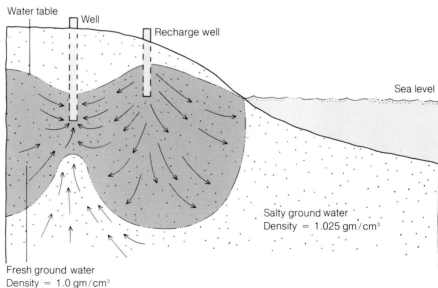

water intrusion to rise toward the base of the well. If the cone of depression is 1 m deep, the cone of saltwater intrusion will be 40 m high. Obviously, such wells must be carefully managed lest excessive production pull the cone of depression down far enough to draw the cone of saltwater intrusion into the well.

Numerous communities along all coasts of the United States as well as in the Canadian maritime provinces have mined much of their freshwater thus permitting seawater to encroach into the coastal aquifer. In some cases saltwater encroachment has been arrested by injecting large quantities of freshwater into the aquifer to form a freshwater ridge that acts as a barrier.

Heat Storage

Water is capable of holding large quantities of heat and rock is an excellent insulator so the ground-water reservoir can store heat as well as water. Large volumes of hot or cold water pumped into an aquifer through an injection well will retain their temperature for many months and will remain nearly in place unless the rock is uncommonly permeable. In areas where suitable aquifers exist, it is possible to store hot or cold water in them during one season and retrieve it during another for use in space heating or air conditioning. In many areas aquifer storage will be the least expensive and most efficient basis for solar heating systems.

Numerous industrial plants discharge large volumes of water that is hot but otherwise uncontaminated after having been used for purposes of cooling. Hot water disposal often presents serious environmental problems and many factories have installed expensive equipment to cool such effluents before dumping them into a stream. Where suitable aquifers exist, hot water effluents could be stored underground and then sold to local consumers for space heating. Such systems could convert expensive problems into profitable opportunities while conserving large quantities of valuable heat energy.

Natural Hot Water

Most springs and shallow wells produce water at the mean annual temperature of the region. Wells deeper than about 100 m produce warmer water, which increases in temperature at a rate of about 3°C for every 100 m increase in depth. Geologists call that change in temperature with depth the **geothermal gradient.** Most hot springs produce water that was heated as it circulated to considerable depth along fractures (Figure 8–14) although in some areas, such as Yellowstone Park, hot springs reflect the existence of an abnormally high geothermal gradient. That is commonplace in regions that have seen volcanic activity within the last few million years.

Where large masses of hot rock and a generous flow of ground water combine to produce abundant reservoirs of natural steam at shallow depth, they can provide the energy to generate electricity inexpensively and with minimum pollution. One such area, the Geysers in northern California, has been generating electricity with natural steam since 1927 and now provides San Francisco with approximately half its supply. The total reserves of **geothermal energy** at the geysers are unknown but evidently ample to drive the turbines at their present rate for several centuries at least.

Many much larger geothermal areas, including some on the Atlantic and Gulf Coasts, produce hot water instead of steam. This hot water is useful for space heating even if not for generating electricity. Several fortunate communities in western North America, such as Klamath Falls, Oregon and Boise, Idaho already use natural hot water on a modest scale and enjoy the potential for greatly increased production.

Most current estimates suggest that full development of the geothermal energy potential of the United States may satisfy as much as 5 percent of the national demand for energy. However, facilities using geothermal energy operate with almost twice the efficiency of those using coal or petroleum so that could actually reduce national consumption of fossil fuels by as much as 10 percent.

GROUND-WATER QUALITY

Septic Systems

Most people who have domestic wells also have septic tanks that recycle their sewage effluents back into the ground-water reservoir thus helping to keep the well productive. Distasteful though that recycling may seem at first

FIGURE 8-14

One type of natural hot water system. Cold surface water percolates down through the fractured rocks in a fault zone, becoming hot as it penetrates to great depth. It then returns to the surface along another fault zone to emerge in a series of hot springs. The well will produce natural hot water or steam if its developers drill deeply enough to intersect the fault zone.

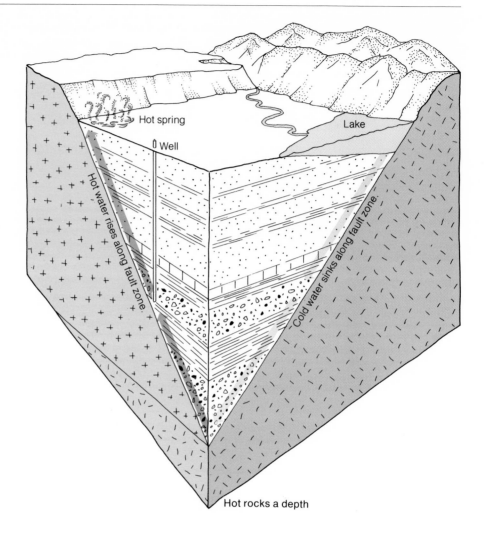

Hot spring

Well

Lake

Hot water rises along fault zone.

Cold water sinks along fault zone.

Hot rocks a depth

thought, it is basically a natural and wholesome process because the soil contains bacteria adapted through millions of years of evolution to consume organic wastes. They do their work well with the help of the simple mechanical filtration that occurs as sewage effluent percolates through the soil. The clarity and purity of most well and spring water testify to the efficiency of soil bacteria and natural filtration as cleansing agents. In fact, most cases of sewage contamination in well water turn out upon

investigation to be due to surface water seeping into the hole.

Septic systems may fail if the ground has extremely high or extremely low permeability (Figure 8–15). In places where ground permeability is very high, sewage effluent gurgles through to the water table too rapidly for soil bacteria to act and without benefit of mechanical filtration. In places where ground permeability is very low, sewage effluent does not infiltrate the soil but instead runs off the surface. It is customary to check ground permeability before installing a septic system by performing a *percolation test*. This test is simply a matter of digging a small hole, filling it with several liters of water, and watching to see how fast the water seeps into the ground.

Chemical Contamination

The gravest threat to ground-water quality comes from chemicals that pass through the soil and into aquifers because bacteria do not consume them and mechanical filtration cannot catch them. During the early 1960s, for example, wells all over North America began producing sudsy water that foamed in the drinking glass. Analyses showed that it contained dishwashing and laundry detergents that soil bacteria could not consume. Obviously, that foaming water included septic tank effluents that still con-

tained residues of detergent even though it had been thoroughly cleansed of ordinary organic wastes. Fortunately, the detergents were not poisonous and soon made their presence offensively obvious. Aroused public opinion, backed by legislation, forced manufacturers to switch to biodegradable products that bacteria could destroy and the problem began to disappear. However, other potential chemical contaminants include substances that are much less obvious and far more dangerous.

During the Second World War, an aircraft factory on Long Island dumped solutions containing dissolved cadmium and chromium, both extremely poisonous, into a pond whence they soaked into the ground and contaminated a shallow aquifer. Measurements made 20 years later revealed that the chemicals had percolated through the aquifer to spread over an area almost 1,300 m long and about 300 m wide—that kind of observation gives an im-

FIGURE 8-15

Locating septic tank and drainfield sewage disposal sites. The people who live on the floodplain may contaminate their own water supply because the water table is less than 2 m deep and does not provide adequate percolation distance through aerated soil. Those who live in the house on the river terrace should have no problem because the sands and gravels in the terrace provide good percolation and the water table is more than 2 m deep. The owners of the house on the cavernous limestone may contaminate their aquifer because open fractures in the bedrock provide no filtration. Those who live in the house built on impermeable shale may find their sewage running off the surface instead of soaking into the ground.

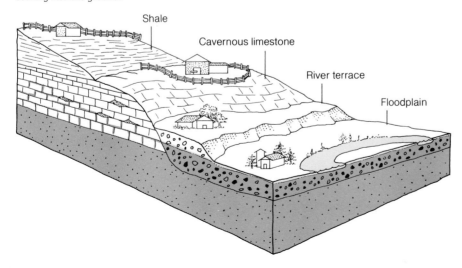

Shale

Cavernous limestone

River terrace

Floodplain

pression of how fast ground water moves. Discharge from the aquifer is now contaminating a nearby creek and will doubtless continue to do so for many decades until the aquifer has finally been flushed.

During the mid-1930s, enormous plagues of grasshoppers swarmed over the western plains, devouring any crops that had survived the drought. Many control methods were tried, including the use of poisoned bait consisting of bran heavily laced with arsenic. In 1972 most of the people who worked for one firm in a small town in western Minnesota became ill shortly after their employer had drilled a new well. Their illness was finally diagnosed as arsenic poisoning, which was finally traced to the new well after some weeks of uncertainty, during which they continued to drink the tainted water. Investigation revealed that the well had been drilled where unused grasshopper bait had been buried in 1934. The people who buried the bait must have thought they were disposing of it in a safe way and surely did not dream that they were, in fact, poisoning their grandchildren. Many modern insecticides, herbicides, industrial chemicals, and especially radioactive wastes, are far more poisonous than arsenic, more insidious in their effects, and equally immune to bacterial degradation. It is tempting to wonder how many mysterious illnesses may be due to undetected contaminants in the water supply.

KARST LANDSCAPES

Caves and Sinkholes

Ground water dissolving carbonate rocks may create caverns that eventually collapse. Some caverns break through to the surface to become **sinkholes.** In some areas underlain by large carbonate rock formations, cave and sinkhole development becomes the dominant process of erosion primarily responsible for shaping the landscape. Such distinctively pitted countryside with numerous caverns and underground drainage is usually called **karst topography** after a region of Yugoslavia where it is classically developed. Karst landscapes exist in large regions of the central and southeastern United States, most notably in peninsular Florida but also in the hills of Missouri, Arkansas, Tennessee, Kentucky, southern Indiana, and Alabama. Smaller but equally spectacular and interesting areas of karst landscape are widely scattered elsewhere in North America.

Rainwater dissolves atmospheric carbon dioxide to become a weak solution of carbonic acid capable of dissolving carbonate rocks. If the rock is porous, the slightly acidic rainwater will soak evenly into it and dissolve the entire surface uniformly without creating distinctive landforms. Karst landscapes do not develop on porous carbonate rocks even though they erode by going into solution. If, on the other hand, the carbonate rock is dense, compact, and lacking in porosity, the water can enter it only along fractures that become caves as solution erosion enlarges them. Maps of cave systems typically show them following the fracture pattern in the rock. There are a few instances, such as the Eocene Ocala formation of Florida, in which caves develop in porous limestones. However, in those cases the formation develops an armor of chert on the weathering surface that channels entry of water into the limestone below.

Although most limestone formations consist of mechanically strong rock easily capable of spanning large openings, caves are nevertheless inherently unstable and tend to collapse. When that happens, the ceiling falls along fractures that break upward in sweeping curves, outlining a form like that of a gothic arch. The new ceiling vaults above a conical pile of fallen debris that slopes at its angle of repose (Figure 8–16). Fallen rubble on the cave floor occupies about one-third more volume than it did as intact rock in the ceiling because of the void spaces between the fragments. Therefore, the cave progressively diminishes in volume as each successive ceiling collapse displaces it closer to the ground surface.

If the original cave was either very large or very shallow, its collapsing roof may eventually break through to the surface to form a sinkhole, such as that shown in Figure 8–17. New sinkholes commonly appear with dramatic suddenness as a cave collapses to open a gaping hole in what had appeared to be solid ground. That most often happens when heavy rains follow a long drought because the lowered water table will have withdrawn buoyant support from cave roofs to which the newly fallen rain adds an extra burden.

An especially striking example of wholesale sinkhole formation occurred near Birmingham, Alabama after drainage in two quarries lowered the water table beneath a large area during the late 1950s. Hundreds of new sinkholes appeared during the next decade as caves collapsed because their ceilings had lost buoyant support. In a parallel example more than 100 new sinkholes appeared within a year after drainage of a deep quarry lowered the water table beneath an area of several square kilometers near Hershey, Pennsylvania in 1949. In that case the problem was alleviated by injecting cement into the rock surrounding the quarry. That created an impermeable barrier to ground-water movement, which permitted the operators to drain the quarry without lowering the regional water table.

Many new sinkholes are shaped like jugs with a narrow surface opening leading to a much wider cavity below. Their shallowest depth is typically beneath the middle of the hole because it generally opens directly above the top of the debris cone standing on the cave floor beneath. Large surviving remnants of the cave may lie concealed beneath the overhanging edges of the new sinkhole, making it nearly impossible to estimate the amount of fill it might absorb. As the years pass and the edges of the sinkhole collapse, it slowly enlarges and softens its outlines until it finally becomes a broad dimple in the landscape. Other sinkholes may start with that shape if they form through collapse of a broad cave at shallow depth.

Sinkholes also develop without benefit of cavern collapse in places where solution of the carbonate rock is locally concentrated at the surface. There is no sure way to distinguish the two kinds of sinkholes from the surface; so it is difficult to estimate their relative abundance. As Figure 8–18 shows, the issue could be resolved if it were possible to see the hole in cross-section: Those formed by cavern collapse would have rubble beneath them whereas those that developed through locally concentrated solution would not. Unfortunately, very few roadcuts section sinkholes so the appropriate observations are rarely feasible.

Karst landscapes typically contain few surface streams because water drains into sinkholes and flows underground through caverns. In many karst landscapes, including some in very wet climates, small and widely scattered sinkhole ponds contain the only surface water. However, the underground streams commonly surface through large

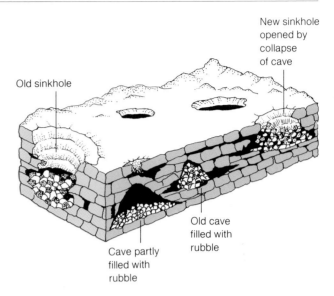

FIGURE 8-16
Caverns migrate upward as their roofs collapse and bury their floors beneath broken rock. Small caverns may completely backfill themselves with rubble and only the larger or shallower ones finally break through to the surface to become sinkholes.

FIGURE 8-17
A gaping sinkhole opened by collapse of a cavern in Wyoming. (N. H. Darton, U.S. Geological Survey)

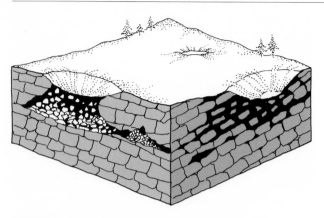

FIGURE 8-18
Although they form in quite different ways, collapse sinkholes, left, and sinkholes created by locally concentrated solution of the bedrock, right, may look alike. It is sometimes difficult to distinguish one from the other without knowing what lies beneath the surface.

springs, many of which produce millions of tons of water daily—an entire stream surging out of a single spring.

Unfortunately, the beautifully clear and inviting water that emerges from most big limestone springs is quite likely to be contaminated. Sinkholes make inviting places to dump refuse and the limestone caverns that drain them provide neither filtration nor effective bacterial action.

Nevertheless, limestone springwater is desirable for distilling because it is absolutely free of dissolved iron which would, if it were present, make the whiskey black. That is true because the water becomes slightly alkaline as it reacts with the limestone. Dissolved iron precipitates from alkaline solutions as the oxide mineral goethite.

Dripstone Formations

Bizarre and grotesquely beautiful dripstone decorations adorn many caves. Most are composed of **travertine,** a rock that consists of calcite precipitated from water. A few caves, the drier ones, also contain gypsum ornaments, which form as sulfate created by oxidation of pyrite combines with calcium released by solution of calcite.

Water seeping into a cave contains calcite, which it dissolved from the carbonate rock formation. If the water evaporates or loses part of its dissolved carbon dioxide content after it enters the cave, it will precipitate part of its dissolved calcite. Evaporation causes precipitation simply by making the solution too concentrated. Loss of carbon dioxide deposits calcite because it was the water's content of carbon dioxide that made it acidic enough to dissolve the calcite originally.

The most familiar forms of travertine dripstone are stalactites, which hang like icicles from the roof of the cave and stalagmites, which rise like pedestals from the floor. Successive drops of water create these formations by depositing part of their load of dissolved calcite as they hang from the cave ceiling and another part as they splash onto the floor. Therefore, stalagmites form directly beneath stalactites and the two may eventually join to form a column. In some caves such formations may develop so lavishly that they actually fill openings and help support the roof.

Small stalactites are tubular, like soda straws, because the hanging drops deposit calcite at the water surface where they lose carbon dioxide. Each drop leaves a minute ring of freshly precipitated calcite on the cave roof before it falls and successive rings merge to form a tube (Figure 8–19). If the drops of water trickle down the cave ceiling instead of falling to the floor, they deposit a thin curtain of travertine that hangs like drapery.

Thin films of water flowing across a smooth surface tend to lose water vapor or carbon dioxide in places where some irregularity stretches the water surface, providing more area for the escaping gases to pass through. The initial irregularity grows into a ridge, which finally develops into a dam capable of impounding a pool of water such as those shown in Figure 8–20. Many caves and hot springs develop sets of such cascading pools, which are called *travertine terraces.* Once the travertine dams have formed, the water spilling over their lips will build them higher or repair them if they are damaged.

Broken or sawed travertine surfaces almost invariably display fascinating patterns of curving bands and irregularly shaped void spaces. Numerous quarries produce travertine from old cave or hot spring deposits and sell the rock for use as an ornamental building stone. Excellent displays exist in almost every town either as rough outside facing stone or as sawed and polished stone on interior walls or floors.

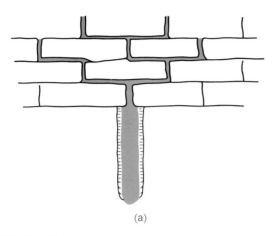

(a)

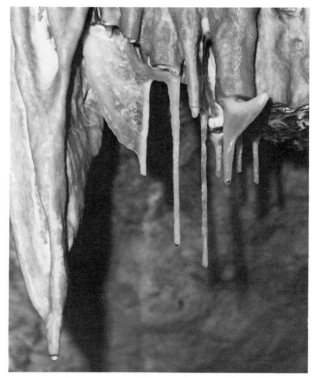

(b)

FIGURE 8-19

(a) Water seeping down through fractures in limestone dissolves calcite, which it later redeposits as cave dripstone formations. Soda straw stalactites grow as each successive drop of water deposits calcite on its surface, adding a thin ring to the tip of the straw. (b) These stalactites were broken during the 1930s and have since grown new soda straw tips, some of which are now several centimeters long. The hanging drops of water suggest their size. Lewis and Clark Cavern, Montana.

FIGURE 8-20

Unusually large travertine terraces at Mammoth Hot Springs in Yellowstone National Park. The series of walls in the right foreground are about 20 cm high.

Carbon dioxide escapes where water passes over the lip of the pool, causing calcite to deposit there.

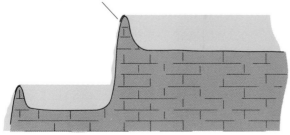

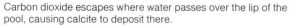

SUMMARY

Surface water may seep into the ground to become ground water and eventually reach the water table where the ground becomes saturated. Water fills all pore spaces below the water table and some water may be free to seep into wells or springs. The water table is the upper surface of the reservoir of available ground water and it fluctuates in level as the amount of water stored in the ground changes.

Some rocks—aquifers—are porous and permeable enough to contain and yield useable quantities of water. Others lack either the porosity or permeability, or both, to function as aquifers and some actually form barriers to ground-water seepage. The problem in ground-water geology is not to locate the water table, which exists nearly everywhere, but to find a good aquifer at a reasonable depth. Artesian aquifers contain water under pressure, which causes it to rise in the well bore and even in some cases to flow to the surface.

Water within the ground-water reservoir flows in the direction the water table slopes, tending always to flatten it. Production from a well creates a funnel-shaped dimple in the water table—a cone of depression—toward which water flows to replace that drawn from the well.

If ground-water production exceeds annual recharge, the water table will gradually drop as the ground-water reservoir empties. When the reserves of freshwater are gone, wells will either go dry or produce brine and centuries may pass before natural recharge can replenish the reservoir.

Therefore, production of water at a rate greater than natural recharge is tantamount to mining and constitutes exploitation of the ground-water reservoir as a nonrenewable resource.

Natural hot water and steam exist at shallow depth in some areas, usually those that have seen relatively recent volcanic activity. They provide a potential source of relatively inexpensive and virtually nonpolluting geothermal energy. The high heat capacity of water combined with the excellent insulating qualities of rock make the ground-water reservoir an excellent place to store hot or cold water for later use in space heating or air conditioning.

Bacteria and natural mechanical filtration efficiently remove organic wastes from ground water under most circumstances, making the ground-water reservoir relatively resistant to sewage pollution. However, nothing removes many of the numerous potential chemical contaminants and they are likely to enter the ground-water reservoir. Ground water moves so slowly that it may take centuries to flush a contaminated aquifer.

Ground water dissolves carbonate rocks, opening caves in them. Many of these caves will eventually collapse to form sinkholes. Locally concentrated solution at the surface may also produce sinkholes. Solution processes locally create a distinctive landscape—a karst landscape—which is heavily pocked with sinkholes and deficient in surface streams. Such landscapes typically contain large springs where underground streams emerge from caves.

KEY TERMS

aquiclude
aquifer
artesian aquifer
capillary fringe
cone of depression
connate water
drawdown
geothermal energy
geothermal gradient

karst topography
perched ground water
permeability
piezometric surface
porosity
sinkholes
travertine
water table

REVIEW QUESTIONS

1. How is it possible for a well to be dry even though the bottom of the hole is some distance below the water table?
2. Why is the height of the cone of saltwater intrusion below a coastal well much greater than the depth of the corresponding cone of depression in the water table?
3. Briefly explain why extreme permeability or lack of permeability make soil unsuitable for treating sewage effluent.
4. Under what circumstances of use is ground water a renewable or nonrenewable resource?

5. Why are new collapse sinkholes most likely to appear during a long drought?

6. Suppose that you and your neighbor both have wells producing at the same rate but that yours has the greater drawdown. Whose well is producing from the better aquifer?

7. Why does soil treatment of sewage effluent generally remove common organic wastes while leaving many chemical contaminants untouched and free to enter the ground-water reservoir?

SUGGESTED READINGS

Davis, S. N. and R. J. M. De Wiest. *Hydrogeology.* New York: John Wiley, 1966.

An excellent general text on ground water.

Freeze, R. A. and J. A. Cherry. *Groundwater.* Englewood Cliffs, N.J.: Prentice-Hall, 1979.

An outstanding text covering all major aspects of ground water.

Jennings, J. N. *Karst.* Cambridge, Mass.: The M.I.T. Press, 1971.

A text dealing entirely with karst landscapes. Thorough coverage of the subject and easy to read.

Kazmann, R. G. *Modern Hydrogeology.* New York: Harper & Row, 1965.

A superb treatment of both surface and ground water. Fairly technical but well written.

Langbein, W. B. and W. G. Hoyt. *Water Facts for the Nation's Future.* New York: Ronald Press, 1959.

A general interest book dealing with the sources, supply, and management of surface and ground water.

Leopold, L. B. *Water: A Primer.* San Francisco: W. H. Freeman & Co., 1974.

Exactly what the title implies—a broad discussion of water resources. Authoritative and readable.

Stevens, Leonard A. *Clean Water: Nature's Way to Stop Pollution.* New York: E. P. Dutton & Co., 1974.

A nontechnical book dealing mostly with soil treatment of sewage effluents. Includes discussion of several large-scale projects.

Sweeting, M. M. *Karst Landforms.* New York: Columbia University Press, 1973.

An outstanding book on karst landscapes.

CHAPTER NINE

Sea stacks on the Oregon coast.

COASTAL PROCESSES

Waves pick up energy from the wind—ultimately from the sun—and use it in doing the work of shaping the coastline. They commonly travel thousands of kilometers across the open ocean efficiently carrying energy from distant storms to quiet coasts that never felt their winds.

WAVES AND BREAKERS

Deepwater Waves

The first waves to arrive from a distant storm are long swells that rise and fall with monotonous predictability. However, waves more often come in choppy crests and troughs that appear to rise and sink chaotically. In most cases those more complex waves can be resolved into several sets of simpler waves that cross each other at angles. When that happens, some crests arrive at the same place at the same time and reinforce each other to make higher waves whereas other crests will meet troughs and the two will cancel each other (Figure 9–1). Such interference of two or more sets of simple waves creates a complex pattern in which the underlying order appears only upon careful analysis.

Deepwater waves transport energy, not water; otherwise, no ship could make seaway against them. A stick tossed out of a boat drifting in deep water illustrates this by bobbing up and down as the waves pass, just as the boat does, instead of moving along with them. If the stick is weighted at one end to make it float vertically, it will reflect the internal movement of the water in the waves by waggling back and forth as it bobs up and down.

It is possible to observe wave motion more precisely in the laboratory: Put neutral density floats adjusted so they hang in the water without rising or sinking in a long tank and watch them move as waves pass. They go through circular orbits that diminish in size with increasing depth (Figure 9–2). Somewhat parallel observations can be made more casually by watching schools of tiny fish too weak to swim against the motion of the water. Such schools make natural neutral density floats that pitch forward with the wave as they bob upward in its crest and then move back again as they sink with the succeeding trough.

In theoretical principle deepwater wave motion extends in some degree to infinite depth. However, the extremely small water motion at depths greater than about one-half the crest-to-crest wave length moves little or no sediment on the bottom so geologists call that depth the **effective wave base.** Of course, the depth of the effective wave base also depends upon the size of the bottom sediment because it takes much more vigorous water movement to transport gravel than sand or silt.

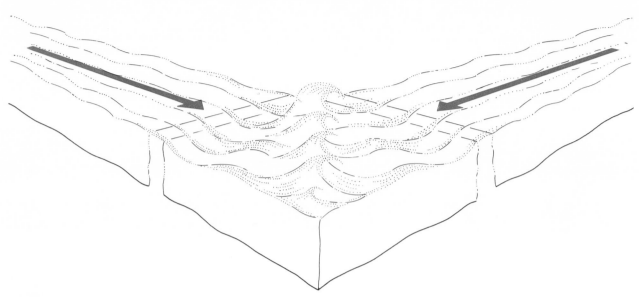

FIGURE 9-1

Choppy seas can generally be shown upon analysis to consist of two or more superimposed sets of simple waves. Where two crests or troughs arrive at the same place at the same time, they reinforce each other to make a much higher crest or deeper trough. Crests and troughs tend to cancel each other where they coincide.

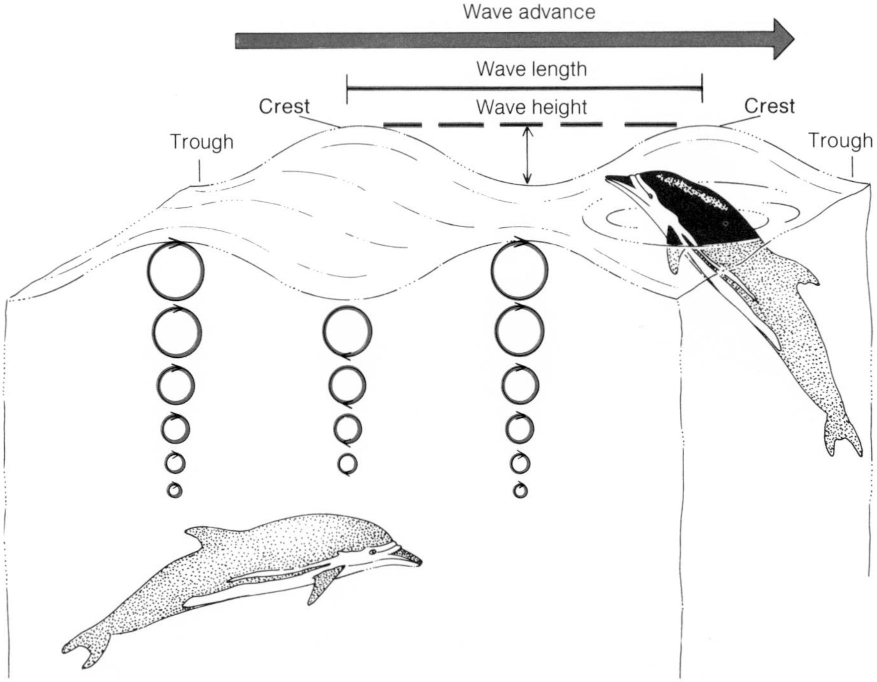

FIGURE 9-2

The anatomy of a deepwater wave. As the wave form passes, the water moves in circular orbits that rise to their highest points beneath the wave crests and sink to their lowest beneath the troughs. The size of the orbits diminishes with increasing depth.

Wave advance

Wave length

Wave height

Crest

Crest

Trough

Trough

Breakers

Waves slow, and their crests rise higher as they begin to drag on a shallow bottom that restricts the orbital water motion typical of deepwater waves. The rising crest rushes ahead of the lagging base of the wave until it finally loses support, curls forward, and breaks (Figure 9–3). That happens when the wave gets into water approximately 1.3 times as deep as the trough-to-crest height it had while still in deep water. Therefore, the small waves that arrive during mild weather get much closer inshore before they break than do the heavy seas that roll in during stormy weather.

Waves do not break if they run against a cliff or seawall founded in water significantly more than 1.3 times as deep as their height. Instead, they reflect harmlessly from its face, merely washing up and down the steep surface as

they would the side of a boat. That effect protects many cliffs and a few seawalls against wave erosion.

Waves change character completely as they break and become rushing masses of water that wash onto the coast and then drain back to sea. Sticks thrown into the surf illustrate that point by returning with the next wave instead of merely bobbing up and down with their passage as they would in deepwater waves offshore.

Rip Currents

Most of the water breakers cast onto the shore returns with their backwash. Some, however, return through a series of **rip currents,** narrow streams of water that flow seaward into the area just beyond the line of breakers where they dissipate (Figure 9–4). When a heavy sea is heaving large amounts of water onto the beach, the rip currents are

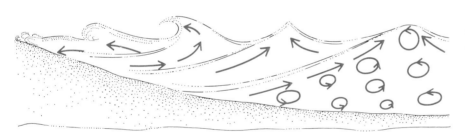

FIGURE 9–3
Waves pile up, curl over, and break as they get into shallow water.

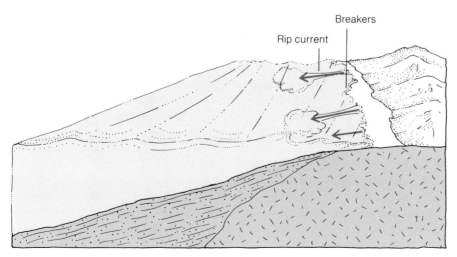

Breakers

Rip current

FIGURE 9–4
Water cast onto the beach by breaking waves returns to the sea through a series of more or less regularly spaced rip currents. They race out from the beach as narrow streams of water that dissipate as they pass seaward of the line of breakers.

closely spaced, perhaps at intervals as short as 100 m. They spread out to much wider spacings when lighter surf casts less water onto the shore.

Rip currents are easy to see from the air or from a high cliff. They appear as more or less regularly spaced lines of muddy water trending out through the surf. They are much more difficult to see from the beach where they appear as narrow zones in which the water may be slightly muddy and incoming waves shape into a nasty chop. Rip currents generally flow swiftly enough to sweep a strong swimmer out past the line of breakers. The best method of escape is to swim across the current in a direction parallel to the beach. Unfortunately, people who escape from a rip current generally misinterpret their experience as an encounter with an apparently mythical current called the "undertow." It has so far eluded all attempts at scientific observation and appears to exist only in the minds of people who post warning signs on beaches.

Wave Refraction

Every visitor to a beach, even a complete novice, confidently expects to find the waves rolling in from the open ocean regardless of the wind direction. And they are never disappointed. Waves always bring surfers onto the beach and never carry them along parallel to the coast. Evidently, the coastline must somehow turn the waves to make them approach it more or less directly no matter which way they may be running offshore.

Consider a series of waves approaching a coast obliquely. Their shoreward ends will slow down as they drag on the shallow bottom while the rest of the wave continues to race ahead in the deep water offshore. The part of the wave still in deep water catches up with the part already in shallow water so the entire wave tends to arrive on the coast almost at the same time. In other words the wave pivots about its slow-moving shoreward end and thus turns to align itself parallel to the coast. Marching bands pivot around corners in essentially the same way by having the members on one side take longer steps than those on the other. The tendency of incoming waves to anticipate and then to conform to the map outline of the shoreline they are approaching is called **refraction.** It explains much of what happens along the coast.

Smoothing the Coast

Where waves approach a projecting headland, they get into shallow water off its tip farther offshore than along the rest of the coastline. The part of the wave destined to crash against the headland slows as it gets into shallow water off its tip while the rest races ahead in deeper water on either side. That makes the wave pivot in both directions about the tip of the headland to embrace it. The part of the wave front that wraps around the point of the headland decreases in length as it molds its shape to fit that of the coast.

A measured length of wave front of given height represents a definite quantity of energy, specifically the amount of work necessary to stack water to the height of the wave. If the length of the wave front shortens, then it must simultaneously rise higher to contain the same amount of energy. Therefore, the effect of the wave's drawing shorter and rising higher as it wraps itself tightly around the top of the headland is to focus its energy at that point. That explains why we commonly see great breakers crashing against an exposed point while the same waves break as a moderate surf on straight beaches nearby. The opposite effect occurs where waves refract into an embayment. They must stretch to extend themselves along the entire shore of the bay (Figure 9–5) thus distributing their energy along a much longer and correspondingly lower length of wave front. That is why bays receive much lower waves than nearby sections of straight coast even if their mouths are wide open to the sea.

Therefore, an irregular shoreline refracts waves in ways that concentrate the main fury of their attack on the tips of exposed headlands and maintain the bays as havens of relative calm. As the incoming surf molds itself to fit the outlines of the coast, it preferentially erodes the headlands and sweeps their debris into the bays thus destroying both simultaneously. The end result is a smooth shoreline unimpeded by either projecting headlands or open bays and generally lacking good harbors.

The resistance of the bedrock largely dictates the extent to which waves have been able to reshape irregular coasts since sea level rose to its present stand as the last Ice Age ended. For example, long stretches of the Atlantic Coast of the United States between Long Island and Boston have been rather thoroughly remodelled by surf attacking vulnerable deposits of unconsolidated glacial debris. In strik-

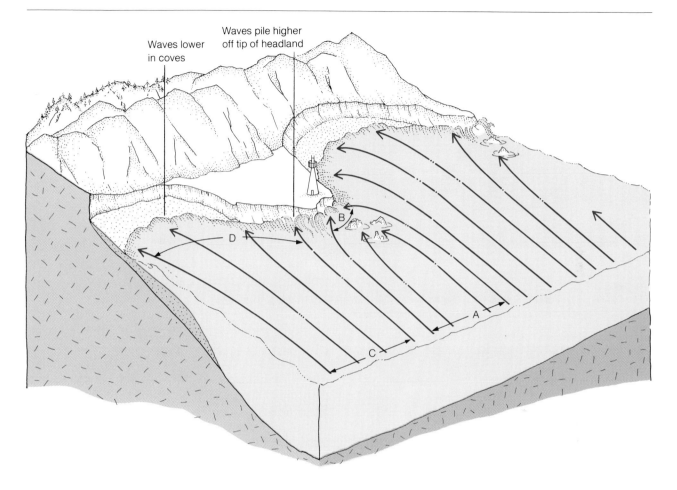

Waves lower in coves

Waves pile higher off tip of headland

FIGURE 9–5

Refraction focuses the energy of wave attack on the tips of exposed headlands and disperses it in bays. The length of wave front A, destined to crash on the tip of the headland, contracts to the much shorter and correspondingly higher length of wave front B before it does so. Conversely, the length of wave front C extends to the much longer and lower wave front D as it stretches to fit the longer shoreline in the cove.

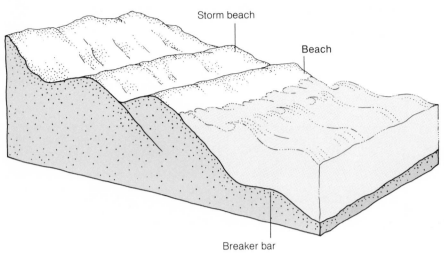

FIGURE 9–6

Along most coasts the beach consists of a complex of features, including the regular beach, an offshore breaker bar, and a storm beach, which is active only in extremely heavy weather.

FIGURE 9–7

Cross-bedded sand layers exposed in a trench dug in the modern beach near Ocean Springs, Mississippi.

ing contrast, the rocky coasts of northern New England and the Canadian maritime provinces show relatively little effect of wave attack.

THE SURF ZONE

Breaking waves shape and reshape the sand and gravel within their reach into a beach, which extends from the zone where waves reach at high tide to that where they break offshore and constantly changes form to reflect the shifting moods of the surf. The beach in its turn refracts the incoming waves, shaping them to conform to the outline of the coast. Interaction between the beach and the surf creates a smoothly continuous coast that conducts a steady stream of moving sediment along the shore.

Beaches

A simple beach created during a long period of uniform wave action has a shape suggesting that of a roof with its two sides pitched at different angles (Figure 9–6). In general, beach surfaces tend to slope more steeply when the surf is heavy than when it is light and they also tend to slope more steeply on beaches made of gravel than on those made of sand. Breaking waves wash up and down the seaward side of the beach and occasionally spill over its landward side, building a deposit of sand or gravel with internal layering such as that shown in Figure 9–7, which clearly reflects the external form of the beach. Ancient sandstones originally deposited as beaches also contain similar internal structures, which help to identify them.

Waders commonly find the bottom shelving gently downward and then shoaling again farther offshore before beginning its continuous descent into deep water. In most cases that offshore shoal is a **breaker bar** that forms where large waves crash during heavy weather. It forms a submerged ridge that follows the shore, trending generally parallel to the beach. Most shorelines have at least one breaker bar and many have several (Figure 9–6).

During severe storms heavy surf erodes the beach landward of its normal position and builds it into a much higher and steeper form. During milder weather smaller waves restore the original beach, leaving the crest of the storm beach as a low ridge trending parallel to the shore (Figure 9–6) and in some cases as much as a few hundred meters inland from it. Most coastlines have one or more such storm beach ridges along them and they give an excellent indication of what the most severe storms can do.

Longshore Sediment Transport

Surf watchers commonly see breaking waves curl over and crash from one end to the other instead of all at once as they would if they were approaching the coast squarely. That happens if the waves approach the coast obliquely so that one end gets into water shallow enough to make it break before the other. Evidently, waves commonly refract enough to approach the coast almost, but not quite, at right angles. That small difference explains wave transport of sediment along the beach.

When waves approach the coast obliquely, the rush of water from each breaker swashes up the beach at a slight angle and then its momentum carries the backwash farther along in the same direction as it drains off the beach (Figure 9–8). Therefore, each breaking wave washes water

FIGURE 9–8
The swash of breakers approaching at a slightly oblique angle sweeps sand down the beach. Then the momentum of the rushing water carries both the backwash of the wave and the sand still further in the same direction. Waves build sand spits across the mouths of rivers and bays, deflecting them in the direction of the longshore transport.

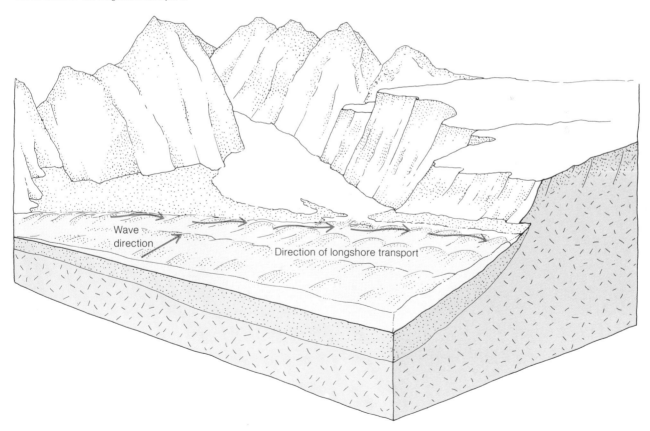

along the length of the beach as well as back and forth across it. That generates a slow drift of water called a **longshore current** (Figure 9–8) along the coast. This current moves suspended sediment in the same direction the waves are running.

Trying to snatch an attractive seashell out of the surf can be frustrating because the incoming breakers commonly wash the treasure almost within reach and then suck it back out with them as they drain away. In most cases they also wash it along the beach (Figure 9–9), forcing the collector to walk a few steps with each breaker to keep pace. It is less obvious but far more important to consider that the same breakers also move every grain of sand along the beach just as they do the shell. The wave-driven drift of sediment down the beach is generally called **longshore transport** although some authors have expressed the point more vividly by describing the beach as a "river of sand" along the coast.

That metaphorical river flows erratically; on some days it moves one way, on other days the opposite way, and on some days it does not flow at all. Like conventional rivers, it tends to move sediment almost catastrophically during occasional heavy storms. It is not possible to discover what is happening on a beach simply by watching seashells wash along in the surf on any particular day.

FIGURE 9–9
Waves breaking obliquely on a beach wash sand along in the direction of their approach.

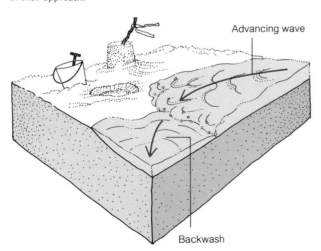

Advancing wave

Backwash

However, waves do move sand more persistently and more voluminously in one direction than in the other, commonly in the direction closest to that of the prevailing winds. That dominant direction of longshore transport is clearly expressed in the general shape of the coastline. For example, much larger beaches and tracts of coastal sand dunes generally form on one side of a river mouth than on the other. The reason for this is that the waves carry more of the sediment the river brings to the coast in one direction than in the other. Similarly, longshore transport lengthens beaches into sand spits that extend across one side of open bays and the mouths of rivers as they grow in the direction of sand movement. Continued growth of such spits generally deflects river mouths, making them jog in the direction of the longshore transport before they finally enter the ocean.

THE COASTAL BUDGET

Beaches continually adjust themselves to maintain an equilibrium between the rates at which they gain and lose material. Streams bring sediment to the beach and the waves add more by eroding the coastline although theirs is a self-limiting contribution because a wide beach armors the coast against further wave attack. Beaches lose some of their substance to the waves that move sediment offshore beyond the reach of the surf and to the wind that blows sand off the upper beach into the tracts of coastal dunes. The amount of sand and gravel on a beach represents a balance between the gains and losses so the beach, like a river, retains its form as material flows through it.

Although the pounding surf certainly ranks among the elemental forces of nature, it is amazingly easy to interfere with its workings to alter and in some cases transform the beachscape. Human intervention in beaches generally interrupts the steady flow of sand and commonly increases the rate of sand loss with effects that vividly illustrate how beach processes work.

Walls on the Beach

It is easily possible to dam the river of sand on the coast by building a wall called a groin across the beach and into the surf. People build such walls because they trap sand, making the beach grow wider on the side facing toward the direction of longshore transport (Figure 9–10). However, that also stops the supply of sand to the beach on the opposite side of the groin, causing it to erode through starvation, not through any increase in wave activity. People who find their beach property eroding because a neighbor has built a groin rarely respond, as they logically might, by demanding the immediate removal of the offending structure. Instead, they build a wall of their own to salvage their beach thus passing the erosion problem down to their neighbors who defend themselves by building still another wall. One after the other, they build a long sequence of defensive groins that eventually ribs many kilometers of beach, giving it an outline that suggests—when seen from the air—the teeth of a ripsaw. Examples of such ravaged beaches abound along almost every stretch of heavily populated sandy coastline.

Short groins may actually serve their intended purpose of holding sand on the beach but long ones have the opposite effect. As trapped sand fills the space behind a short groin, it eventually begins to leak past the tip where the breaking waves pick it up and sweep it back onto the next part of the beach. However, longer groins reach a depth at which the waves move sand seaward onto the continental shelf where it is permanently lost to the beach. The numer-

ous long groins at Miami Beach, Florida, to cite an especially notorious example, have bled enough sand out to sea to virtually destroy the beach they were supposed to enlarge. Waves now break directly against seawalls in many places where they formerly lapped onto beautiful expanses of soft sand. The current ''solution'' to the problem is to truck sand in from old beaches a few meters above sea level in the nearby everglades and use it to ''nourish'' the modern beach. Similar solutions have been attempted in many other places, depending in some instances upon sand hauled in overland and in others upon sand dredged from the bottom offshore. Unfortunately, both approaches are expensive and neither can solve the problem as long as waves stir the ocean and groins rib the beach.

Breakwaters, which are walls built offshore more or less parallel to the beach, have much the same effect as groins because they also obstruct wave action. Islands close offshore cast a wave shadow on the coast in much the same way that a breakwater does and they also trap sand. In many such cases the accumulating sand builds a long spit called a **tombolo.** That ties the island to the shore, converting it into a peninsula of sand—the tombolo—with a rocky knob, the former island, at its tip.

Jetties are long walls that extend the mouths of streams or tidal inlets to maintain an open passage for navigation.

FIGURE 9–10

A groin obstructs passage of sand, enlarging the beach on the side facing into the direction of longshore transport. This situation causes corresponding erosion on the opposite side by depriving the beach there of its normal sand supply. (Photo, U.S. Coastal Engineering Research Center)

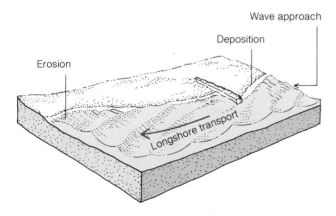

They also disrupt the coastal budget partly by obstructing the normal passage of sand down the beach and partly by channeling the stream's contribution of new sediment beyond the reach of the surf. Like groins and breakwaters, they trap sand on the side that faces the direction of longshore transport while starving the beach on the other side. Many jetties have caused enough erosion of sand spits guarding the mouths of harbors to open new inlets, generally in places where none were wanted.

TYPES OF COASTLINES

Waves crash and break in the same way everywhere but with results that differ from one coast to another, depending upon the kind of rock they attack, the depth of the water offshore, the history of the coastline, and even the climate. Let us begin by considering what happens where the water is shallow enough to make waves break offshore.

Plains Coasts

Waves breaking offshore create breaker bars that grow if the waves continue to break in the same place until they finally emerge above sea level to form exposed beaches called **barrier islands** (Figure 9–11). Most of the eastern seaboard of the United States has a continuous fringe of barrier islands as does most of the Gulf Coast west of the Florida peninsula. They exist only locally along the Pacific

Coast of the United States and Canada and along the coasts of northern New England and the maritime provinces because the water is generally too deep to make waves break offshore.

The body of water enclosed between the offshore barrier island and the original shoreline is a coastal *lagoon.* Those along rainy coastlines, such as the south Atlantic Coast of the United States, are generally brackish because they receive large amounts of freshwater runoff from the land. They discharge to the ocean through numerous inlets. Lagoons along dry coastlines, such as that of south Texas, receive very little freshwater. They become extremely salty and maintain relatively few inlet connections to the ocean.

Barrier islands grow as the waves add sand to their seaward sides and the wind blows it back into the tracts of dunes on their lagoon sides. During severe storms the waves wash enormous quantities of sediment over the crest of the island into the lagoon where it makes fan-shaped deposits that give the landward margins of barrier islands a scalloped outline in map view. Some barrier islands are large enough to provide room for such cities as Jersey City, New Jersey and Galveston, Texas. However, cities so located are never beyond the reach of the surf that built the island. Galveston, in particular, has suffered greatly from flooding when hurricanes have driven high tides across the island.

Coastal lagoons have a short life expectancy in the geologic sense because they catch sediment brought in from

FIGURE 9–11

Waves breaking offshore along a plains coast build barrier islands that fringe the original coast and isolate shallow lagoons from the sea. Eventually, the lagoons will fill with sediment and the barrier islands become the mainland coast. Successive barrier islands commonly step slightly seaward in the direction of longshore transport because deposition of sediment coming out of the inlet makes the water slightly shallower on that side.

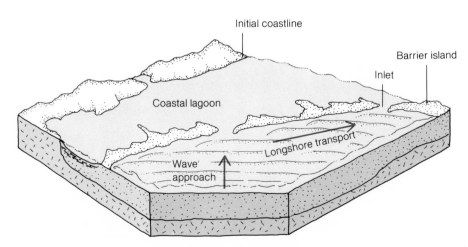

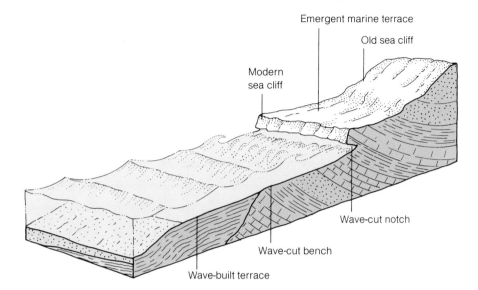

Emergent marine terrace

Old sea cliff

Modern
sea cliff

Wave-cut notch

Wave-cut bench

Wave-built terrace

FIGURE 9–12

A deepwater coast in cross-section. The wave-cut bench and wave-built terrace make a continuous underwater surface. This surface slopes gently seaward and extends itself landward as the waves undercut the notch in the base of the sea cliff and cause it to slide. If sea level drops or the land rises, the old wave-shaped surface is exposed as an emergent marine terrace—similar terraces also form on plains coasts for the same reasons. A series of emergent marine terraces rises like giant stair steps along many coastlines.

both the sea and the land. Filling lagoons become shallow marshes threaded with tidal channels and then fill completely thus converting the offshore barrier island to an onshore beach. Numerous unfilled coastal lagoons now exist because most modern coastlines began their development quite recently, with the worldwide rise in sea level that accompanied the melting of the great glaciers at the end of the last Ice Age.

Deepwater Coasts

Waves breaking against a deepwater coastline erode the rocks partly through the sheer watery mass of their assault, partly through a process of compressed air plucking, and to some extent by abrading rocks against each other. Consider first the enormous weight and momentum of even a moderate wave that slams many tons of water against the shore with a force that would instantly level most buildings and can easily dislodge loose rocks. Even the soundest bedrock is full of fractures and every pounding wave sends a pulse of compressed air into them patiently working blocks loose and then popping them out into the water. As the loose blocks of rock shift back and forth in the surf, they grind and batter each other and the solid bedrock.

All the processes of wave erosion concentrate their attack in a narrow vertical zone ranging from a level slightly below that of the lowest tides to one slightly above the highest. The waves erode a notch in the base of the sea cliff undermining it until it finally collapses into the surf. Then the waves clear the rubble and start a new notch, repeating the process as they saw the edge of the sea into the land. Such wave erosion creates a smoothly planed bedrock surface called a **wave-cut bench** (Figure 9–12) that slices indiscriminately across hard and soft rocks. Meanwhile, the surf works much of the debris torn from the retreating sea cliff back across the wave-cut bench to construct a **wave-built terrace.** As Figure 9–12 shows, the wave-cut bench and wave-built terrace make one smoothly continuous underwater surface that slopes gently seaward. A drop in sea level or uplift of the land exposes that surface as an *emergent marine terrace,* a smooth topographic bench that slopes gently seaward from an escarpment that marks the position of the old sea cliff. Such exposed benches are especially prominent along the West Coast of the United States where they provide level sites for numerous communities and for long stretches of highway. Figure 9–13 shows an excellent example.

Waves battering a sea cliff exploit every natural weakness in the rock, picking relentlessly at the softer or more

FIGURE 9-13
The little town of Mendocino, California sprawls across the surface of an old wave-cut bench now raised above sea level to become an emergent marine terrace.

fractured material and commonly hollowing them into caves or tunnels. The more resistant rocks remain standing in the surf seaward of the retreating sea cliff as small islands or rocky outposts called **stacks,** such as those shown in Figure 9–14. Even those stalwarts finally yield to the surf and vanish.

The ultimate form of deepwater coasts, like that of plains coasts, is a smoothly sweeping series of curves lined with beaches (Figure 9–15). The relatively few such coastlines that exist in that finished form today are those composed of rocks too soft to resist wave attack. Deepwater coasts in resistant rock generally have a ragged map outline that reflects the original form of the coast modified to some extent by wave action.

Mangrove Coasts

Certain tropical trees, most notably the mangroves, grow in seawater and thrive so luxuriantly along some plains coasts that they dominate their development. Mangroves grow on long prop roots that hold the tree as though it were standing on a thicket of stilts. Waves entering a mangrove forest dissipate their energy in those root thick-

ets instead of expending it on a beach. Fine sediment accumulates in the quiet water among the mangrove roots. This sediment builds an island beneath the forest that eventually rises high enough to support other kinds of vegetation while the mangroves continue to extend its margins. Meanwhile, mangrove seedlings aggressively take root wherever they find a muddy shoal and start new islands. Such coasts become mazes of growing islands in which the sea melts so imperceptibly into the land that it is difficult to draw a sharp boundary between them. Florida is the only part of the United States with the combination of warm climate and low tidal range favorable to such growths.

Mangrove coasts tend to accumulate deep deposits of peat, the partially decayed plant material that lithifies into coal. Most ancient coal seams are closely associated with marine sediments and appear to have been deposited in sedimentary environments closely resembling modern mangrove coasts.

Coral Coasts

Corals are small sedentary animals that strain their food from the water. A number of the tropical species form colo-

FIGURE 9–14
An armada of sea stacks, all that remains of an old headland on the Oregon coast.

FIGURE 9–15
Development of an initially irregular deepwater coast. The waves use debris derived from rapid erosion of the projecting headlands to fill the bays and build bars across their mouths. After the process has gone a bit further than shown here, the final result will be a smooth coast lacking both headlands and bays.

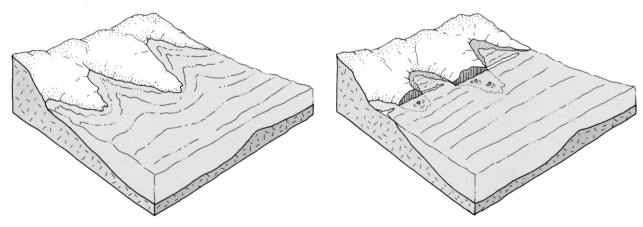

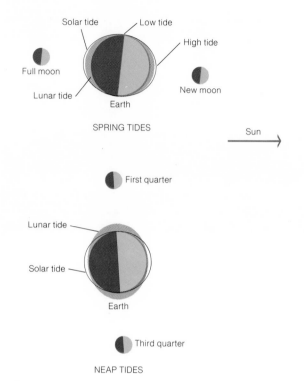

SPRING TIDES

NEAP TIDES

FIGURE 9–16
The greatest tidal ranges come during the new and full moon when the solar and lunar tidal bulges are superimposed; the least come when the moon is in its first and third quarters and the solar and lunar tidal bulges are as far apart as possible.

nies that live in close association with algae and build massive calcium carbonate reefs that fringe the coastline and dominate its development by absorbing the energy of the breaking waves. In addition to warm water they require abundant light and oxygen so they thrive in shallow, wave-torn waters where they grow right into the teeth of the surf. The mainland United States has only one coast, that of the Florida Keys, where the water is warm enough for reef-building corals and even there conditions are precarious.

THE TIDES

Gravitational attraction raises bulges on the parts of the earth's surface directly beneath the moon and the sun. Figure 9–16 shows them as well as the corresponding bulges that rise on the opposite sides of the earth because

they are farthest from the moon and sun and therefore subject to the least gravitational attraction. Tidal bulges always remain fixed in position relative to the moon and the sun that raise them. They therefore appear to move around the earth as it rotates, creating tides that vary in height according to whether solar bulge is in or out of phase with the much higher lunar bulge.

When the sun, the moon, and the earth are in a straight line—as happens every two weeks when the moon is full and when it is new—the solar and lunar tidal bulges are superimposed and add to each other. Those are the periods of the spring tides which have the greatest range. When the sun and the moon are at right angles to each other as seen from the earth, the solar high tide coincides with the lunar low tide and they tend to cancel each other. Those are the periods of the *neap tides* which have the lowest daily range. Although the pattern of the tides is fairly simple when considered broadly on a global scale, the detailed local schedules of their ebb and flow along the coast are generally complex. Every bay and lagoon must fill and empty as the tides come and go at rates that vary according to their volumes and to the sizes of the inlets that connect them to the open ocean.

SUMMARY

The wind transfers part of its energy to water surfaces, which carry it onto the shore in the form of waves that do the work of shaping the coastline.

Waves slow down and rise as they drag on shallow bottoms until they finally break onto the beach. The parts of a wave moving through shallow water slow down while other parts still in deep water race ahead thus shaping the map outline of the wave to conform to the contours of the bottom and therefore to the map outline of the coast. Nevertheless, most waves approach coasts at slightly oblique angles, which cause them to sweep sediment along the beach.

Waves approaching a headland mold themselves around its tip, focussing their energy there. Waves entering a bay extend themselves around its periphery and thus distribute their energy thinly along a great length of shoreline. Therefore, wave attack erodes headlands and fills bays, eradicating both to create a smoothly continuous shoreline.

Waves breaking along plains coasts build offshore barrier islands that isolate lagoons, which eventually fill with sediment. Waves breaking onshore on deepwater coasts carve wave-cut benches and build wave-built terraces offshore while converting the initially irregular coastline into a series of smooth curves.

Most modern coastlines are incompletely developed because wave action along them started in the recent geologic past when sea level rose as the last Ice Age ended.

KEY TERMS

barrier island	rip current
breaker bar	stacks
effective wave base	tombolo
longshore current	wave-built terrace
longshore transport	wave-cut bench
refraction	

REVIEW QUESTIONS

1. Why do waves increase in height as they approach headlands and become lower as they enter bays?

2. Briefly explain why waves break far offshore along some coasts, directly onshore along others, and not at all against cliffs founded in deep water or against the sides of ships at sea.

3. Why does wave attack tend to smooth rather than to enhance initial irregularities in the coast?

4. Why does the dominant direction of longshore sediment movement change seasonally along many coasts?

5. How might construction of dams on rivers eventually cause wave erosion of the coast near their mouths?

6. Explain why deepwater waves merely cause floating objects to bob up and down whereas breakers carry them onto the shore.

SUGGESTED READINGS

Bascom, W. *Waves and Beaches: The Dynamics of the Ocean Surface*. 2d ed. New York: Doubleday Anchor, 1980.

An outstanding book, as exciting and readable as a novel. Filled with good information.

Bird, E. C. F. *Coastal Landforms*. Canberra, Australia: Australian National University, 1964.

An excellent general text on coastal processes and landforms written for students. Well illustrated.

Barnes, R. S. K., ed. *The Coastline*. London: John Wiley, 1977.

A compendium of articles on various aspects of the coastline, most of which emphasize ecological considerations.

King, C. A. M. *Beaches and Coasts*. London: Edward Arnold Ltd., 1959.

A massive, comprehensive, and fairly technical text.

Komar, P. D. *Beach Processes and Sedimentation*. Englewood Cliffs, N.J.: Prentice-Hall, 1976.

A thorough treatment of the subject at the advanced level. Includes a comprehensive review of the literature.

Russell, R. J. *River Plains and Sea Coasts*. Berkeley, Ca.: University of California Press, 1967.

An informal discussion full of personal reminiscence and observation.

Steers, J. A. *Coasts and Beaches*. Edinburgh, Scotland: Oliver & Boyd, 1969.

A brief treatment of coastal landforms. Eminently readable and nicely illustrated. Full of specific descriptions.

CHAPTER TEN

Footprints on a wind-rippled sand surface.

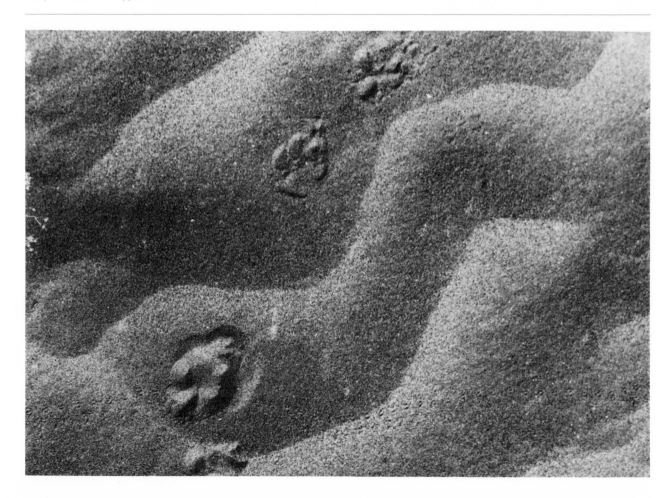

WIND

Because sand dunes are common in deserts, we tend to associate wind erosion with arid climates. However, sand dunes are also conspicuous in many other landscapes, including some with wet climates, such as the coasts of Oregon and Florida. In fact, the wind operates wherever there is barren ground—in deserts, along beaches, around glaciers, even in city streets where gusts of wind raise miniature dust storms and send sand grains skittering over the pavement. And even though they do not suffer from wind erosion, well-vegetated areas may collect thick blankets of dust blown in from elsewhere.

Despite its widespread activity, the wind ranks among the minor geologic agents. It rarely shapes entire landscapes and generally only adds detail to scenes created primarily by other processes (Figure 10–1). Nevertheless, those details are invariably interesting, commonly beautiful, and in some cases economically important.

DUST AND SAND

To analyze how the wind works, we must first distinguish dust from sand. For our purpose here, we can define *dust* as those particles so small and light that they travel in suspension because the upward air velocities in turbulent wind exceed their settling velocities. In general, dust particles are similar in size to silt particles, which travel in suspension in streams for the same reason. Sand grains, on the other hand, are heavy enough to settle through the strongest air currents. The wind bounces and rolls sand grains along the ground in much the same way that water moves them along streambeds. The distinction between dust and sand can easily be observed by tossing a handful of soil into a strong wind. The sand will settle to the ground while the dust blows away.

Even if the wind should raise both sand and dust simultaneously, the two will quickly separate: The sand will lag behind near the ground surface while the dust will swirl away in a cloud. Sand storms rise with strong wind gusts, whip along close to the ground surface leaving the air above them clear, and sink back to the ground as the wind subsides. Very few sand grains reach heights of more than about a meter and most remain below knee height. Dust storms, on the other hand, may rise as high as the tallest thunderheads and travel many hundreds of kilometers, moving as fast as the wind itself.

Anyone who watches clouds of dust towering behind a tractor must wonder why the wind needs the tractor's help since the wind itself is so obviously capable of carrying dust. Close observation has shown that dust particles are small enough to nestle within a thin surface film of stationary air (Figure 10–2). They are kept still by friction between the wind and the ground just as particles of silt are protected by a similar film of quiet water on the bottom of a stream. Unless something such as a tractor stirs soil into the air, the dust particles are safe from all but the strongest

FIGURE 10–1

Wind-rippled dune surfaces in southern California. (W. C. Mendenhall, U.S. Geological Survey)

FIGURE 10–2

Friction between the air and the ground surface maintains a paper-thin film of quiet air next to the ground while light or moderate winds blow. Sand grains project into the turbulent wind above and may roll along the surface while the much smaller dust particles nestle undisturbed within the quiet air. Strong winds destroy the ground film of quiet air and move both sand and dust.

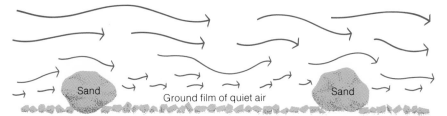

winds. However, sand grains are large enough to project through that quiet surface film into the turbulent air above. Thus, a moderate wind can move sand more readily than dust for the same reason that a moderate stream flow can move sand more readily than silt.

WINDBLOWN DUST

Wind Erosion

Aerial views of rising dust storms typically, suggest columns of smoke billowing up from numerous small grass fires. Each marks a place where the wind is exploiting a break in the plant cover or an irregularity in the ground surface to create a hole called a **blowout.** Thousands of blowouts pock the semi-arid high plains of the United States and Canada where many people call them "buffalo wallows." They typically have flat bottoms and irregular outlines and may sport little rims of sand on their downwind sides. Most blowouts are small enough that one can imagine them as buffalo wallows but some are much larger and a few are enormous. One oversized example near Laramie, Wyoming is approximately 5 km wide, nearly 15 km long, and almost 100 m deep. Although the wind is responsible for excavating blowouts, it is true that cattle wallow in them as the buffalo must have done in their time. Cattle do break up the ground surface in blowouts and that probably makes it more vulnerable to wind erosion.

Loess

Severe wind storms raise dust clouds that commonly contain hundreds of tons of dust per cubic kilometer, millions of tons in the entire storm. After such storms travel into more humid regions the bulk of the dust tends to settle along the downwind sides of major rivers. The more humid air there presumably moistens the dust particles, sticking them together into heavier aggregates that sink. The hills fringing the eastern sides of the larger rivers in the midwestern United States generally bear mantles of dust many meters thick. Some of that dust certainly came from the high plains and some consists of glacial rock flour that blew off the river floodplain during the ice ages.

Thick deposits of wind-blown dust make a distinctive

kind of soil called **loess.** Loess is typically yellowish and generally lacks obvious internal structure except for a distinct pattern of vertical fractures probably initiated by plant roots. Small amounts of calcite lightly cement most loess deposits, making them far more coherent than ordinary soils and capable—as Figure 10–3 shows—of standing in a vertical cut without slumping. People in parts of central China take advantage of that stability by digging cave homes in loess. Despite their structural stability loess deposits are extremely vulnerable to wind erosion. Machines raise great clouds of yellow dust when they work loess fields during dry weather. Paths and unpaved roads quickly etch themselves deeply into the ground as every

FIGURE 10–3

Loess exposed in the old Natchez Trace near Natchez, Mississippi. Like most unpaved roads in loess, this one was deeply eroded as the wind blew away puffs of dust raised by passing feet and wheels. The banks remain steep after many decades.

passing foot or wheel kicks up puffs of dust that drift away in the breeze.

In picking up dust in dry regions and carrying it into humid ones, the wind provides the best of all possible agricultural combinations: The fertile soil of the desert with its abundant mineral nutrients imported into a region with ample water. That is the opposite of the human strategy of bringing irrigation water to the desert. Fertile loess soils provide the base for many of the world's most productive agricultural regions.

The United States is fortunate in having large areas of excellent loess soils. The states of the Mississippi Valley contain rather sizeable patches of relatively thin loess in addition to the thick deposits along the eastern sides of the major rivers. There is another large region of deep loess soils in the Palouse Hills of eastern Washington and western Idaho. Those hills are simply large dust dunes upon which grow enormous crops of wheat.

WINDBLOWN SAND

Sand Transport

Wind moves sand mostly by bouncing it along the ground surface in a series of short hops, a style of movement often called **saltation** (Figure 10–4). Sand also moves through **surface creep,** which involves grains that roll along the ground surface or are kicked along by impacts of bouncing grains. Both processes are slow so the wind rarely moves sand very far from its source.

Desert Pavement

Where wind erodes bare ground it blows the dust away, urges the sand slowly along, and leaves behind a lag concentrate of pebbles too heavy to move. The wind searches out every gap between the pebbles, working them down into a continuous mosaic called a **desert pavement** (Figure 10–5) which effectively armors the ground against further wind erosion. Desert pavements similar to that shown in Figure 10–6 cover vast areas of deserts with a stony armor. However, they also develop on a much smaller scale in many other situations, such as river sandbars, urban vacant lots, and the upper parts of beaches where they may consist of seashells instead of pebbles.

Sand travels rapidly across desert pavements because the pebbles provide a hard surface on which the grains bounce high and far, perhaps as much as a meter in a single hop. The effect is comparable to that of bouncing golf balls across a hard pavement. In both cases, the bouncing object conserves most of its energy of motion in the rebound. Soft surfaces, on the other hand, absorb the energy of bouncing objects.

Sand Dunes

The very existence of sand dunes is amazing! Why should the wind sweep loose sand into neat piles instead of scattering it across the countryside?

If you were bouncing golf balls across a hard floor that had a scrap of shag carpet lying on it, the balls landing on

FIGURE 10–4

The wind drives a dense cloud of bouncing sand grains across the ground. Where they cross a loose sand surface, the individual grains tend to follow roughly parabolic paths of similar shape and length, giving the cloud of blowing sand a well-defined upper surface only a few centimeters high. Where they cross desert pavement, the individual grains follow much less predictable paths and often bounce two or more meters high.

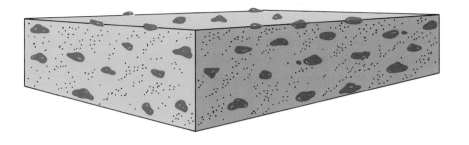

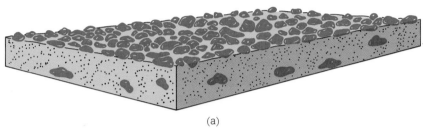

(a)

FIGURE 10-5
(a) Wind erosion of dust and sand deflates the ground surface, leaving the larger pebbles behind as a lag concentrate that forms the desert pavement and effectively armors the ground. (b) Desert pavement. (W. H. Bradley, U.S. Geological Survey)

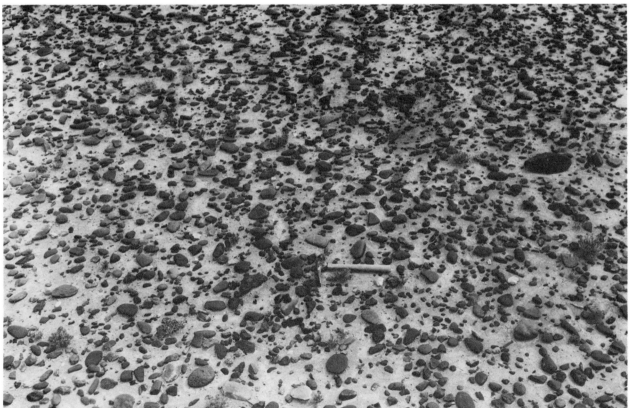

(b)

FIGURE 10–6
View across a smooth and gently sloping desert plain surface to the Furnace Mountains in the
California desert. The stony surface in the foreground is a desert pavement. (D. W. Hyndman)

the carpet would collect there. The reason for this is that
the carpet would absorb the golf balls' energy of motion
upon impact and thus prevent their bouncing back onto
the floor. Similarly, sand grains bouncing rapidly across a
hard desert pavement tend to stick if they strike a soft
place and accumulate there (Figure 10–7), making a pile
of loose sand that is itself soft and will therefore trap more
sand. That is why the wind sweeps sand off the desert
pavement into neat piles that become **dunes** in the prop-

erly defined sense of the word when they begin to march
before the wind.

Anything hard on the surface of a dune violates its reason for existence and therefore affects its behavior. For example, a loose scattering of pebbles on the windward surface of a dune will function as partial desert pavement and
speed sand movement, causing the dune to move rapidly
downwind until it leaves them behind. Similarly, we commonly find empty bottles and other hard debris lying in the

FIGURE 10-7
Sand grains bounce higher and farther on the hard surface of a desert pavement than on the soft surface of a pile of sand. Therefore, the pile of sand traps more sand and eventually grows into a dune.

FIGURE 10-8
The sharp crest of a small dune on an Oregon beach. Ripples cover the surfaces on which sand moves before the wind but not the steep lee slope where it simply slides.

bottom of a hole in a dune. In most cases, the hard debris caused the wind to excavate the hollow by accelerating sand movement.

Dunes move as the wind transfers sand from their exposed windward sides to their sheltered leeward sides. Sand grains bounce and creep up the gently sloping windward side of the dune to its sharp crest (Figure 10–8). There, many accumulate to build a small cornice that eventually breaks and slides down the steep leeward side

(Figure 10–9). Other sand grains simply whip over the dune crest and fall into the wind shadow on the sheltered leeward slope. The depositional leeward sides of dunes always stand at the angle of repose of sand—about 30°—whereas the eroded windward sides slope at much flatter and more variable angles.

The internal structures of sand dunes clearly reflect their style of movement. A trench dug along the length of a dune reveals layers of sand slanting down in the direction of

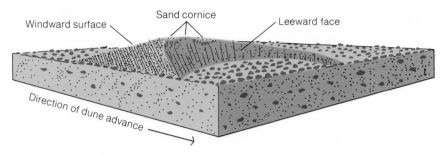

FIGURE 10-9

A dune migrates across the desert pavement as the wind erodes sand from its gently sloping windward surface and deposits it on the steeply sloping leeward face. Individual grains bounce up the windward surface of the dune and then settle near the crest to make small cornices that slide down the leeward face. Steeply dipping cross beds within the dune record former positions of the leeward face buried as the dune advances.

FIGURE 10-10

Dune crossbedding in sand—Osoyoos, British Columbia. (D. W. Hyndman)

movement, former lee surfaces of the dune that were buried as it advanced over them. Such distinctive patterns of steeply dipping **cross beds** as those shown in Figure 10–10 make it possible to identify inactive dunes and to establish that certain ancient sandstones were deposited as dune sands.

Thick deposits of dune sand typically display a more complex internal structure that consists of layers containing the steep cross beds separated by much flatter bedding surfaces. The two kinds of layering combine to form a distinctive pattern on roadcuts or weathered cliff faces, such as that shown in Figure 10–11. Some of the flatter bedding surfaces probably formed as advancing dunes buried the gently sloping windward faces of older dunes; others may have formed as the wind whisked sand off the drying surface of a dune left damp by a heavy rain.

The wind tends to gather sand dunes into vast **dune fields,** such as that shown in Figure 10–12. As a result, most deserts consist of vast expanses of windswept rock and desert pavement locally interrupted by dune fields which seem, when one is within them, to be veritable seas of sand. The actual amount of sand-covered ground varies greatly from one desert to another, averaging perhaps one third of the area. Most of the North American deserts contain relatively little sand, much less than those in North Africa and Arabia.

FIGURE 10–11
The herringbone pattern of steeply dipping cross beds, which formed on the lee slopes of dunes, is clear evidence that the Navajo sandstone was deposited by the wind. (J. K. Hillers, U.S. Geological Survey)

FIGURE 10–12
Oblique aerial view of a dune field in the Imperial Valley, California. The steep slopes face in the direction of sand movement. (J. R. Balsley, U.S. Geological Survey)

Drifts

Streamlined heaps of sand trapped in the wind shadows behind obstacles are not dunes in the proper sense of the word but **drifts** (Figure 10–13). Unlike dunes, drifts do not move but remain tied to the obstacle that trapped the sand. However, many drifts eventually outgrow the wind shadow that trapped them and then spawn dunes that march away before the wind.

Dune Stabilization

The vast majority of sand dunes exist harmlessly in unpopulated regions where they move a few meters each year and menace nothing; a few, however, do threaten to bury valuable property as they advance. In those cases, establishment of even a thin cover of plants generally stops dune movement.

Several kinds of plants have adapted to the rigors of life in loose sand (Figure 10–14) where they thrive even as the shifting surface partially buries their stems or exposes their roots. The leaves of such plants efficiently arrest sand movement. They do so in part by breaking the force of the wind and in part by intercepting flying sand grains thus breaking the force of their impacts and preventing them from splashing other grains into the air stream. Plant roots have little influence because they grow within the dune where they can hardly affect sand movement along the surface and through the air.

Wherever something breaks the plant cover on a stabilized dune, the wind resumes its former role and carves a blowout, moving the sand into the nearest downwind clump of plants. A long succession of such random episodes of erosion and deposition produces a disorderly landscape of small hollows and humps that contrasts sharply with the cleanly sculptured lines of actively moving dunes.

Sand Ripples

The wind ruffles a moving sand surface into trains of remarkably straight and regular ripples (Figure 10–15) that change their direction and spacing with every shift in the

FIGURE 10–13
Sand accumulating behind drift fences on Cape Cod. (John Herrlin)

FIGURE 10-14
Dunes partially stabilized by beach grass on the Oregon coast. (D. W. Hyndman)

FIGURE 10-15
Straight ripples rule a windswept sand surface on the California coast.

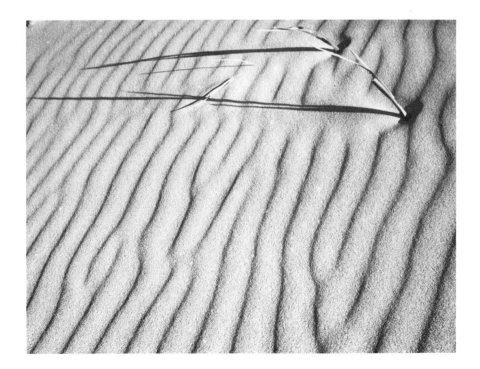

breeze. Careful observation of their behavior reveals many details of how sand grains move before the wind. Begin by imagining a situation such as that shown in Figure 10–16, an irregularity in a windswept sand surface.

Incoming sand grains will tend to skim over the upwind side of a depression and strike the side that faces into the wind. Therefore, incoming sand grains will splash more grains into the air from the windward than from the sheltered side of the depression and surface creep will likewise be more active on the windward side. The difference in the intensity of sand motion between surfaces that face into and away from the wind is obvious to anyone who watches a sand surface rustling beneath the breeze.

At this point it becomes important to note that nearly all the sand grains in any particular part of a dune are nearly the same size. Such is the case because the wind sorts sand by driving the smaller grains ahead while letting the larger ones lag behind. Dune sand always has a much narrower range of grain sizes than beach or river sand, a property that often helps determine whether an ancient sandstone was deposited by the wind. The same fact plays a significant role in ripple formation.

Because all the grains in a cloud of blowing dune sand are nearly the same size, they all tend to bounce along similar trajectories. Therefore, the larger number of grains splashed into the air stream from the upwind side of an irregularity in the sand surface will all land about the same distance downwind. That area will likewise become one of more intense sand movement, which will in turn generate another such area yet another bounce step downwind and so on indefinitely. Furthermore, sand splashes and creeps out of the areas of more intense movement and into the quieter zones. Those become the troughs and ridges of a series of slowly moving sand ripples spaced according to the average distance the grains bounce. An increase in the wind speed will immediately increase the average bounce distance of the sand grains and the spacing of the ripples will likewise increase within a few minutes.

As the wind ripples a sand surface it sorts the larger grains into the crests and the smaller ones into the troughs (Figure 10–17). That happens because a fine sand grain perched in the crest of a ripple is likely to blow into the wind. Then, being smaller than the average, it will bounce a slightly greater than average distance, overshoot the crest of the next ripple, and land in the trough beyond. Coarser grains, on the other hand, are harder to move and therefore likely to lag behind in the ripple crests.

Sand ripples similar in size to those that form on dunes also form beneath running water. However, the two kinds of ripples differ somewhat in general appearance. Particularly, those formed under running water have their coarser grains sorted into the troughs instead of onto the

FIGURE 10–16

Incoming sand grains tend to miss the leeward side of an irregularity in the sand surface and strike the windward slope where they kick more grains up into the wind. Since sand grains in a dune are approximately the same size, they all bounce approximately the same distance. They thus propagate the initial irregularity downwind in a series of ripples spaced according to the average bounce distance of the grains.

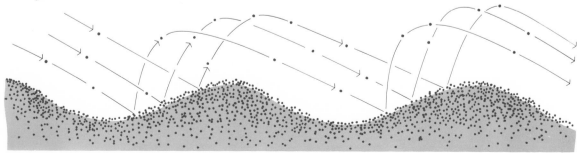

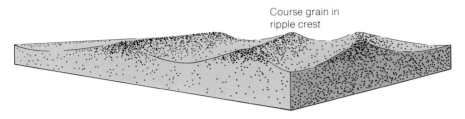

Course grain in
ripple crest

FIGURE 10-17

The larger grains concentrate in the crests of wind ripples and the smaller ones in the troughs.
Ripples in water-transported sand are sorted oppositely.

crests. Geologists working in the field carefully observe ripples in ancient sandstones to determine whether they formed beneath running water or the wind. Such observations help to determine in what kind of sedimentary environment the original sand accumulated. If the rocks are tightly folded, the sorting of the coarser sand grains into the crests and troughs of ripples can also help distinguish between tops and bottoms of sedimentary layers. Such observations may be necessary in determining whether the layers of rock are right side up or upside down.

Wind-Carved Rock

Wind sculpture of solid rock is a minor geologic process that normally operates on a small scale near the ground surface. Most wind-carved rocks are merely pebbles on desert pavements, occurring only locally.

Windborne particles of sand, dust, and possibly snow abrade rocks to carve fluted surfaces that meet in sharp edges and corners, and commonly have a smooth polish and a texture resembling an orange peel. Figure 10-18 illustrates such a rock. Its elegantly sculptured form contrasts sharply to that of stream-rounded rocks, which tend to have convex surfaces and to lack sharp edges or corners. Such rocks are called **ventifacts,** a latinate term meaning "wind made" that emphasizes their origin as wind carvings and their superficial resemblance to human-made artifacts.

Ventifacts tend to be more abundant in cold regions; some of the finest examples have been collected in Antarctica. That climatic relationship is doubtless due in large part to the greater density of cold air that increases its ca-

FIGURE 10-18

Windblown sand, dust, and possibly snow sculpture rocks into distinctive forms consisting of slightly concave surfaces that meet in sharp edges and corners. This unusually large specimen, which is about 35 cm across, came from the Jefferson Valley of western Montana.

pacity to carry abrasive sediment. However, it is true that extremely cold snow becomes as hard as many common rock-forming minerals and could therefore act as an effective abrasive. So it is conceivable that blizzards, as well as sand and dust storms, may contribute to forming ventifacts.

SUMMARY

The wind excavates basin-shaped blowouts as it erodes dust, which it carries long distances in suspension. Thick deposits of windblown dust in humid regions form fertile soils that nourish large human populations. Windblown sand, on the other hand, moves along the ground surface rather than in suspension and rarely travels long distances from its source.

Wind erosion creates a desert pavement consisting of a lag concentrate of pebbles that armor the ground against further erosion and provide a hard surface on which sand grains travel in long hops. Dunes form where sand collects in soft places on the desert pavement and grow because their softness enables them to trap more sand. Dunes migrate as the wind blows sand off their windward surfaces, across the sharp crest, and onto the sheltered leeward side. They tend to congregate into dune fields, vast seas of sand. Ripples on dune surfaces are spaced according to the average bounce distance of the sand grains and change their spacing with variations in wind velocity. Unlike ripples that form under water, the coarser grains concentrate in their crests and the finer ones in the troughs.

Windblown sand, dust, and possibly snow locally carve bedrock, pebbles, and boulders into distinctively sculptured forms composed of shallowly concave flutes that meet in sharp edges and corners.

KEY TERMS

blowout	dune field
cross bed	loess
desert pavement	saltation
drift	surface creep
dune	ventifact

REVIEW QUESTIONS

1. How would a change in temperature or wind velocity affect the maximum size of the sediment particles that the wind carries in suspension as dust?

2. What properties distinguish loess from other sedimentary deposits?

3. The grass-covered Sand Hills of western Nebraska are an enormous and poorly understood sea of naturally stabilized dunes. Briefly suggest at least three different kinds of observation that might indicate the directions in which the

dunes formerly moved and at least two that might shed some light on their age.

4. What field observations might enable you to distinguish between deposits of sand laid down by the wind and by streams?

5. Distinguish between the types of sediment transport that shape the gently sloping windward face of a dune and its more steeply sloping leeward surface.

6. Why is windblown sand more likely to abrade the paint on cars that are parked on paved lots than it is those on unpaved lots?

SUGGESTED READINGS

Bagnold, R. A. *The Physics of Blown Sand and Desert Dunes.* London: Methuen, 1941.

The classic reference on its subject since first published and the principal reference today. Extraordinary. There is no better source on the activity of the wind as a geologic agent.

Cooke, R. U. and A. Warren. *Geomorphology in Deserts.* London: Batsford, Ltd., 1973.
Contains an excellent section on the effects of wind erosion.

Cooper, W. S. *Coastal Sand Dunes of Oregon and Washington. Washington, D.C.:* Geological Society of America, Memoir 72, 1958. Describes many large tracts of coastal dunes. Excellent.

Schultz, C. B. and J. C. Frye. *Loess and Related Eolian Deposits of the World.* Lincoln, Neb.: University of Nebraska Press, 1968.
A comprehensive discussion of loess deposits.

CHAPTER ELEVEN

A group of glacially carved peaks in
Waterton Park, Alberta.

GLACIERS AND ICE AGES

Relatively few people ever see a glacier and most tend to think of them as strange objects that exist only in remote places. Yet, a large proportion of North Americans live in landscapes shaped in one way or another by ice age glaciers and include glaciated mountains among their ideals of scenic beauty. Whether we see them or not, glaciers are part of our daily lives.

GLACIERS AND GLACIATION

Glaciers consist essentially of thick masses of ice that flow under their own weight. Some fill mountain valleys and flow downhill. Others form large ice caps that flow downhill where they lie on slopes and spread like pancake batter where they cover flat terrain. All form, move, and shape the landscape in essentially similar ways although the landforms they create vary greatly in appearance.

From Snow to Ice

People who neglect to shovel snow from their walks have the opportunity to watch soft snow turn into hard ice through a series of diagenetic processes. Snow falls as delicate crystals that shrivel as they age and turn into grains of ice. That process converts the fluffy, powder snow of winter into the granular corn snow of spring. Water trickles into the spaces between grains of ice on warm days, rainy days, and days in which moisture condenses directly out of the air into the snowpack. Then cold weather freezes the water and cements the soggy slush of half-melted corn snow into solid ice. If snow and ice consistently survive from one winter to the next, each successive layer adds to the weight on those below.

Ice Movement

As ice reaches a thickness of about 50 m, depending upon the temperature and the slope, it begins to flow under its own weight (Figure 11–1). To some extent, movement is due to recrystallization as grains melt where they are under pressure and the water freezes in open pores where the pressure is lower. Another process called *crystal gliding*—in which layers of atoms slip past each other as though they were cards in a deck—probably contributes to ice movement. Some of the movement may occur as ice grains slip past each other within the glacier. However, it is difficult to estimate the relative importance of those processes.

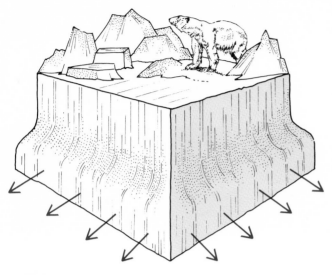

FIGURE 11-1
As ice accumulates to depths greater than about 50 m, the lower part begins to flow beneath the weight of that above.

Glaciers rarely contain either the elongated grains we would expect to form through crystal gliding or the granulated grain boundaries that would tell of grain slippage. Presumably, recrystallization obliterates much of the ice texture created by other processes.

The brittle ice in the upper part of a glacier rides passively on the zone of the plastically flowing ice below, breaking into blocks that creak and groan as they shift past each other along the fractures separating them. Yawning **crevasses** open between blocks of brittle surface ice where movement stretches the glacier and close where further movement eases the tension (Figure 11-2). Crevasses close where they enter the zone of plastic flowage at a depth of 50 m or more. However, plastic flowage is not the only way that glaciers move.

Figure 11-3 shows a common sight in glaciated regions—a bedrock surface covered with parallel grooves called **striations.** Such grooves appear to have formed as skidding ice rasped the bedrock with particles of sediment embedded in the sole of a glacier. Many such surfaces are polished (Figure 11-4), evidently by fine-grained sediment in the ice, and some contain rows of evenly spaced chatter marks that appear to have formed as the sliding ice vibrated, gouging a new hole with every bounce. Test holes drilled to the bases of glaciers reveal that many rest on a film of water that may encourage skidding by lubricating the surface between ice and bedrock.

Rate of Ice Movement

Glaciers usually move quite slowly, so slowly that debate about whether they move at all persisted well into the last century. One experiment that helped resolve that debate,

FIGURE 11-2
The basic anatomy of a valley glacier.

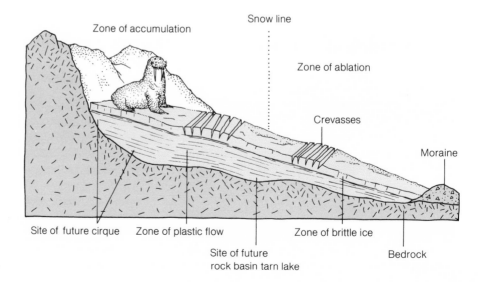

and remains a standard method of observation, involved driving a line of stakes across the glacier and then surveying its movement over a period of months. Such studies show that valley glaciers, like rivers, flow most rapidly near their centers and more slowly along the valley wall where friction restrains them. Vertical holes drilled through glaciers likewise show that the rate of ice movement increases with distance above the bed.

Glaciers generally move at speeds of between several centimeters and several meters per day although some occasionally move much faster. The rate varies seasonally, increasing to a maximum during spring and early summer when a heavy burden of new snow weights the ice, and decreasing to a minimum during fall and winter. For reasons not well understood, many glaciers occasionally move rapidly in **surges**, such as the one that advanced the Black Rapids Glacier of Alaska some 6.5 km during 1936 and 1937. Many Alaskan glaciers surged in the year or two after the Good Friday earthquake of 1964 triggered numerous large rockfalls that spread enormous loads of debris, such as that shown in Figure 11–5, across their surfaces. That experience might lead us to suspect that surges reflect seasons of heavy snow were it not true that individual glaciers commonly surge while their neighbors—that share the same climatic history—do not. It seems more likely that surging reflects episodic slippage of the glacier over its bedrock bed.

FIGURE 11–3

Long striations and rows of chattermarks gouged into the smooth surface of a glacially scoured outcrop of slate. The pebbles in the foreground are about the size of walnuts.

FIGURE 11–4

Remnants of a glacially polished surface on granite in the Sierra Nevada, California. The roughly pitted areas reveal the extent of weathering since the ice melted, presumably about 10,000 years ago. (G. K. Gilbert, U.S. Geological Survey)

FIGURE 11–5
A fast-moving rockfall spread this longitudinally grooved sheet of debris across a glacier in Alaska. The added load will accelerate ice movement. (T. L. Pewe, U.S. Geological Survey)

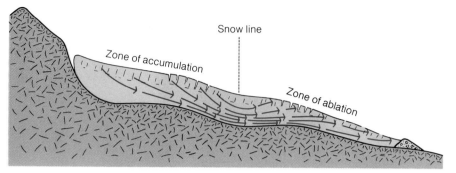

FIGURE 11-6

The arrows describe the paths of individual particles of ice from where they fall as snowflakes on the zone of accumulation to where they finally melt or evaporate on the zone of ablation. The snow line is the only part of the glacier beneath which every particle of ice must pass. Therefore, it marks the area of fastest flow and of greatest capacity to erode.

The Ice Budget

Glaciers, like business firms, have budgets and reflect changes in the balance between their gains and losses by growing or shrinking. Glaciers gain ice in a region called the **zone of accumulation** in which winter snowfall exceeds summer melting. They lose ice in a region called the **zone of ablation** in which the reverse is true (Figure 11-2). Although glaciers sensitively reflect any change in their budgetary balance by growing or shrinking, it is difficult to interpret their behavior. For example, does the dramatic shrinkage of most North American glaciers since about 1900 reflect a decrease in snowfall, a warming of the climate, or some combination of the two?

In more or less temperate regions glaciers flow away from their zones of accumulation until they reach an area in which the ice front melts at the same rate as it advances. That balance dictates the positions of their leading edges. Polar glaciers, such as the one that covers Antarctica, do not melt significantly and tend to flow until they reach the ocean.

Glaciers reveal their budgetary status most clearly during the last weeks of summer when the seasonal melt is virtually complete and the next winter's snows are still to come. Then the zone of ablation is an expanse of blue ice cut by numerous crevasses and littered with rocks and mud melted free during the summer. The zone of accumulation still lies beneath a blanket of white corn snow left from the previous winter. The sharp, late summer snowline that separates the zones of accumulation and ablation marks the point where summer melting exactly balances winter snowfall.

Figure 11-6 shows the paths of several snowflakes from the points where they fall on the zone of accumulation to those where they finally melt in the zone of ablation. They sink into the ice as each winter's snowfall buries them deeper while they move through the zone of accumulation. They then rise toward the surface as each summer's melt strips more ice off the glacier in the zone of ablation. Clearly, the snowline is the only point on the glacier's surface beneath which all the ice passes. The power of the glacier as an agent of erosion must increase as it moves through its zone of accumulation and then decrease as it shrivels in its zone of ablation. That explains why melted glaciers leave landscapes dominated by erosional landforms in the area that was the zone of accumulation, and by depositional landforms in the former area of the zone of ablation.

Glacial Erosion

Glaciers erode bedrock in two ways: (1) by abrading it where sediment-laden ice rasps across exposed surfaces; and (2) by quarrying it where ice freezes onto loose blocks of rock and plucks them free. Figure 11-7 shows how the two processes combine to sculpture bedrock knobs into glacially streamlined **whalebacks,** which are also called **roches moutonees.** Those ice-sculptured knobs show the effects of abrasion on the side that faced into the direction of flow and those of quarrying on the downflow side. That difference is especially obvious in the glacially scoured lakes of Ontario and northern Minnesota. There, the north sides of bedrock islands generally show a streamlined form

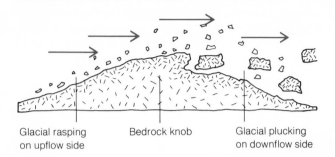

Glacial rasping   Bedrock knob   Glacial plucking
on upflow side   on downflow side

FIGURE 11–7
Debris embedded in the sole of a glacier abrades the upflow sides of bedrock knobs while the moving ice plucks blocks away from the downflow side. After the ice melts, the eroded bedrock knob remains as a whaleback with one side rasped smooth and the other roughly quarried.

FIGURE 11–8
Glacial till—an unsorted and unstratified deposit of all sizes of sediment mixed indiscriminately.

and striated bedrock surfaces whereas the south sides are ragged. Canoeists determine direction simply by looking at the shapes of the whaleback islands; they do not need a compass.

Glacial Deposition

Glaciers deposit a characteristic sediment called **till** in a variety of landforms called **moraines.** Glacial till, which is sometimes called *boulder clay,* consists of an unlayered deposit of all sizes of material mixed indiscriminately together (Figure 11–8). If the margin of a glacier remains in the same position for a considerable time, it deposits a ridge of till that remains after the ice melts to record its former position. Such ridges are called **terminal moraines** if they mark the farthest advance of the ice, and **recessional moraines** if they record a position where the ice margin remained stationary for some time during its retreat. Glaciers also plaster till onto the ground surface to form **ground moraines,** which generally have no distinctive form but locally become streamlined hills called **drumlins** (Figure 11–9). Moraines of all kinds typically contain many lakes and ponds. Some of these bodies exist merely because irregularly deposed till left undrained depressions in the landscape whereas others fill holes called **kettles** (Figure 11–10), which form as large blocks of ice incorporated in the till melt. Surfaces of till deposits are generally littered with boulders called **erratics** because they differ from the

underlying bedrock and must therefore have been transported. Figure 5–3 shows a large glacial erratic.

Melting glaciers produce torrents of water that carry enormous quantities of sediment and deposit them both on and beyond the ice. Those water-transported sediments form layered accumulations of well-sorted clay, sand, and gravel called **glacial outwash,** which occurs in a wide variety of distinctive landforms. Outwash may accumulate on

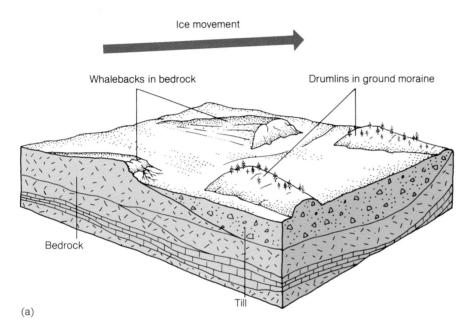

(a)

FIGURE 11–9

(a) Glaciers streamline bedrock whalebacks with their steep surfaces facing in the direction of ice movement and drumlins with their blunt heads facing in the opposite direction. It would be unusual to find whalebacks and drumlins in such close proximity as they appear in this schematic diagram. (b) A small drumlin near Eureka, Montana. The ice moved from right to left. (W. C. Alden, U.S. Geological Survey)

(b)

FIGURE 11-10

A kettle pond in a moraine in Yellowstone National Park. Erratic boulders dropped from melting ice stand in the water and litter the low hummock of till beyond.

or within the ice as fillings in crevasses and in the beds of meltwater streams. After the glacier melts the crevasse fillings remain as isolated small hills or short ridges called **kames** and the stream channel sediments become long, sinuous ridges called **eskers.** Outwash carried beyond the ice may fill stream valleys with deep accumulations of sand and gravel or form vast alluvial fans called **outwash plains** that spread away from the moraine. Like moraines, glacial outwash deposits typically contain numerous kettle ponds and lakes.

Glacial meltwater also forms lakes that accumulate sediments in their floors, which record the seasons of the lake's existence in dark and light layers called **varves.**

Summer meltwater carries large quantities of finely pulverized rock called **rock flour,** which make the water look almost like milk and settle to the bottom of the lake to form a layer of light-colored silt. During the cold winter months the influx of meltwater and rock flour virtually ceases and suspended clay mixed with the remains of minute plants and animals that flourished during the summer settles to the bottom of the lake to form a layer of dark sediment. Each pair of light and dark layers not only records a year of the lake's existence but also tells something about the seasons of that year. It can tell, for instance, whether the summer was long or short and whether any midwinter thaws dumped thin streaks of pale rock flour into the dark winter sediments. The layers of glacial lake sediments may be correlated from one exposure to another by matching their characteristic sequences of seasonal patterns in much the same way that sequences of tree rings can be matched from one log to another. Geologists have pieced together remarkably complete records of the waning stages of the last Ice Age by studying glacial lake sediments.

Although geologists occasionally refer to all glacial sediments collectively as **drift,** they nevertheless distinguish carefully between till and outwash because the difference has great practical importance. Lack of sorting makes most till deposits almost completely impermeable and therefore poor sources of ground water and unsatisfactory locations for septic tank systems. Till is also unsuitable for use as construction aggregate. The well-sorted and layered deposits of outwash, on the other hand, generally contain good aquifers, excellent septic tank sites, and provide abundant construction aggregate.

Valley Glaciers

Alpine glaciers fill old stream valleys and gouge them into long troughs with steep side walls and sharp ridges between them. They create many of our most picturesque mountain landscapes capped by craggy peaks and sparkling with lakes.

The head of a large valley glacier freezes tightly onto the bedrock and then plucks out blocks as the ice moves downslope, opening a gap between the upper end of the glacier and the rock wall behind it (Figure 11–2). Snow and meltwater refill the gap and freeze once again, attaching the glacier to the rock wall at its head thus enabling it to pluck out more blocks of bedrock as it continues to move away. After the glacier melts the deeply quarried bedrock headwall remains as a cliff overlooking a deep hollow called a **cirque,** which commonly holds a lake in its floor. Figure 11–11 shows a typical large cirque. From a distance cirques look like basins scooped out of the mountaintop by a giant with an ice cream scooper. If several glaciers bite their cirques into a mountaintop, they leave nothing but a craggy remnant called a **horn peak** (Figure 11–12).

Large valley glaciers generally gouge their courses into deep troughs that are nearly semicircular in cross section.

FIGURE 11–11

A headwall cliff towers above a lake in a cirque on Red Castle Peak in the Unitah Mountains of northeastern Utah. The talus cones at the base of the cliff must have formed since the ice melted, presumably within the last 10,000 years. (W. R. Hansen, U.S. Geological Survey)

After the glaciers melt, the upper portions of their valleys retain that cross section because relatively little sediment accumulates there. Those upper parts of glaciated valleys commonly contain numerous small lakes called **tarns,** which fill basins carved into solid bedrock. Many glaciated

FIGURE 11–12

A horn peak in Glacier National Park, Montana.

valleys contain long chains of such lakes strung like beads along a stream that tumbles over rapids and waterfalls as it flows from one to another.

Large alpine glaciers also straighten their valleys (Figure 11–13), presumably because ice flows around sharp bends less readily than water. Hikers commonly walk up winding stream valleys in which the view extends no far-

ther than the next bend until they reach the lower limit of glaciation where the valley abruptly widens and straightens, providing a long view of distant peaks. Adjacent glaciers commonly reduce the ridge between them to a thin wall with a serrated crest called an *arete,* which may contain low passes marking places where the glaciers eroded through the ridge and coalesced.

FIGURE 11-13

The heavily glaciated eastern front of the Northern Rocky Mountains in Glacier National Park, Montana. During the last Ice Age, only the mountain peaks and the crests of the higher ridges stood above the glaciers. The two large lakes in the floor of the long, straight valley fill basins formed where large masses of stagnant ice prevented accumulation of outwash as the glaciers melted. (National Park Service)

Big glaciers erode their troughs much more deeply than do their smaller tributaries. That difference is not visible while the glaciers exist because their ice surfaces meet at the same level. However, after the ice melts, the floors of the larger valleys lie at much lower levels than do those of their smaller tributaries. Therefore, tributary streams typically enter large glaciated valleys over waterfalls or cas-

cades. Such *hanging* tributary valleys, as they are called, are common features of glaciated mountain landscapes.

Valley glaciers build hairpin-shaped moraines (Figure 11–14) around their lower ends. The tip of the hairpin is called a terminal or **recessional moraine** whereas the parts that follow the valley wall, the arms of the hairpin, are called **lateral moraines.** In most cases, lateral moraines

FIGURE 11-14

The low ridge in the center of the picture is a terminal moraine neatly cross-sectioned by a postglacial stream. Waterton Park, Alberta. (D. W. Hyndman)

are largest on the north and east sides of valleys because those are the sides that face the warm afternoon sun. More rapid melting makes the ice surface lower there, and that causes both ice and sediment to move toward the warm side of the valley. Meltwater also tends to flow along the warmer side of the valley where it may erode an ice-marginal stream channel in the bedrock valley wall or in the top of the lateral moraine.

Melting valley glaciers leave deep deposits of ground moraine and outwash in the lower parts of their valleys, building broad valley floors (Figure 11–15) dotted with lakes and meadows. In some cases, terminal or recessional moraines impound lakes. However, lakes of that type are not common because their outlet streams generally drain them by eroding a channel through the moraine. Many of the lakes in the lower parts of glaciated mountain valleys fill basins formed where stranded masses of stagnant ice were left behind as the glacier melted and then were partially buried in outwash. Many other lakes are impounded behind alluvial fans of outwash swept in from tributary valleys.

Piedmont Glaciers

Streams of ice pouring out of mountain valleys may spread along the base of the range to form broad sheets called

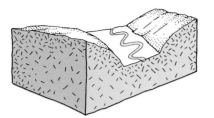 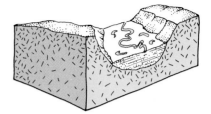

Pre-glacial stream valley

Ice-filled valley

Post-glacial valley floored with deposits of till and outwash.

(a)

(b)

FIGURE 11-15
(a) Glaciers gouge stream valleys to an approximately semicircular cross section and then leave them partially filled with debris as they melt. (b) The glacially gouged valley of Rock Creek at the eastern edge of the Beartooth Plateau, Montana.

piedmont glaciers, which commonly coalesce into a continuous mass. Such ice sheets formed along much of the eastern front of the northern Rocky Mountains in the United States as well as along the front of the Canadian Rockies (Figure 11–16) where the piedmont sheet spreading east from the mountains met the great continental glacier spreading west from central Canada. In that region, distinguishing deposits left by the piedmont and continental glaciers is difficult except by examining the boulders they contain to determine whether they came from bedrock exposed in the mountains or in central Canada.

Rock Glaciers

High and cold mountainous regions commonly contain numerous streams of angular rubble called **rock glaciers** (Figure 11–17), which have lobate forms, steep fronts, and broadly corrugated surfaces that make them look as though they are moving as extremely viscous masses. Some rock glaciers dribble out of cirques and down the floors of the upper parts of glaciated valleys, looking plausibly like the lingering remnants of large glaciers. Others have no association with glaciated valleys and start from large talus slopes or fill long draws in the mountainside.

Geologists long debated whether rock glaciers actually contain ice and move or simply remain as inactive relics of the Ice Age. The question was at least partially resolved when highway construction in the Rocky Mountains revealed ice deep within rock glaciers sectioned in roadcuts. Continuing maintenance problems requiring frequent removal of rocks from the road have erased any doubt that they move. It seems reasonable that they should move because most common rocks are about 2.7 times as dense as ice so a burden of only 18 m of rock, instead of 50 m of ice, would exert enough pressure to make ice flow. Therefore, a relatively thin rock glacier would be heavy enough to flow if its lower part contained an ice matrix.

A pile of angular rubble, such as a rock glacier, can easily maintain a filling of ice even in regions where the mean annual temperature is well above freezing. New air can sink into the spaces between the rocks only if it is colder and therefore denser than the air already there. Rocks pro-

FIGURE 11–16

Hummocky morainal topography along the eastern front of the Canadian Rockies near Waterton, Alberta. This moraine was left by a piedmont glacier but it looks like those left by continental glaciers. (D. W. Hyndman)

FIGURE 11–17
A rock glacier in the San Juan Mountains of Colorado. The broadly wrinkled surface is typical of
extremely viscous masses that flow very slowly. (W. Cross, U.S. Geological Survey)

vide such good insulation that the interior of the pile can
remain cold through the summer and absorb new air only
during the coldest days of winter—ice caves work the same
way. Rock glaciers normally consist of light colored rocks
because they neither absorb nor radiate heat as readily as
would dark rocks and therefore maintain the interior of the
mass at a more uniform temperature.

Continental Glaciers

As the name implies, **continental glaciers** cover areas of
regional or even continental extent. Nearly all of Antarctica
lies beneath a continental glacier as does most of Greenland and parts of several large islands in the Canadian Arctic. Extensive regions of high mountains may acquire a
continuous cover of ice that behaves very much like a continental glacier. Such ice caps exist today in Alaska where
the Juneau ice field is the best-known example and in the

Canadian Rockies where many people visit the Columbia
ice field. During the ice ages, continental glaciers similar to
that which now covers Greenland to a depth of about
4,000 m spread across most of Canada and the northern
part of the United States, and mountain ice caps existed as
far south as the central Rocky Mountains.

Continental glaciers and alpine ice caps generally
spread outwards in all directions from their zones of accumulation. Most alpine ice caps and some continental glaciers spawn large valley glaciers that drain ice away to
lower elevations.

The great continental glacier of the last Ice Age left a
deeply eroded landscape in most of Ontario and Quebec,
in much of Manitoba and the maritime provinces, and in
parts of the northern tier of the United States from northeastern Minnesota eastward. That is a region of glacially
sculptured whaleback hills and countless lakes and bogs
that fill basins carved into solid bedrock. Soil is generally

scarce and consists largely of local deposits of till or outwash.

A peripheral region—in which landforms created by processes of glacial deposition dominate much of the landscape—embraces the glacially eroded central area. It includes virtually all of Saskatchewan, large regions of Alberta and Manitoba, and much of the northern United States from central Montana eastward. Ground moraine and outwash cover much of the region so deeply that many large areas completely lack bedrock outcrops. Most of the ground moraine is virtually formless but in some areas the ice moulded it into large fields of drumlins oriented parallel to the direction of ice flow. When seen from the air, they suggest schools of tadpoles with their blunt heads facing into the direction of ice movement and their long tails trailing off downstream.

Terminal moraines that extend hundreds of kilometers in great looping curves define the margins of the area covered by continental ice and a complex of recessional moraines records the stages in its final retreat. Geologists recognize the moraines as low ridges as much as several

kilometers wide in which the landscape is a hummocky maze of little hills and kettle ponds. Detailed mapping of those moraines reveals a complex pattern recording the advance and retreat of great lobes of ice that suggest the "arms" of an amoeba. Glacial outwash extends far beyond the terminal moraines, filling the valleys of many streams to form prominent sets of terraces and spreading across expanses of flat countryside as outwash plains (Figure 11–18).

The great continental glaciers obliterated the original streams in the regions they covered and the numerous lakes and swamps in those regions are evidence that they remain without effective drainage today. The present courses of the Missouri and Ohio Rivers rather closely trace the boundaries of maximum glaciation, presumably because they established those channels along the edges of large continental glaciers. Most of the streams that formerly drained the big ice sheets carried much more water during the ice ages than they do today. Now, they flow through greatly enlarged valleys inappropriate to the size of the modern stream.

FIGURE 11–18

Schematic view of the margin of a continental glacier showing the relationships between the ice and its associated deposits of till and outwash.

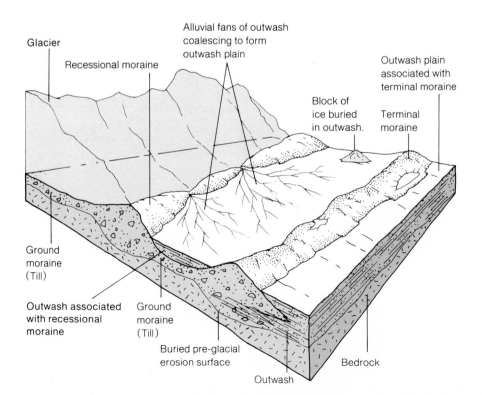

THE ICE AGES

During the past 2.5 million or so years—the period geologists call the Pleistocene—the earth has experienced a long succession of great ice ages, perhaps as many as 20. Enormous continental glaciers spread from the general region of Hudson Bay to cover, at one time or another, most of North America north of the line of the Missouri and Ohio Rivers and from the Canadian Rockies to the Atlantic. Ice also spread from the Baltic Sea region to cover most of northern Europe from the Atlantic almost to the Urals and from the Arctic as far south as central Germany and Poland. Valley glaciers flourished in high mountains everywhere and numerous ranges developed extensive ice caps along their crests. However, vast regions of Alaska, northwestern Canada, and Siberia remained ice-free, presumably because they received too little snowfall to form glaciers.

The Weight of the Ice

The great ice sheets depressed the lithosphere beneath their weight just as Greenland and Antarctica are now depressed beneath their loads of ice. If the Greenland ice were to melt—and it shows no sign of doing so—its disappearance would leave an inland sea where the center of the island is now depressed below sea level. The only land to appear immediately would be a chain of islands corresponding to the present coastal regions. The lithosphere would then rise as it floated back into isostatic equilibrium and the center of the island would emerge above sea level after some thousands of years. Large regions surrounding Hudson Bay in North America and the Baltic Sea in Europe are still rising after having been unburdened as the great continental ice sheets melted at the end of the last Ice Age about 10,000 years ago. Maps of old beaches show that the Baltic coast of Sweden has risen about 200 m since the end of the last Ice Age, an average rate of about 2 cm each year, and that concentrically surrounding regions have risen lesser distances at a slower rate. Raised beaches in Canada reveal a similar pattern of uplift in the regions surrounding Hudson Bay. In both continents archeologists find sites once inhabited by coastal peoples far inland. Viking ship graves, for example, and the remains of old ports now exist in the midst of farm fields many kilometers from the modern coast of Sweden.

Meanwhile, the low countries sink as mantle rock flows out from beneath the lithosphere there into the region farther north. The Dutch often find such things as old Roman roads and mileposts when they dike and drain coastal areas to create new farms, evidence that those areas were still above sea level less than 2,000 years ago. Precise measurements of the earth's gravity field in the Hudson Bay and Baltic Sea regions reveal deficiencies in mass, which suggest that the postglacial isostatic rebound is now about half complete. Another several thousand years of continuing uplift will drain Hudson Bay and the Baltic Sea, leaving only layers of marine sediments and raised beaches as mementos of their former existence.

Changing Sea Level

Water to make glaciers must come from the oceans. Thus, sea level drops as land ice accumulates and rises as it melts. Profiles of drowned shorelines appear at depths as great as 130 m on recording fathometer traces across many coastlines. They show that vast areas of continental shelf were above sea level at the same time that large regions of the continents were buried under ice. The ice now on Greenland and Antarctica would raise sea level approximately 65 m if it were to melt. However, melting of the pack ice in the Arctic Ocean would not affect sea level because it is already floating. Although the subject is a matter of considerable dispute, many geologists believe that the highest Pleistocene shorelines on stable coasts are only about 10 m above present sea level. If they are correct, then it follows that neither the Greenland nor the Antarctic ice cap could have melted during any of the interglacial periods between ice ages.

Pleistocene Chronology

Although the archives of Pleistocene time surround us in many of our familiar landscapes, geologists nevertheless have great difficulty assembling the record into a coherent historical account. One reason for their problem is that Pleistocene events occurred at a pace much faster than that of organic evolution, which makes using fossils to date the events a bit like using a sundial to time a foot race. Radioactive age-dating methods fail to solve the problem because most Pleistocene materials are too old to date by the carbon-14 method and too young to date by the other standard techniques. The nature of the evidence poses further problems because advancing glaciers generally obliterate the record left by their predecessors—bulldoze it off the landscape. Nevertheless, the surviving record does contain a richly detailed history of the later phases of the last Ice Age and a partial account of at least several earlier and more extensive glaciations. In a few especially informative exposures we find glacial drift deposited during two or more ice ages stacked to create a sequence of layers. Old soil profiles on the buried layers convey an impression of the durations and climates of the interglacial episodes that separated the ice ages.

In recent years, study of deepsea cores has added a new dimension to our knowledge of Pleistocene time and events. Layers of sediment accumulate fairly continuously on the ocean floor to preserve a record far more complete than any known to exist on the continents. Analysis of deepsea cores has already shown that there were far more than the four ice ages that geologists working on land had been able to recognize. Study of deepsea cores may also resolve the long-debated question of whether the Pleistocene epoch was a time of long interglacial episodes punctuated by relatively brief ice ages or a time dominated by ice ages with relatively brief interglacial intermissions. They may even reveal more information about the climates that prevailed during ice ages and interglacial times.

Pleistocene Climates

Obviously, the climates of the glaciated regions must have been considerably colder during the ice ages than they are now. Clear evidence of the former operation of tundra erosion processes in the soils and landscapes of the regions immediately surrounding the big glaciers shows that they too must have cooled considerably. There is also evidence—in the form of fossils—that the ranges of many northern plants and animals shifted hundreds of kilometers south during the ice ages. However, there is no reason to believe that the ice ages chilled the entire world, and temperatures may have changed little in the warmer regions.

The most precise information about ice age temperatures comes from analysis of the ratios of the isotopes oxygen-16 and oxygen-18 in fossil seashells. Those ratios reveal the temperature of the water in which the shell grew to within about 1°C. The results of many such analyses show that ocean surface water temperatures dropped approximately 4° to 5°C at middle and high latitudes during the ice ages. Tropical surface water temperatures changed very little.

Abundant evidence from many regions of the northern hemisphere shows that the ice ages must have been wet as well as cool. Northern Hemisphere deserts contain old shorelines, such as those shown in Figure 11–19, which record lakes that filled undrained desert basins during the last Ice Age. Furthermore, archeologists find artifacts left by peoples who lived during the last Ice Age in regions much too arid to support them now. However, it is difficult to know how much of that wetness actually reflects increased rainfall and how much is due to reduced evaporation rates caused by climatic cooling. On the other hand, it is also possible to imagine increased precipitation causing climatic cooling; the cause and effect relationships are not clear.

Presumably, our modern climate more nearly resembles those of the interglacial periods than of the ice ages. However, ours is far from being a stable climate; the last several thousand years have seen considerable fluctuations in both temperature and rainfall. Some of these have dictated the course of historic events, such as the Viking colonization of Greenland, which began and ended with a warm climatic period. There is no apparent reason to assume that the present state of the earth differs significantly from that of the last 2 million or so years. It thus seems reasonable to expect more ice ages although we still cannot predict when or how the next one may start. However, there is good evidence that the climatic change ending the last Ice Age about 10,000 years ago came rather suddenly so it is at least conceivable that the next one could start abruptly.

A new ice age will almost certainly make the climates of the formerly glaciated regions much wetter and colder. However, thousands of years would pass before new glaciers growing in the old centers of ice accumulation could begin their slow spread into milder regions. Meanwhile, many of the world's dry regions would become much wetter and vast expanses of land now too arid for agricultural use would become productive. And as the growing ice sheets lowered sea level, large regions of continental shelf now shallowly submerged would become productive coastal plains. It is difficult to be sure whether the onset of a new ice age would increase or decrease the world inventory of arable land.

FIGURE 11-19
The gravel bench on the lower slope of the hill is an old beach deposit left by a pluvial lake that filled this desert valley during the last Ice Age. Tucker Hill near Paisley, Oregon.

Causes of Ice Ages

Many geologists jokingly say that the number of theories attempting to explain the cause of ice ages is the same as the number of theoreticians. Although somewhat exaggerated, that statement does convey a useful impression of a field inhabited by numerous competing ideas all in some way at odds with the evidence. Nevertheless, some of those theories may contain part of the truth and many provide interesting insights into the workings of our planet.

The simple statement that glaciers form where winter snowfall exceeds summer melting embraces a superficial explanation of ice ages. Under modern conditions the snowpack lasts well into the summer in large parts of the former areas of ice accumulation and new snow comes early in the fall. Thus, a relatively minor climatic change involving some combination of cooler summer and wetter winters might tilt the balance in favor of ice age conditions. Most of the existing theories attempt to explain ice ages as the result of a shift toward either a wetter or a cooler climate, regarding one as the primary cause and the other as a secondary effect of glacial conditions.

The Indonesian volcano Krakatoa inspired one theory to explain ice ages after an eruption in 1883 injected immense quantities of ash into the stratosphere where the upper-level winds circulated it throughout the northern hemisphere. The suspended ash reflected enough solar radiation to cause distinctly cooler and wetter weather during the two years that passed before it finally settled. That

led many geologists of the time to suggest that great volcanic eruptions may have been responsible for the ice ages. However, there is little evidence of unusual volcanic activity associated with ice ages and volcanoes have erupted throughout geologic time whereas large-scale glaciation has been most infrequent.

An entirely different theory proposes that ice ages occur after the Antarctic ice cap surges and spreads a greatly expanded shelf of floating ice across the surrounding ocean. The enlarged expanse of ice would reflect sunlight thus reducing the earth's heat absorption and therefore causing a cooler ice age climate. That theory could explain the apparently sudden ending of the last Ice Age through breakup of the floating ice shelf. However, it fails to explain why Northern Hemisphere glaciation began only about 2.5 million years ago although evidence from deepsea cores suggests that the Antarctic ice cap has existed for at least 12 million years. Further study of deepsea cores collected off Antarctica will test that theory by revealing whether the ice ages corresponded to periods of greatly expanded shelf ice.

A few geologists have suggested that the ice ages may correspond to periods when the sun was slightly cooler than it now is. The idea is appealing in its simplicity and has a certain plausibility because the sun does vary in brightness during the 11-year sunspot cycle. However, the theory does not explain why continental glaciation has occurred so infrequently in the earth's history without resorting to special assumptions about the sun's behavior for which there is no independent evidence. The greatest obstacle to seriously considering the idea is the lack of any obvious way to test it—the rocks contain no known record of the sun's brightness.

One of the more interesting theories depends upon the small fluctuations in the amount of sunlight that reaches the earth because of the way it wobbles in its orbit. Analysis of deepsea cores has shown that the periodic fluctuations in solar radiation seem to correspond to the timing of the last major Ice Age and perhaps to that of some earlier ones. However, it is difficult to imagine how such small variations in solar input could cause such large climatic effects and why they should have done so only during Pleistocene time. Nevertheless, many geologists find the coincidence in timing persuasive and suggest that the small orbital fluctuations in the amount of solar radiation the earth receives may trigger large changes in an already unstable climatic situation.

One interesting theory attributes the primary cause of ice ages to wetter rather than to colder conditions. That approach requires a scenario that would increase the amount of water evaporating from the oceans to explain the increase in precipitation. However, the temperatures of oceanic surface waters at temperate latitudes were demonstrably colder during the ice ages than during interglacial episodes so the rate of evaporation at those latitudes must have been correspondingly lower. Nevertheless, it would be possible to increase the total amount of water evaporating from the oceans by diverting the warm waters of the Gulf Stream through Denmark Strait between Greenland and Iceland and thence into the Arctic. That would melt the Arctic pack ice and almost certainly cause much wetter climates throughout much of the Northern Hemisphere. There is archeological evidence suggesting that the Arctic coasts were indeed warmer during the last Ice Age than they now are.

Although the idea has its shortcomings, it does explain several of the more puzzling aspects of Pleistocene climatic history. The Gulf Stream is a winddriven current so it could presumably be deflected into a different course by an unusual storm season and that could also explain the apparent suddenness of the climatic change that ended the last Ice Age. Because the mechanism depends upon the present geographic arrangement of continents and oceans, it becomes easy to imagine the Pleistocene pattern of alternating glacial and interglacial episodes; the

pattern beginning as the continents moved into their present positions and eventually ending as their continued movement reshapes the Arctic and North Atlantic Oceans. Unfortunately, the theory does not explain the large-scale glaciation of New Zealand and southern South America, which apparently coincided in time with the Northern Hemisphere ice ages. Furthermore, it is possible to trace the former paths of ocean currents through study of deepsea cores and data gathered so far from the North Atlantic. Such study does not reveal the expected pattern.

Whatever its merits, the Arctic pack ice theory does emphasize the importance of the arrangement of continents and oceans in determining climatic patterns. Antarctica evidently acquired its ice cap when it moved into its south polar location during Miocene time and will presumably retain it until it eventually moves on into more temperate latitudes. The tight clustering of the northern continents around the Arctic Ocean certainly contributes to the present climatic pattern by restricting exchange of heat between the tropics and the Arctic. If the poles were in unenclosed oceans—as they must have been during most of geologic time—global heat exchange would improve and the climatic pattern would become more uniform.

SUMMARY

Glaciers form where the annual snowfall exceeds melting, permitting ice to accumulate to depths greater than 50 m, at which it begins to flow under its own weight. Although glaciers assume various forms according to the terrain in which they develop, all erode rock and deposit sediment in basically similar ways and leave comparable landforms when they melt.

Glaciers erode bedrock partly by freezing onto blocks and plucking them loose and partly by rasping it with pieces of rock embedded in the sole of the ice. They shape knobs of bedrock into streamlined forms rasped smooth on their upflow sides and plucked into roughly quarried surfaces on their downflow sides. Glaciers also tend to erode bedrock basins, which become ponds and lakes after the ice melts. Valley glaciers gouge straight channels separated by thin bedrock ridges and reduce mountain peaks to craggy spires.

Glaciers thin and deposit part of their sediment load as they pass from the area of ice accumulation into that in which annual melting exceeds snowfall. Much of the ice-deposited sediment forms morainal ridges that mark former positions of the glacier margin and the rest is plastered onto the ground surface in deposits that locally develop into drumlins streamlined in the direction of ice flow.

Glacial meltwater flowing on or within the ice and beyond its margins deposits layered accumulations of well-sorted silt, sand, and gravel, which remain after the ice melts to record the former positions of meltwater streams and lakes. Other meltwater deposits fill stream valleys and form large plains stretching away from the former ice margin.

Large regions of glaciated landscape testify to the occurrence of ice ages. The exact nature of their climates and the factors that caused them remain in vigorous dispute, although most of the evidence suggests that ice age climates were generally colder and wetter than those of the modern world. None of the competing theories to explain ice ages has gained general acceptance because all fail to account for significant elements of the geologic record. However, there is no known reason to assume that the operative mechanism, whatever it may have been, has ceased to function. Most geologists agree that the pattern of recurrent ice ages characteristic of the recent geologic past will probably continue into the near geologic future and that another ice age is probable at an unknown time.

KEY TERMS

arete	hanging valley	rock glacier
cirque	horn peak	striation
continental	kame	surge
glacier	kettle	tarn
crevasse	lateral moraine	terminal moraine
drift	moraine	till
drumlin	outwash plain	varve
erratic	piedmont glacier	whaleback
esker	recessional moraine	zone of ablation
glacial outwash	roche moutonee	zone of
ground moraine	rock flour	accumulation

REVIEW QUESTIONS

1. What would you look for to determine whether an undrained depression is a tarn, a kettle, a wind blowout, or a limestone sinkhole?

2. What properties make most glacial tills poor aquifers and poor sites for ground treatment of sewage effluents? Why does glacial outwash generally serve better for both purposes?

3. Glacial tills and mudflow deposits make virtually identical sediments. What would you look for in the field to determine whether an exposure of such material is a mudflow or a moraine?

4. How would an increase or a decrease in either mean annual temperature or precipitation affect the position of the snowline on a glacier?

5. What landforms would you look for to determine whether you were in the lower part of a glaciated valley and how would you expect them to change as you ascended to higher elevations in the same valley?

6. What landforms would you expect to see in (1) the area beyond the terminal moraine left by a large continental glacier, (2) the area beneath the former zone of ablation of the same glacier, and (3) the area formerly beneath the zone of accumulation?

SUGGESTED READINGS

Embleton, C. and C. A. M. King. *Glacial and Periglacial Geomorphology.* New York: St. Martin's Press, 1968.
A massive and technical treatment of glacial landforms.

Dyson, J. L. *The World of Ice.* New York: Alfred A. Knopf, 1963.
An interesting review of ice in general and glaciers in particular for the nontechnical reader.

Flint, R. F. *Glacial and Quaternary Geology.* New York: John Wiley, 1971.
An authoritative reference.

Imbrie, J. and K. P. Imbrie. *Ice Ages: Solving the Mystery.* Short Hills, N.J.: Enslow Publishers, 1979.
An outstanding nontechnical account of efforts to understand ice ages and their causes.

Matsch, C. L. *North America and the Great Ice Age.* New York: McGraw-Hill Book Co., 1976.
An excellent brief treatment.

Sharp, R. P. *Glaciers.* Portland, Ore.: Oregon State System of Higher Education, Condon Lectures, 1960.
An extremely readable and authoritative treatise on glaciers. Brief.

CHAPTER TWELVE

A reef composed of stromatolites,
distinctive mound-shaped structures built
by blue-green algae. These are in
Precambrian limestone approximately 1.2
billion years old but blue-green algae still
create similar structures today. These
stromatolites are about the size of
cabbages.

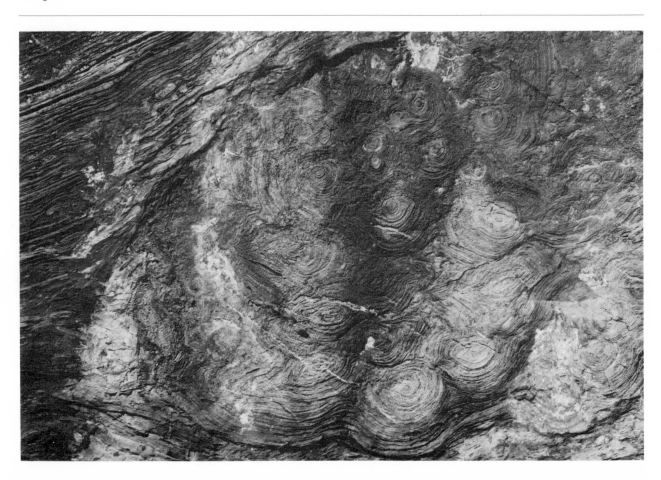

AIR, IRON ORE, AND FOSSIL FUELS

Figure 12–1 shows the approximate composition of the earth's modern atmosphere. It is a wildly improbable combination of gases maintained in its present state by the complex interplay of processes briefly summarized in Figure 12–2 and discussed in the following pages. There is no reason to believe that the earth's atmosphere has always been as it now is—it appears to have evolved to its present state through processes that also created most of our iron ore and all of our fossil fuels.

We have no certain way of knowing when or how the earth acquired its atmosphere. Neither do we know what the early atmosphere contained. However, there is some rather circumstantial evidence that provides the basis for interesting conjecture.

AIR

Origin of the Atmosphere

Spectroscopic analyses of light from the sun and other stars reveal that the inert gases helium, argon, and neon are more abundant in other parts of the universe than in our atmosphere. That presents a serious problem because argon is heavier than the other atmospheric gases, including water vapor, and therefore could not have escaped from the earth unless the others did likewise. That could conceivably have happened if the earth became very hot at

some early stage of its development but the escape of the inert gases would have left the planet with neither an atmosphere nor surface water. Some geologists argue that such an event did indeed happen and that the present atmosphere and surface water were later derived from the mantle through a process called **outgassing.** Some argue in favor of rapid outgassing early in geologic time—the "big burp" hypothesis—and others prefer to think of a long series of lesser burps—volcanic eruptions. For many years most geologists regarded gases escaping from volcanoes as new contributions from the mantle, as evidence that outgassing is a continuing process. However, modern isotopic analyses of those gases show that less than 1 percent are indeed new contributions from the mantle. The remainder clearly consists of material recycled from the surface. Nevertheless, the mantle does appear to be releasing small quantities of new gases and may have done so on a much larger scale during the past.

Other geologists prefer to argue that the scarcity of inert gases in the earth's atmosphere may simply reflect a peculiarity of its original composition rather than a major event early in the history of the planet. They point out that the terrestrial planets have densities ranging from 3.4 to 5.5 times that of water, which must reflect differences in their bulk compositions. If the compositions of the solid parts of planets can vary so greatly, then why shouldn't the compositions of their atmospheres also vary?

FIGURE 12-1

The composition of the atmosphere. Green plants account for the large concentration of oxygen and the small amount of carbon dioxide. The argon consists almost entirely of argon-40 produced by radioactive decay of potassium-40 and released from rocks as they weather. The origin and history of the nitrogen is unclear.

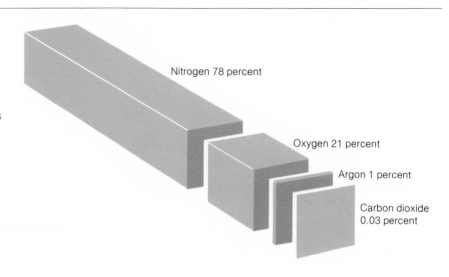

Nitrogen 78 percent

Oxygen 21 percent

Argon 1 percent

Carbon dioxide 0.03 percent

FIGURE 12-2 (below)

Green plants play the central role in the atmospheric balance of both carbon dioxide and oxygen and therefore in the creation and maintenance of the air we breathe.

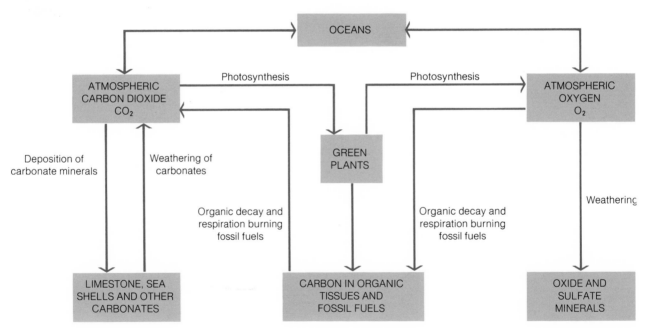

Composition of the Earth's Early Atmosphere

Our knowledge of the composition of the earth's early atmosphere is almost as conjectural as our ideas about its origin. Some of the evidence is clear but much is highly circumstantial and subject to widely varying interpretations.

It does seem reasonably certain that the earth's atmosphere contained virtually no free oxygen until about 2 billion years ago and very little for a long time thereafter. One of the main items of evidence exists in certain sandstones and conglomerates deposited more than 2 billion years ago. They contain stream-rounded grains of pyrite, which would certainly have rusted into the iron oxide minerals goethite or hematite had they been carried in a stream flowing beneath an oxygenated atmosphere.

There is fairly good evidence that the earth's early atmosphere may have had large quantities of carbon dioxide. Sedimentary rocks contain large amounts of coal, petroleum, disseminated organic matter, and limestone. All of those substances contain carbon that must have been derived from atmospheric carbon dioxide. There is no other way in which carbon could become incorporated in a sedimentary rock.

We have relatively little evidence of what other gases the earth's early atmosphere may have contained. Considerable speculation has revolved around the possible presence of methane and ammonia because those gases could have provided an environment chemically congenial to the development of early forms of life. However, no direct evidence for early atmospheric methane and ammonia exists in the rocks. Nor does nitrogen—which comprises some 79 percent of the modern atmosphere—leave any evidence of its existence in the rocks.

If the earth's early atmosphere did indeed contain large amounts of carbon dioxide, that must have had a considerable climatic effect. Carbon dioxide molecules absorb heat radiation—a phenomenon called the **greenhouse effect** (Figure 12–3)—so large concentrations of carbon dioxide in the early Precambrian atmosphere must have made it very warm. High temperatures would have maintained a high water vapor content in the atmosphere and therefore presumably a dense cloud cover. Geologists still

have no way to measure the temperature of the early Precambrian atmosphere. They can only speculate that it must have been tolerable to blue-green algae and other equally primitive lifeforms, which thrive today at temperatures throughout the entire range between the freezing and boiling points of water. Had blue-green algae not been able to start growing on earth, our planet might well have gone the way of Venus.

Geologists who study Precambrian sedimentary rocks generally notice a striking absence of evidence that wind played a significant role in their deposition. Even though Precambrian sandstones are numerous and many appear to have been deposited on land, none show the kind of crossbedding typical of dune sands. Perhaps the wind did not blow strongly enough then to move sand. It may be significant that the Venus probes detected very little surface wind on that planet with its dense atmosphere whereas violent winds blow immense clouds of dust high into the extremely thin atmosphere of Mars.

Whatever the early Precambrian atmosphere may have been, we can safely assume that it and the environments it created were quite unlike anything we know today. If we could somehow transport ourselves back to that remote time, equipped of course with suitable breathing apparatus, we might find a shadeless planet devoid of familiar animal and land plant life but abundantly supplied with scummy growths of blue-green algae. Those algae were already beginning the long process of removing carbon dioxide from the atmosphere and adding free oxygen. Green plants continue that process today.

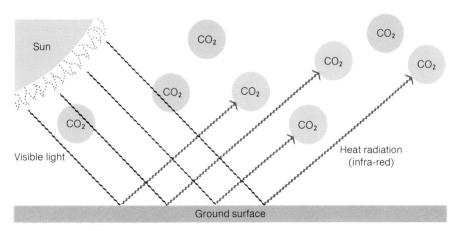

Figure 12–3
The greenhouse effect. Carbon dioxide transmits visible light, which warms the earth but absorbs infra-red heat radiation thus trapping the warmth.

Atmospheric Evolution

Many Precambrian carbonate sedimentary rocks contain abundant remains of blue-green algae, such as those shown in Figure 12–4 as well as large quantities of organic matter. Some rocks contain enough organic matter to stain them black and cause them to emit a foul odor when broken. The algae, the organic matter, and the carbonate rock itself tell the basic story of atmospheric evolution.

The algae, like all photosynthetic green plants, absorbed carbon dioxide from the atmosphere, used energy from the sun to break it down to obtain carbon for their tissues, and released the oxygen as a by-product. Some of the organic matter in their tissues was incorporated in the accumulating sediments, now become rocks, and therefore did not decompose and recombine with oxygen to return to the atmosphere as carbon dioxide. Every molecule of free oxygen in the atmosphere must correspond to an atom of carbon either tied up in organic matter, such as wood, or buried in the earth's crust.

Green plants also remove carbon dioxide from the atmosphere by assisting in the deposition of limestone, a rock type that formed in small quantities during early Precambrian time but did not begin to appear in large quantities until about 2.6 billion years ago. Any growing aquatic plant extracts carbon dioxide from the water in its immediate vicinity, making the chemical environment there slightly alkaline. Calcite is insoluble under such conditions so it precipitates as a white crust on the plant—not part of the plant but an accidental by-product of its metabolism, which is inconvenient because plants cannot grow well under a crust of calcite. Blue-green algae solve that problem by sending thin filaments through the calcite crust, which then grow into a new mat of fibers on its surface. They in turn cover themselves with a new crust of calcite. Continued repetition of that process builds thick sections of limestone, both Precambrian and younger, which consist of countless thin laminae of calcite stacked like sheets of paper. Under certain conditions the mats of algae grow into humps called **stromatolites** (Figure 12–4b), which are composed of numerous thin laminae that wrap around each other almost like the leaves of a cabbage. Modern stromatolites are essentially indistinguishable from those that grew 3 billion years ago.

Geologists still have no certain way of knowing when the atmosphere finally evolved to resemble the air we breathe.

Some argue that the abrupt appearance of abundant animal fossils at the beginning of Cambrian time may mark the development of a breathable atmosphere. However, experiments on the closest modern relatives of those earliest animals show that they can survive in an atmosphere containing only a small fraction of the oxygen in our modern air. Thus, the appearance of their ancestors does not imply the existence then of a fully oxygenated atmosphere—it may have no atmospheric implications whatsoever. Oxygen must have played another quite different role in permitting development of the upper atmospheric ozone layer, which screens out ultraviolet radiation that would otherwise kill animals. However, that too could happen with only a small fraction of the present atmospheric oxygen. Presumably, the atmosphere must have reached something very like its present condition by the middle of Paleozoic time when advanced fishes and land vertebrates appeared and sandstones with dune cross bedding became common. The active warm-blooded land animals that began to appear during early Mesozoic time surely required air that we would regard as breathable.

It is difficult to avoid the impression that development of the atmosphere may have progressed somewhat unevenly as plants and animals evolved and placed new demands on its gases. For example, rocks deposited during Mississippian and Pennsylvanian time contain enormous amounts of coal that must have depleted the atmosphere in carbon dioxide as they formed while simultaneously enriching it in oxygen. Rocks deposited during the succeeding Permian and Triassic periods contain little coal but enormous volumes of red mudstone and sandstone, prompting some geologists to speculate that the atmosphere may have been very rich in oxygen during those periods. Limestone also appears to have formed more abundantly during some periods than in others.

IRON ORE

Sedimentary Iron Formation

Anyone who has left a screwdriver outside to rust has observed the great affinity of iron for atmospheric oxygen. Iron and oxygen, like carbon and oxygen, combine readily. Therefore, it was impossible for significant amounts of oxygen to accumulate in the atmosphere until nearly all the iron at the earth's surface had oxidized.

(a)

(b)

FIGURE 12–4

(a) Towering cliffs of Precambrian Siyeh limestone exposed in the headwall of a cirque in Glacier National Park, Montana. The dark band high on the cliff face is a basalt sill. (b) A detailed view of algal stromatolites about the size of brussels sprouts exposed in the same cliffs. Deposition of such massive limestones withdrew vast quantities of carbon dioxide from the atmosphere.

If the early Precambrian atmosphere did indeed contain large amounts of carbon dioxide, then the surface water in contact with it would have been slightly acidic, probably more so than modern rainwater. Slightly acidic water can dissolve **ferrous iron** (Fe^{++}) and it seems reasonable to suppose that the early Precambrian oceans may have contained large inventories of dissolved ferrous iron. It evidently reacted with free oxygen to form insoluble ferric (Fe^{+++}) iron oxides that precipitate as rust, the mineral **goethite.** That seems to have happened on a massive scale during early Precambrian time as algae released free oxygen, which reacted with dissolved iron.

Precambrian sedimentary sections more than about 2 billion years old commonly contain thick formations of a striking rock composed of alternating thin layers of hematite and quartz, the so-called **banded iron formation.** The thin bands of hematite commonly continue for distances of many kilometers and must, when deposited, have covered areas of many hundreds of square kilometers. Some geologists envision a scenario in which vast blooms of algae flourishing in the sunlit surface waters periodically raised the concentration of dissolved oxygen to levels that would precipitate iron oxide. That, or whatever happened, evidently continued until virtually the entire oceanic inventory of dissolved ferrous (Fe++) iron was exhausted about 2 billion years ago. Only then did it become possible for free oxygen to accumulate in the atmosphere.

FOSSIL FUELS

Atmospheric Carbon Dioxide

All **fossil fuels** consist of organic matter that was buried in the earth's crust where the carbon it contains could not react with atmospheric oxygen. When we burn those fuels, we complete the natural cycle that was interrupted by their burial. The energy they yield is solar energy that plants stored millions of years ago as they converted carbon dioxide into reduced carbon and free atmospheric oxygen. We reverse the process that added oxygen to the atmosphere when we burn fossil fuels.

However, green plants continue to use carbon dioxide and some of that released by consumption of fossil fuels dissolves in seawater. Therefore, the rate of increase in the atmospheric content of carbon dioxide does not nearly match the rate at which it is released from smokestacks and exhaust pipes. Nevertheless, the carbon dioxide content of the atmosphere has been increasing at an easily measurable rate throughout the present century. Many current estimates now project an increase of about 25 percent during the present century. That sounds alarming but it is important to remember that the atmosphere contained only about 300 parts per million carbon dioxide when the century began. It will still contain very little even if the figure does rise to about 375 parts per million by the year 2000.

Whether the current slow increase in atmospheric carbon dioxide will enhance the greenhouse effect enough to significantly warm the earth's climate is a difficult question. If the climate does become warmer, that will increase evaporation from the oceans, which will in turn increase the cloud cover thus reducing the amount of sunlight reaching the earth. Therefore, the climatic effects of a small increase in atmospheric carbon dioxide are to some extent self-cancelling. That makes the net result difficult to predict.

Many estimates suggest that burning all known fossil fuel reserves would reduce the oxygen content of the atmo-sphere by about 2 percent, perhaps a bit more. The figure is so small because the overwhelming bulk of all the carbon stored in rocks is too widely disseminated to be economically mineable. Only a relatively small proportion exists as useable deposits of fossil fuel. The reserves are now dwindling at an alarming but quite unsurprising rate. Let us consider the basic arithmetic of growing consumption.

The Implications of Growth

Since 1973 the increasingly short supply and rapidly rising price of petroleum have made its discovery, production, and consumption the subject of widespread discussion. The conversation could easily be extended because the pattern of petroleum use is basically similar to that of most nonrenewable resources and could therefore provide a good general model. We can expect to see fairly similar situations develop in the cases of many other mineral commodities, especially those which, like fossil fuels, cannot be recycled.

Figure 12–5 shows the historic pattern of petroleum consumption in the United States. This pattern provides the basic information necessary to explain the current energy shortages and project the probable future course of the problem. The curve in Figure 12–5 has the general form of an exponential function, the mathematical expression that describes the growth of anything that increases at a constant rate. Compound interest, for example, increases the value of a savings account exponentially because the interest from last year is added to the principal on which this year's interest will be computed. Energy consumption in the United States, and therefore the demand for fossil fuels, has historically shown precisely the same kind of **exponential growth.**

All exponential curves have the property of doubling their height at fixed intervals of time, which depends upon the percentage growth rate. The **doubling time** can easily be computed as

$$\text{Doubling time} = \frac{70}{\text{percentage growth rate}}$$

In other words, if growth is at an annual rate of 10 percent, the quantity will double in 7 years, double again in another 7 years, and so on. A growth rate of only 5 percent will double the amount in 14 years. A town growing at an annual rate of 2 percent—which most chambers of commerce regard as tantamount to no growth at all—will double in size within 35 years and quadruple within 70. Such a modest growth rate would mushroom a community of 50,000 into a bustling city of 200,000 within one human lifetime.

Consider now that every doubling period increases the total to more than that achieved in all previous doubling periods combined. For example, when our community of 50,000 doubles into one of 100,000, it will contain twice as many people as it had reached in all of the doubling periods that took it to 50,000. To cite a slightly racier example, all gamblers are familiar with a scheme whereby they continue to bet the same game at the same odds but double the stake with every loss. When they finally win, as they eventually will if the game is not too badly rigged, their take must exceed all previous losses combined. The only defense gambling houses have against that scheme is to place a lower and upper limit on bets to restrict the number of times a player can double the stake. Otherwise, any gambler with enough financial backing to play the game to its inevitable end would finally break the house. A general familiarity with the properties of exponential curves eliminates any need for nerve because there is no element of luck.

Figure 12–5 shows that petroleum consumption in the United States has grown historically at an annual rate that varies between about 6 and 9 percent, a doubling rate that averages approximately 11 years. In other words, during every 11-year period the United States has used more petroleum than during its entire previous history. World consumption has followed a very similar trend, staying close to an annual growth rate of 7 percent throughout the twentieth century. Obviously, there must be some limit to how long that kind of steady increase in consumption can continue.

The Limits of Consumption

Since all nonrenewable resources are finite, the time must presumably come when the last bit has been produced and consumption drops to zero. Presumably, in most cases that will happen after a long period of decline following a curve generally similar to that shown in Figure 12–6. The main obstacle to plotting such a curve into the future is the

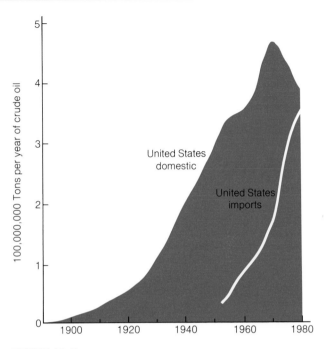

FIGURE 12–5

United States crude oil consumption has steadily increased at a rate that doubles demand about every 11 years. As domestic production began to lag behind demand, imports increased to maintain the steady growth in consumption. The trend is finally beginning to flatten.

difficulty in estimating the amount of still undiscovered reserves. The problem is somewhat similar to that of estimating the number of fish in a lake because in both cases the quarry remains invisible until actually in hand.

Anglers generally conclude that a pond is about fished out when hours spent at their sport begin to produce consistently more disappointing results. If fishermen were as statistically inclined as some other sportsmen, they could probably plot curves showing numbers of fish caught per hour spent fishing, which would fairly accurately reflect the changing status of the fish population. It is similarly possible to arrive at an estimate of the amount of oil remaining to be discovered in a region by considering the proportion of dry holes to successful oil wells. In the United States the failure rate of exploratory oil wells drilled in search of new petroleum deposits rose dramatically from about 33 percent dry holes in 1950 to about 42 percent by 1970. The

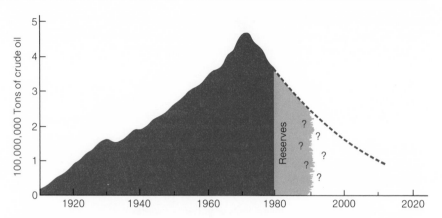

FIGURE 12-6

The past history and probable future course of United States domestic oil production, which
has been declining steadily since 1970. The projection is based on known reserves until about
1990. Beyond that date it depends increasingly upon hypothetical reserves still awaiting
discovery. Authorities differ but the range between the optimistic and pessimistic estimates
will make relatively little difference in the form of the curve. In any case, it seems that we are
now into the second half of the resource.

trend has continued since then and strongly suggests that
the pond may be about fished out.

Obviously, we run the risk of being seriously in error if
we use such methods to estimate the reserves of either fish
or oil. It is always possible that the fish just are not biting
and that the wells were drilled in the wrong places. However, even large mistakes in estimating future discoveries
make little difference in the time it will take to use up all the
remaining oil. Let us return to the characteristics of the exponential growth curve.

Some estimates place the amount of oil remaining to be
discovered in the lower 48 states at about 12.7 billion tons,
approximately as much as has already been produced.
However, those estimates do not imply a long future for the
current scene; those undiscovered reserves could maintain domestic production at 1970 levels for about 29 years
but would last only 16 years if production continued to rise
at the 7 percent annual rate. If we include the estimated
potential reserves of Alaskan oil, which some people have
optimistically described as a "glut of crude oil," the total
rises to some 14.1 billion tons, enough to last about 31
years at the 1970 rate but only 17 years if the 7 percent
growth rate were maintained. If we accept the estimates

that place the total reserves of shale oil equal to the remaining supply of conventional crude oil, the total of all
those supplies could maintain domestic consumption for
63 years at the 1970 rate and for only 24 years if production were to continue its historic 7 percent annual increase. Therefore, doubling the estimate of total undevel-
oped reserves by adding the shale oil increases the life of
the resource by only 7 years if the 7 percent growth rate
continues.

World petroleum prospects are just as bleak in the long
run as those in the United States are now. Most estimates
suggest that world production will pass the peak of its
curve sometime in the early 1990s and decline thereafter.
Many countries, including the Soviet Union, which now ex-
port large amounts of oil will pass their own production
peaks long before then. Even major discoveries of new reserves will make little difference if the present growth in
demand continues. By the year 2000 the present growth
rate would bring world consumption to a level that would
consume all of the Mexican reserves, which may rival
those of the Persian Gulf region, in less than two years.
Obviously, consumption cannot continue to increase at a
rate that will bring it to such a devastating level. How can

FIGURE 12-7
The Bighorn Coal Mine about 10 km north of Sheridan, Wyoming produces from one of the thick coal seams in the northern high plains. (Kirk Badgley)

we avoid concluding that we must shortly see the end of economic growth based on continually expanding use of petroleum?

Coal

The United States has enormous deposits of coal distributed through several large regions: High and medium quality coals of Pennsylvanian age in the Appalachian and mid-western states and lower quality coals in the northern high plains (Figure 12-7). Lesser deposits exist in other regions. As Figure 12-8 shows, coal production in the United States increased at an average annual rate of just under 7 percent from the time of the Civil War until about 1920 when it leveled as oil took command of the market. Then it began to climb again as the market position of oil weakened in the 1970s. Had coal production not leveled off in 1920, we would be facing a coal crisis today.

Coal now provides approximately 20 percent of the energy consumed in the United States. Known reserves amount to approximately 1,500 billion tons, about one-quarter of the world total. That is enough coal to sustain United States production at the present rate for about 2,500 years but those same reserves will last only about 75 years if production increases at an annual rate of 7 percent. If production were increased at an annual rate of 10 percent, those same reserves would last only 56 years. The U.S. Geological Survey estimates that discovery of new coal beds in the United States will eventually more than double the reserves to a total in the neighborhood of 3,500 billion tons. If we actually do find those hypothetical reserves, they will suffice to sustain production at its current rate for almost 6,000 years but will last only 86 years if growth continues at an average annual rate of 7 percent. More than doubling the reserves would add only about 11 years to the estimated life of the resource if constant growth continues. Once again, the arithmetic of exponential growth shows that the total reserves of a nonrenewable commodity make relatively little difference to its life expectancy; what matters is the rate at which consumption increases.

Oil Shale and Heavy Crude

As the name suggests, **oil shale** is a fine-grained sedimentary rock that contains large amounts of oil. It is possible to

FIGURE 12-8

United States domestic coal production increased at an average rate of slightly less than 7 percent annually until about 1920 when it began to fluctuate about a more nearly constant figure. Had production continued to increase at its former steady annual rate, we would by now be experiencing a coal crisis. If production now resumes the growth it experienced before 1920, we can expect a coal shortage crisis to occur by about 2020.

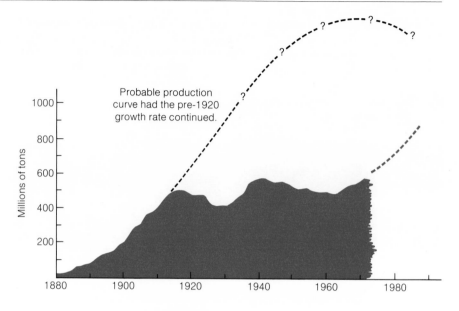

light small chips of the rock with a match and watch them burn with a smoky yellow flame that leaves large residues of ash. However, it is not possible to pump oil out of shale through a well. Instead, the oil must be distilled out of the rock at high temperatures.

Shale oil, most of which is in northwestern Colorado and nearby areas of Wyoming, probably constitutes approximately 23 percent of the total reserves of fossil fuels in the United States. Most of the production so far has been experimental work directed to development of finding some means of extracting the oil, preferably without having to strip mine and then reclaim a large corner of Colorado. The environmental costs of using shale oil will almost certainly be prohibitive unless some means is found of extracting it from the shale through wells, perhaps by using steam or solvents.

Heavy crude oil is basically tar, which is petroleum too viscous to be produced from a conventional well. The largest deposits exist in Canada in the Athabaska Tar Sand and the United States also has substantial reserves. As in the case of shale oil the problem lies in finding some means of extracting the tar from the sand. Improved methods have been developed and we can expect some contribution from heavy crude but it will not solve the problem posed by expanding consumption. Nothing except reduced consumption and development of a renewable energy source can solve the basic problem.

SUMMARY

Although the origin and composition of the earth's early atmosphere are unknown and subject to some debate, it seems reasonable to believe that it lacked oxygen and contained large amounts of carbon dioxide as well as other gases that left little or no record in the rocks. During the course of geologic time most of that early atmosphere was transferred into sedimentary rocks and the remainder converted into the oxygen-rich air we know. Green plants were the major agents of atmospheric evolution and they still maintain our modern atmosphere.

During early Precambrian time, green plants began absorbing carbon dioxide and converting it to carbon and atmospheric oxygen. To the extent that the carbon was incorporated into sediments, it could not recombine with the oxygen and return to the atmosphere as carbon dioxide. However, oxygen could not accumulate in the atmosphere until all the available ferrous iron (Fe^{++}) had been converted into ferric (Fe^{+++}) iron oxides, which accumulated as deposits of Precambrian iron formation until about 2 billion years ago. Sometime after then, green plants increased the atmospheric oxygen content to its present level. Green plants also contribute to deposition of limestone, which withdraws carbon dioxide from the atmosphere without liberating oxygen. The composition of the modern atmosphere expresses an equilibrium between the rate at which weathering of limestone and oxidation of carbon release carbon dioxide into the atmosphere and the rate at which plants use carbon dioxide and release oxygen. Large-scale combustion of fossil fuels has now thrown that equilibrium into slight imbalance, the consequences of which remain to be seen.

Constantly growing use of fossil fuels during the past 150 years has driven their consumption upward along an exponential curve which, like all such curves, starts with a gentle slope and then swoops steeply upward. The industrial nations have now reached the steeply sloping part of the curve and cannot therefore continue to maintain economic growth based on increasing consumption of fossil fuels of any kind. No finite and nonrenewable resource, fossil fuels or any other, can indefinitely continue to meet constantly growing demand. In the end, the industrial nations will have no alternative to either curtailing demand for fossil fuels or switching to renewable energy sources.

KEY TERMS

banded iron formation	goethite
doubling time	greenhouse effect
exponential growth	oil shale
fossil fuels	stromatolite

REVIEW QUESTIONS

1. Contrast the effect on the atmosphere of cutting a tree and using the timber to build a table with that of cutting a tree and using it for fuel.

2. If inflation were to continue at an annual rate of 12 percent, how long would it take to halve the value of the dollar?

3. Consider the impacts of (1) smelting iron ore and (2) the eventual rusting of the iron on the atmosphere.

4. Contrast the effects on the atmosphere of weathering limestone and of burning fossil fuels.

5. Why did free oxygen not begin to accumulate in the atmosphere until about 2 billion years ago even though green plants had by then been active for considerably more than a billion years?

6. Why would consumption of all fossil fuel reserves not drastically deplete the oxygen content of the atmosphere?

SUGGESTED READINGS

Averitt, P. *Coal Resources of the United States.* U.S. Geological Survey *Bulletin 1412* (January 1, 1974).
    A review of coal reserves that includes projections of their duration based on various assumptions of rate of use.

Bartlett, A. A. "Forgotten Fundamentals of the Energy Crisis." *American Journal of Physics,* 46 (September, 1978).
    An exceptionally clear review of the basic characteristics of constantly growing consumption. Includes numerous quotations from many sources that illustrate the general lack of comprehension among political and business leaders.

Cloud, P. *Cosmos, Earth and Man: A Short History of the Universe.* New Haven, Conn.: Yale University Press, 1978.
    Brilliant. Written for a nontechnical audience. Deals with the evolution of the atmosphere and the development of life. The author is one of the leading thinkers in the field.

Hubbert, M. K. *Energy Resources: Resources and Man.* San Francisco: W. H. Freeman & Co., 1969.
    A basic reference on resource analysis by a leading authority in the field. The author correctly predicted the onset of the current oil shortages 14 years before they began.

Miller, B. M. *et al. Geological Estimates of Undiscovered Recoverable Oil and Gas Resources in the United States.* Washington, D.C.: U.S. Geological Survey *Circular 725* (1975).
    The title describes it perfectly.

Ruedisli, L. C. and M. W. Firebaugh, eds. *Perspectives on Energy,* 2d ed. New York: Oxford University Press, 1975.
    A series of articles by various authors on energy use and alternative energy sources.

CHAPTER THIRTEEN

The oceans cover more than two-thirds of
the earth's surface.

THE OCEANS

Oceans cover approximately two-thirds of the earth's surface and dominate the planet in many ways that we who live on the continents tend to forget. The oceans and the atmosphere are inseparably linked because water vapor is an atmospheric gas and all the other atmospheric gases dissolve in large quantities in seawater. There is no geologic record of a time when the earth lacked either an atmosphere or oceans—the oldest decipherable sedimentary rocks show evidence of having accumulated in seawater.

SEAWATER

Composition of Seawater

Analyses of seawater show that it contains high concentrations of dissolved sodium and chloride ions along with much lesser amounts of several others (Figure 13–1) and traces of virtually every element known to occur in nature. The total concentration of dissolved salts in normal seawater varies in the range between about 28,000 and 34,000 parts per million. For example, areas such as the Gulf of St. Lawrence—which receive abundant freshwater from rain and rivers—contain more dilute seawater than those such as the Gulf of California, which receives little freshwater to replace that lost to evaporation. However, the proportions of most of the dissolved constituents remain almost constant. Thus, simple dilution will make a sample of the extremely salty water from the Gulf of California virtually identical to one from the Gulf of St. Lawrence. That broad similarity in composition implies that currents keep the oceans of the world fairly well mixed.

Phosphate and silica provide the only important exceptions to the rule that the dissolved substances in seawater vary in concentration but not in their proportions to each other. Both function as fertilizer nutrients in the marine environment. **Diatoms,** the one-celled algae that form the basis of the oceanic food chain, make their delicate shells out of silica. Phosphate is an essential part of bones, teeth, fish scales, and some kinds of invertebrate shells. Diatoms thrive wherever seawater is enriched in silica. The animals requiring phosphate flourish on the rich pasturage diatoms provide. Cold upwellings of deepwater nurture bountiful fisheries because they carry high concentrations of both silica and phosphate.

Residence Times

Generations of scientists regarded the oceans as global sumps where evaporation concentrates the soluble products of rock weathering brought to them by streams. How-

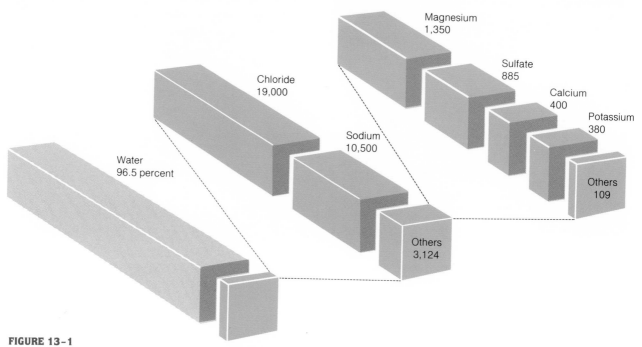

FIGURE 13-1

Seawater is a complex solution containing large amounts of dissolved sodium and chloride ions, lesser amounts of magnesium, sulfate, calcium, and potassium ions, and at least traces of every element known to occur in nature. All the concentrations shown are expressed in parts per million.

ever, consideration of the data in Figure 13–2 shows that seawater cannot possibly owe its composition to simple evaporative concentration of typical riverwater because the proportions of their dissolved constituents differ so greatly. Furthermore, it is now clear that rivers are not the only source of dissolved substances in the ocean.

Oceanographers have recently discovered that seawater sinks into fractures near the crest of the ridge system that winds through all the ocean basins, circulates through very hot rocks deep beneath the ocean floor, and then returns to the ocean through submarine hot springs (Figure 13–3). That circulating water extracts soluble substances from the rocks beneath the oceanic ridge and injects them into the ocean in quantities that are still unknown but are evidently large.

Dividing the total dissolved inventory of an element in seawater by the estimated annual contribution from all sources gives its **residence time** (Figure 13–4), the average period an atom of that element remains in the ocean before being removed from solution. Elements with brief residence times, such as potassium, do not accumulate to great concentration regardless of how much rivers and hot springs may dump into the ocean. Those with long residence times, such as sodium or chloride, are least efficiently removed from solution and therefore concentrate in seawater. The present composition of seawater represents a dynamic balance between the rates at which various elements enter and leave the ocean—an essentially timeless equilibrium rather than a momentary point in a continuously progressive evolution. Because the residence times of all the elements are short compared to the length of geologic time, it seems reasonable to conclude that sea-

RIVER WATER

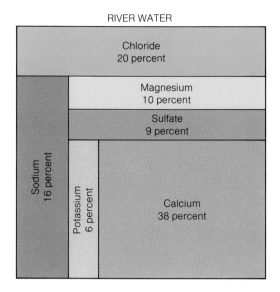

SEA WATER

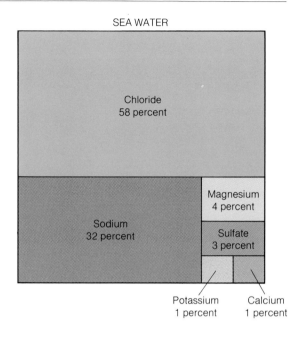

FIGURE 13–2

Because the proportions of their dissolved constituents differ so greatly, we must conclude that seawater does not form through simple evaporative concentration of riverwater.

FIGURE 13–3

Seawater sinks into a rift at the crest of the oceanic ridge and percolates through fractures in the rocks at depth, dissolving material from them as it gets hot. The heated and mineral-laden water returns to the ocean through hot springs on the sea floor that precipitate sulfide sediments in their vicinity.

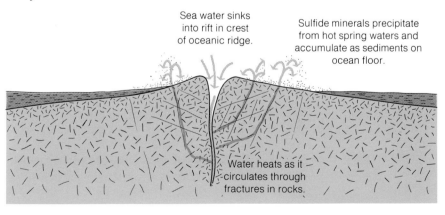

FIGURE 13-4

The concentrations of the dissolved constituents in seawater depend primarily upon their residence times; upon how rapidly they are removed from the oceans. The concentration of sodium is high because few processes remove it efficiently from the ocean; those of potassium and iron are low because both are rapidly incorporated in bottom sediments.

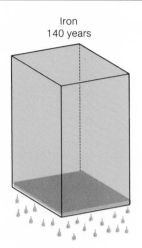
Iron
140 years

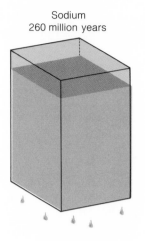
Sodium
260 million years

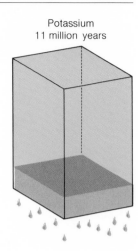
Potassium
11 million years

water must have reached something resembling its present composition rather early in the history of the earth. The major known exception is iron, which must have been abundant in seawater during early Precambrian time before green plants shortened its residence time by releasing free oxygen.

FIGURE 13-5

Although the concentrations of many elements in seawater are very low, the total amount in the oceans is enormous. Nevertheless, the prospects of commercially recovering most of them are very poor.

One cubic kilometer of sea water contains:

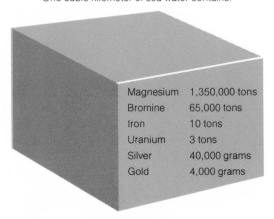

Magnesium	1,350,000 tons
Bromine	65,000 tons
Iron	10 tons
Uranium	3 tons
Silver	40,000 grams
Gold	4,000 grams

The oceans contain approximately 1,350,000,000 cubic kilometers of water.

Mining Seawater

Seawater contains only 0.00001 parts per million of gold, a minuscule concentration by any standard. Nevertheless, that is enough to put some 10,000 grams of gold in every cubic kilometer of seawater, and the oceans of the world contain approximately 1,320,000 km³ of water. Indeed, enormous tonnages of every metallic element exist in solution in the oceans (Figure 13–5).

Those vast inventories tantalize many imaginative people, leading them to devise visionary schemes to recover glittering fortunes from ordinary seawater. Similar schemes often center around such common rocks as granite, which contains about 500 times as much gold as an equal weight of seawater. Most efforts to recover valuable elements from seawater or common rock have failed miserably and the prospects for future success look generally bleak. The major obstacle in most cases is the cost of energy.

A fortune in gold will enrich no one if it remains dissolved in the waters of Long Island Sound or disseminated through the granite of Pike's Peak. It would be necessary to pump all the seawater through an extraction plant or mine and smelt an entire mountain before the gold could be cast into saleable bricks. Either kind of processing requires vastly greater expenditures for energy than the gold could be worth. The same obstacle prevents extraction of most of the many other valuable substances dissolved in seawater or disseminated through common rocks. Nevertheless, it is profitable to recover magnesium and bromine from seawater and the oceans do supply the world with the bulk of both. Prospects for recovering other elements from seawater depend upon development of cheap sources of abundant energy. Solar energy could possibly be used.

Hot brines that issue from springs on the ocean floor and accumulate locally in large pools—such as the several on the bottom of the Red Sea—contain many elements in high enough concentration to make their extraction seem potentially feasible. Furthermore, their large heat content could provide some of the necessary energy. Those waters exist in great volume; a successful process to extract metals from them could prove enormously productive.

Many plants and animals perform dazzling feats of extraction. Diatoms, for example, make their lacy shells of silica that they extract from seawater in which silica concentration varies between 4 and 10 parts per million. Lobsters and other crustaceans extract copper—an essential ingredient in their blood—from seawater in which copper concentration is about 0.003 parts per million. Many people have suggested that such plants or animals could do the preliminary concentrating and then be harvested just as seaweed is harvested for its iodine content. However, few of the natural concentrators are either as abundant or as easy to harvest as seaweed.

OCEAN CURRENTS

The waters of the oceans move in a complex web of surface and deep currents, which flow in different directions at different depths for different reasons. Attempting to visualize the overall pattern is as bewildering as watching an urban freeway intersection with traffic moving in different directions at several levels.

The Surface Circulation

The prevailing trade winds of the tropics blow generally from east to west: out of the northeast in the Northern Hemisphere and from the southeast in the Southern Hemisphere. In temperate latitudes, on the other hand, the prevailing winds blow generally from the west: out of the northwest in the Northern Hemisphere and from the south-

west in the Southern Hemisphere. Polar winds move relatively little water because they blow across pack ice in the Arctic Ocean and across land in the Antarctic.

If there were no continents, the planetary winds would drive the surface waters of the tropics around the world from east to west and those of temperate latitudes from west to east; alternate belts of water moving in opposite directions would girdle the earth. One such continuous current actually does circle the earth in the "roaring forties" between the southern continents and Antarctica where no land obstructs the drive of wind and water. In the rest of the world, continents restrict the circulation of surface water to individual ocean basins.

Figure 13–6 shows the general pattern of the surface water circulation. The tropical trade winds blow surface water away from the continent at the eastern edge of the ocean basin and against the one at the western edge where it piles up to create a slight bulge about a meter high in the ocean surface. That sun-warmed water spills north and south out of the tropics as well-defined currents that flow like rivers along the western edge of the ocean basin. As it reaches higher latitudes the prevailing westerlies of the temperate regions drive the stream of water back across the ocean to the eastern side of its basin. There, the water warms the adjacent continents as it turns and drifts slowly back into the tropics chilled by its sojourn through cold latitudes and ready to pick up another load of heat as it completes the circuit. That circulation turns clockwise in the Northern Hemisphere and counterclockwise in the Southern Hemisphere largely because it follows the wind and partly because it is deflected by the earth's rotation.

The earth's rotation causes all moving objects to tend to drift to the right in the Northern Hemisphere and to the left in the Southern Hemisphere. The phenomenon—called the **Coriolis effect**—is complex if considered in detail but we can partially visualize it by comparing the problem of playing catch on the earth with that of playing the same game on a moving merry-go-round. In either case, an observer standing on the rotating object will see the thrown ball appear to curve in midair. To understand why the deflection is in the opposite direction in the Northern and Southern Hemispheres, it is necessary to imagine playing catch on both the upper and lower surfaces of the merry-go-round. The Coriolis effect deflects surface currents slightly to the right of the wind direction in the Northern Hemisphere and

FIGURE 13-6
Ocean currents circulate clockwise in the Northern Hemisphere and counterclockwise in the Southern Hemisphere as the trade winds of the tropics drive the surface water west and the westerlies of higher latitudes move it to the east.

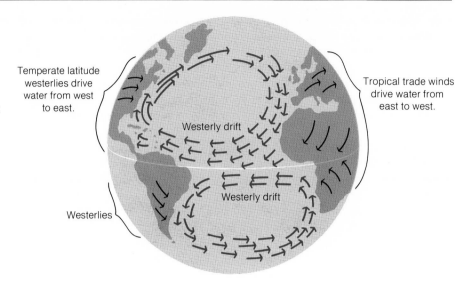

to the left in the Southern Hemisphere because the earth is rotating beneath the moving water.

The Gulf Stream

Warm water piles against the eastern edge of tropical central America and then spills out of the Gulf of Mexico through the strait between Florida, Cuba, and the Bahamas as the **Gulf Stream** (Figure 13–7). This current carries more water than all the terrestrial rivers of the world combined. Most of the flow is at depths of less than 100 m but the current drags the deeper water more slowly along to a total depth somewhat greater than 1,000 m.

The Gulf Stream clings rather closely to the coast of North America in the general vicinity of Cape Hatteras, North Carolina where it begins to edge slowly eastward until it finally heads directly across the North Atlantic off the coast of Newfoundland. Its deep blue waters contrast so sharply with the gray water of the North Atlantic that the boundary between them is obvious from the air, even from the deck of a ship.

In places such as the region off the coast of Newfoundland where the prevailing winds blow surface waters away from the edge of a continent, cold water rises from the depths to replace them. Such cold upwellings support

good fisheries because the water contains large concentrations of dissolved silica and phosphate. Conversely, areas where surface currents meet, pile up, and sink into the depths make extremely poor fishing grounds because the water is deficient in fertilizer nutrients.

Cold upwellings and surface convergences link the wind-driven surface circulation of the ocean to the deep currents. The deep currents are often called the **thermohyaline circulation** because they move in response to density differences caused by variations in water temperature or saltiness.

The Deep Circulation

Where surface water cools or becomes very salty, it may become denser than the water beneath and sink to a depth where it finds water slightly denser than itself. There it spreads horizontally, floating on the denser water beneath as a layer of moving water—a **density current**—that may move thousands of kilometers before it finally loses its identity by mixing with the water above and below itself. The deeper levels of the ocean are a complex web of layers of water sliding past each other as they move in different directions. Much of the bottom water of the North Atlantic, for example, sinks off the east coast of Greenland where the climate is intensely cold and the water slightly saltier than normal. Some of the cold water that sinks off Antarctica moves well north of the equator before it finally loses its identity through mixing.

If water sinks in some places, an equal quantity must rise elsewhere. Therefore, the density-driven deep circulation involves both vertical and horizontal movements that extend to the bottom of the ocean. It is possible to determine when a sample of deep water was last in contact with the atmosphere by running a carbon-14 date on its content of dissolved carbon dioxide. Such data show that virtually all of the water now on the ocean floor was at the surface sometime within the last few centuries.

Since no bottom water is stagnant, we must conclude that no part of the ocean provides a safe dumping ground for radioactive wastes or other soluble contaminants. Neither does the vast size of the ocean necessarily provide safety through dilution. Many plants and animals concentrate minor elements and may introduce dangerous quantities of radioactive isotopes into the food chain.

The Gibraltar Currents

Mediterranean surface water becomes warm and abnormally salty because that sunny region has a high rate of evaporation and receives little freshwater either from rainfall or rivers. A deep current of water salty enough to be considerably denser than normal surface water even though it is also quite warm pours out along the shallow bottom of the Strait of Gibraltar and sinks to a depth of about 1,000 m in the North Atlantic (Figure 13–8). The deep flow from the Mediterranean is clearly identifiable many hundreds of kilometers from its source as a layer of warm and salty water sandwiched within the much colder and less saline waters of the North Atlantic. Meanwhile, a strong counter current of North Atlantic water flows east past Gibraltar into the Mediterranean to replace that exiting at depth and that lost to evaporation. Both currents must pass through the Strait of Gibraltar because it provides the only connection between the Mediterranean and the rest of the world ocean.

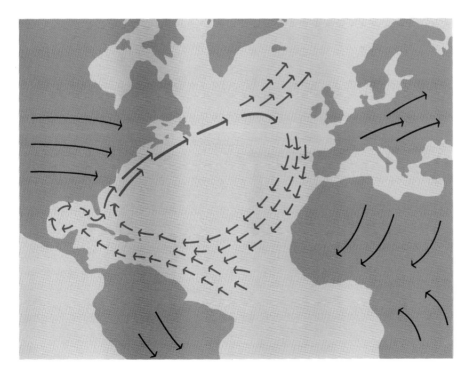

FIGURE 13–7
Water drifts slowly westward before the tropical trade winds and crowds against the western edge of the ocean basin whence it spills into higher latitudes as swiftly flowing and narrowly defined currents.

FIGURE 13–8

Rapid evaporation and scanty rainfall in the Mediterranean region concentrate the seawater there, making it dense enough to sink through the shallow Strait of Gibraltar and into the North Atlantic as a deep density current. A surface countercurrent of less dense seawater flows into the Mediterranean to replace that lost to the deep current.

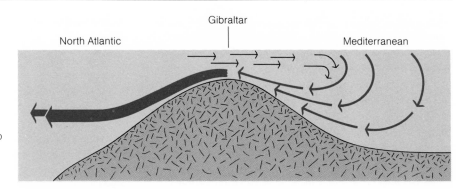

The Gibraltar currents have played a role in history since the ancient mariners discovered that they could slip into the Mediterranean by drifting on the surface current but had to wait for favorable winds to exit into the North Atlantic. During the Second World War, German submarines commuted between their bases in Italy and their prey in the Bay of Biscay by submerging to the appropriate depth and drifting silently past the British defenses.

OCEANIC SEDIMENTS

Clastic Sediments

Waves spread much of the clastic debris that reaches the coast across the surfaces of the continental shelves. Some spills past the edge of the continental shelf and down the face of the steeper continental slope to the deep ocean floor. Another large fraction of clastic sediment pours down submarine canyons in **turbidity flows** that blanket vast expanses of the ocean floor beneath turbidite sediments.

Waves stir the shallower bottom inshore more vigorously than they do the deeper parts of the continental shelf farther offshore. This action tends to work the finer sediments farther offshore while leaving the coarser material, which is less easily moved, lagging behind near the coast. However, detailed maps of sediment distribution on the modern continental shelves reveal a complex pattern superimposed on that broad trend. Local changes occurring in that pattern cause layering within the accumulating sedimentary section.

Consider, for example, a patch of seaweed growing on the bottom and trapping mud because the plants interfere with wave action that might otherwise move the mud farther offshore. If a heavy storm were to bury the patch of seaweed and the mud it had trapped beneath a deposit of sand, the result would be a layer of mud covered by a layer of sand. Numerous repetitions of such events would eventually build a thick section consisting of layers of mudstone and sandstone—the kind of section we commonly see in roadcut and cliff exposures of ancient rocks. However, other processes can also create sedimentary layering. For example, events in the watershed of a stream emptying to the coast may cause it to start carrying sand instead of mud or a change of sea level or bottom elevation may carry the area closer to or farther away from the shore. In most cases, the interpretation of a sequence of layered sediments presents a fairly difficult problem.

Abyssal Oozes

Regions of the ocean floor within reach of the turbidity currents that pour down the submarine canyons acquire thick blankets of turbidite sediments that may extend as far as several thousand kilometers from the nearest land. Beyond the farthest reach of the turbidity currents, the deep ocean floor receives a slow rain of clay mixed with various nonclastic sediments, mostly the remains of small plants and animals that drift in the surface waters.

Most of the small animals secrete calcite shells and in the large regions—primarily tropical regions—where theirs is the dominant contribution the bottom sediment is white **calcareous ooze** that looks and feels almost like toothpaste but lithifies into limestone. That kind of sediment accumulates only in water that is slightly alkaline and therefore incapable of dissolving calcium carbonate—to depths as great as 4,500 m in the Atlantic Ocean and to somewhat lesser depths in the others. However, as anyone experienced in the behavior of carbonated beverages has observed, low temperatures and high pressures both favor the solubility of carbon dioxide in water. At a critical level called the **carbonate compensation depth,** the temperature becomes low enough and the pressure high enough so that seawater absorbs enough carbon dioxide to become slightly acidic and therefore capable of dissolving calcium carbonate. The minute calcium carbonate shells sifting slowly down from the surface dissolve after they pass the carbonate compensation depth and no calcareous ooze can accumulate below it.

Elimination of calcium carbonate shells leaves only clay and minute siliceous shells to accumulate below the carbonate compensation depth as deposits of red clay and siliceous ooze. The siliceous ooze consists mostly of the remains of diatoms and one-celled animals called **radiolaria** and hardens into beds of colorful chert. It accumulates slowly on vast reaches of the ocean floor too remote from land to receive clastic sediment and too deep to permit deposition of calcium carbonate.

Iron-Manganese Nodules

Interesting and potentially valuable black nodules a few centimeters in diameter form in great numbers on large regions of the ocean floor too remote from land to receive significant quantities of turbidite sediment. Bottom photographs show them densely littering the ocean floor like windfall apples after a blustery night in October. Chemical analyses reveal that they differ considerably in composition from one region to another but consist basically of a mixture of the iron and manganese oxide minerals goethite and pyrolusite along with small percentages of copper, nickel, and cobalt as well as traces of many other metals. Most of the nodules contain a hard object, such as a shark's tooth or fragment of volcanic rock at their centers. Sawed sections often reveal a crudely symmetrical pattern of concentric growth rings, which show that the nodules grow more or less uniformly on all sides. Evidently, the iron and manganese oxides precipitate directly from seawater first onto the hard center and then onto the growing nodules. In

laboratory experiments freshly precipitated iron and manganese oxides have often been observed to absorb other metals as impurities. It seems reasonable to assume that the iron-manganese nodules acquire their contents of copper, nickel, cobalt, and other metals in the same way.

Radioactive age dates on volcanic rock fragments from the cores of some iron-manganese nodules have revealed ages as great as 10 million years, which means that they are much older than the siliceous oozes on which they lie. In fact, some data suggest that the siliceous oozes may accumulate as much as 100 times faster than the nodules grow. How, then, do they avoid burial?

It seems that iron-manganese nodules must roll about on the ocean floor, dumping ooze off their upper surfaces and uniformly exposing all sides to the water. Otherwise, they could neither avoid burial nor develop symmetrically concentric growth rings. No one knows what rolls them; perhaps fish do it as they poke about on the bottom, looking for prey. The nodules do eventually end their vagrant careers and disappear into the accumulating ooze, presumably because they get too big to roll or fuse together into awkward shapes that do not turn easily. Meanwhile, new nodules start growing to replace those that get buried.

The iron-manganese nodules are potentially the world's largest and richest ore body. They contain enough manganese, copper, nickel, and cobalt to qualify as high grade ores of all those metals, all of which are critically necessary and in short supply. Furthermore, the nodules are unique in being the world's only renewable metallic mineral resource because their growth rate, although slow for individual nodules, is great enough on the ocean floor as a whole to keep pace with a large-scale mining operation. The technical problems of designing equipment to recover the nodules from beneath 4,000 or more meters of water can certainly be solved as can those of developing processes to separate and refine nodule constituents. However, the diplomatic and legal problems of mining beneath the high seas and dividing the take among interested nations have so far defied solution.

Sulfide Sediments

Numerous mines in many parts of the world exploit spectacular ore bodies that look in most respects like sedimentary rocks but consist mostly of sulfide minerals (Figure

13–9). Until recently, geologists found those rocks as baffling as they are valuable because no such sediments were known to accumulate in modern environments. Many theories proposed that they had been deposited as ordinary sediments and then somehow altered to sulfides, an extraordinary chemical phenomenon that defied rational explanation.

The problem became much simpler with the discovery of abundant sulfide minerals in the sediments accumulating beneath pools of hot brine in the floor of the Red Sea and around submarine hot springs along other parts of the oceanic rift system. It seems clear that they are normal oceanic sediments even though they form only locally near the crests of oceanic ridges. That understanding will certainly help geologists in their search for new deposits of sedimentary sulfide deposits contained in sections of old oceanic crust now incorporated into the continents. It should also be feasible to mine modern sulfide sediments by

dredging them directly off the ocean floor. The problem of separating the different minerals may prove to be much simpler than in the case of the iron-manganese nodules.

Evaporites

Sedimentary deposits consisting of more or less water soluble minerals—such as halite, the mineral form of common table salt, or gypsum—accumulate in basins where seawater flows in to replace water lost to evaporation but none flows out to flush the concentrating brine. Under such conditions, the water entering the basin imports a continuing supply of dissolved salts, which permits the evaporite beds to accumulate to great thickness. Relatively few such basins exist today and most of the modern examples are relatively small. However, numerous large evaporite basins have existed in the past. Formations containing thick sections of evaporite beds occur in many places on the continents—there are salt mines deep beneath the city of Detroit as well as in central Kansas and the Gulf coastal region. Evaporites provide a variety of chemical raw materials as well as gypsum, which is used to make wallboard among other things.

SUMMARY

The oceans cover about 70 percent of the earth's surface to an average depth of about 4,000 m with water that has the same concentration everywhere although it is somewhat concentrated by evaporation in some regions and diluted by freshwater in others. The concentrations of the various dissolved salts in seawater reflect a balance between the rates at which they are added to and subtracted from the ocean. Dividing the total oceanic inventory of a particular dissolved constituent by the annual input gives the average length of time atoms of that constituent remain in the ocean before being removed. Since those times are brief compared to the length of geologic time, it seems likely that seawater reached essentially its present composition early in the history of the earth. Although the total inventories in the world ocean of all substances dissolved in seawater are enormous, efforts to use the oceans as sources of mineral commodities have generally failed and will probably continue to do so because the cost of energy for processing is too high.

FIGURE 13–9

Sulfide minerals glitter brightly on this slab of Precambrian sedimentary sulfide ore from the Sullivan Mine in southern British Columbia. The tightly folded sedimentary layers make an intricate pattern on the sawed surface. (Bernie Rosenblum)

The prevailing winds drive oceanic surface waters from east to west in tropical regions and in the opposite direction at higher latitudes. The result is a general surface circulation that turns clockwise in the Northern Hemisphere and counterclockwise in the Southern Hemisphere. Density differences due to variations in temperature and salinity drive the deeper levels of the ocean in a layered web of currents that move in many directions. The combination of surface and deep circulation moves all the water in the oceans, leaving none stagnant.

Waves distribute clastic sediments across the continental shelves in complex and shifting patterns that create layered sequences of sedimentary rocks. Clastic sediment reaches the deep ocean floor primarily in turbidity currents that carry sand and mud great distances before they finally settle to make turbidite sequences. Large regions of the ocean floor that lie beyond the reach of turbidity currents receive a slow rain of fine clay mixed with the shells of minute plants and animals that sink from the surface. Such sediments make calcareous or siliceous oozes, depending upon whether they accumulate above or below the depth at which calcium carbonate becomes soluble.

Round nodules consisting mostly of iron and manganese oxides along with other elements litter vast reaches of the ocean floor where the oozes accumulate. They may become an important source for several mineral commodities. Small areas of the ocean floor near the crest of the oceanic ridge system accumulate sediments composed mostly of metallic sulfide minerals. Many sulfide ore bodies now on land appear to have formed in such sites and it seems likely that the sediments on the ocean floor will be mined in the future.

KEY TERMS

calcareous ooze	Gulf Stream
carbonate compensation depth	radiolaria
Coriolis effect	residence time
density current	thermohyaline circulation
diatoms	turbidity flows

REVIEW QUESTIONS

1. If estimates of the amount of water moving through the submarine hot spring system increase, how will that affect the figures for the residence times of the elements dissolved in those waters?

2. Why do iron-manganese nodules not form on regions of the ocean floor near continents?

3. Why do warm water creatures such as corals live at higher latitudes along the east coasts of continents than along the west coasts?

4. What could make it possible for the hot brines from submarine hot spring systems to remain on the bottom despite their high temperatures?

5. What factors cause the surface circulation of the ocean to move in a generally clockwise direction in Northern Hemisphere ocean basins and in the opposite direction in the Southern Hemisphere?

6. What factors cause some elements to be more concentrated in seawater than others?

SUGGESTED READINGS

Broecker, W. S. *Chemical Oceanography*. New York: Harcourt Brace Jovanovich, 1974.

An excellent introduction to the chemistry of the oceans.

Carson, R. L. *The Sea Around Us*. New York: Oxford University Press, 1961.

An outstanding general book about the oceans. Regarded as a classic of both science and literature.

Gross, M. G. *Oceanography, a View of the Earth*. Englewood Cliffs, N.J.: Prentice-Hall, 1972.

A superior introductory text covering all aspects of the oceans.

Heezen, B. C. and C. D. Hollister. *The Face of the Deep*. New York: Oxford University Press, 1971.

A beautiful book full of wonderful pictures with an informative and readable text.

CHAPTER FOURTEEN

In the geologic sense of time the edge of
the continent is not so enduring as was
once believed.

ROCK MAGNETISM AND SEAFLOOR SPREADING

Early geologists assumed, since they knew of no contrary evidence, that the present arrangement of continents and ocean basins had always existed. They rather quickly found evidence that the continents move vertically but little reason to suppose that they move laterally—perhaps that was because most geologists confined their attention to one continent.

The basic proposition that continents move vertically while remaining geographically fixed became one of the fundamental tenets of geology. Geologists use the word **tectonics** to refer to large-scale movements of the earth's surface and many now characterize that old view of a crust that moved mostly vertically as ''elevator tectonics.'' This chapter tells how that view of the earth was displaced by new concepts that emphasize large horizontal movements of the earth's surface.

CONTINENTAL DRIFT

Pangaea

Many nineteenth-century geologists observed that the map outlines of the Atlantic coasts of Africa and South America fit like pieces of a jigsaw puzzle (Figure 14–1). Several of those early geologists even ventured to suggest that those two continents had once been joined and then torn apart.

However, none pursued the point in detail until Alfred Wegener, a German meteorologist, began publishing a series of articles on the subject in 1912. He forcefully and cogently argued that an enormous supercontinent, which he chose to call *Pangaea* (Figure 14–2) had broken into fragments that dispersed to become the modern southern continents. That obviously required large horizontal movements of the continents and Wegener's theory came to be known as **continental drift.** Wegener continued to press the point and assembled his arguments into a book originally published in 1924.

Wegener pointed out that a much stronger argument can be made for assembling torn fragments of paper into a single piece if the writing matches across the fitted edges. He then applied that principle to the problem of fitting continents together. He cited several examples of folded mountain trends that now end abruptly at coastlines that would neatly align into continuous linear belts of similar mountains if the continents were fitted together to reassemble Pangaea. He also showed that areas of Africa, India, Australia, and South America that had been glaciated during late Paleozoic time would assemble into a single large region if the continents were fitted back together. Furthermore, the directions of the glacial striations in the various continents would likewise fit into a simple and logical pattern, reflecting ice movement away from a single spreading center (Figure 14–3). Wegener also

233

FIGURE 14-1
The facing coasts of South America and Africa, like those of several other continents, fit almost as though they were pieces of a jigsaw puzzle. The actual edge of the continental crust lies buried somewhere beneath the sediments of the continental shelf, shown here in color, so the exact line of the match is not visible.

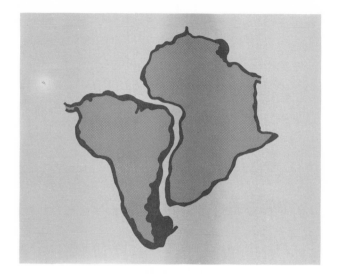

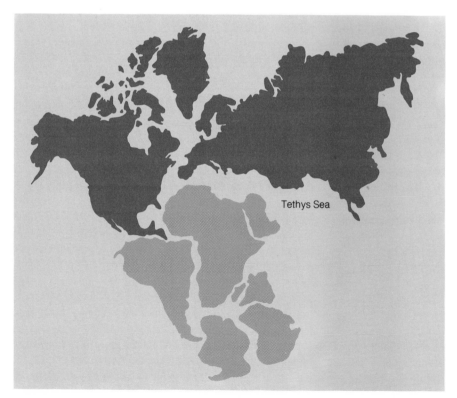

Tethys Sea

FIGURE 14-2
The modern continents can be fitted together like puzzle pieces to reconstruct the supercontinent of Pangaea as it apparently existed about 200 million years ago. Pangaea consisted of two major parts—Gondwanaland in the south and Laurasia in the north—which were nearly separated by the Tethys Sea, parts of which still survive in the Mediterranean, Black, and Caspian Seas. The accuracy with which most of the puzzle pieces fit suggests that they dispersed into their present locations without significantly changing shape.

showed that fossil plants and animals are quite similar in the Paleozoic sedimentary rocks of Africa, India, Australia, and South America but become increasingly dissimilar in sedimentary rocks deposited since Permian time. That observation, among others, led him to infer that Pangaea had fragmented near the end of Paleozoic time.

The Problem of the Mechanism

Wegener envisioned continents plowing through the ocean floor, driven by a small force that should in theory impel them away from the poles of a rotating earth. However, it soon became clear that the oceanic crust is much too strong to permit such a small force to shove continents through it. Nor do we find the continents concentrated along the equator as they should be if they were moving before a force that tended to drive them away from the poles. Furthermore, neither Wegener nor his supporters could show that the continents buckle the ocean floor ahead of them or leave any sort of gap behind; if they moved, it appeared that they must do so without raising a bow wave or leaving a wake. Those problems led the vast majority of geologists to reject the idea of continental drift, abandoning it to a lonely band of advocates most of whom worked in the Southern Hemisphere.

That widespread rejection of Wegener's arguments on grounds that no one could explain how continents move seems a bit curious in retrospect. Nothing in good scientific procedure requires that we fully understand how a phenomenon works before admitting that it exists. If scientists did work under such a constraint, we would have to reserve judgment about the existence of gravity until someone could fully explain how it works and we would have to discover how birds find their way before admitting that they migrate. In fact, advances in scientific knowledge normally begin with observing a phenomenon that demands an explanation, not with the explanation itself.

It is difficult not to suspect that the idea of continental drift was widely rejected for reasons that did not fully meet the test of scientific objectivity. Even today, it is fantastic to think of continents migrating about the face of the earth. We can easily sympathize with the geologists of Wegener's day who felt that they were faced with a proposition more

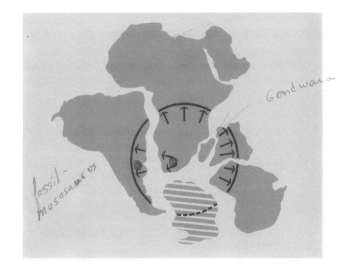

FIGURE 14-3

Wegener's reconstruction of Pangaea neatly assembled the areas of the southern continents that had been glaciated during Paleozoic time into a single large region of the southern supercontinent. The colored arrows show directions of ice movement as indicated by glacial striations on bedrock surfaces. The reconstruction assembled them into a coherent pattern and explained how ice could have moved westward onto South America from Africa instead of from the Atlantic.

nearly akin to science fiction than to science. Furthermore, the idea of continental drift was a truly revolutionary concept that challenged several fundamental assumptions of geology. It threatened to demolish much of the existing edifice of knowledge instead of adding to it in an orderly way, and people rarely welcome such alarming ideas. Unfortunately, the opponents of continental drift put themselves in a logically unsound position by arguing that because the original theory failed in one important respect—that of the mechanism—it must necessarily fail in others. Effective debating points do not always make good scientific arguments.

Nevertheless, data favoring the idea that continents move continued to accumulate and the controversy raged in the background. Many geologists dismissed the matching coastlines and other jigsaw puzzle similarities between continents as mere coincidence. Distributions of fossil plants and animals on widely separated continents were attributed to the former existence of "land bridges." Some of those, such as the one across Bering Strait, were perfectly reasonable whereas others, such as those that supposedly crossed the Atlantic and Indian Oceans, can only be described as wildly fanciful. Despite the general hostility to the idea, firm evidence establishing the fact of continental movement appeared during the 1950s and an adequate explanation of how they moved appeared during the 1960s. Much of the basic data that finally ended the old controversy and set the science of geology on a new course emerged from the study of rock magnetism. This line of inquiry began as an arcane pursuit that seemed unlikely to add more than a few footnotes to our knowledge of the earth.

ROCK MAGNETISM

The Earth's Magnetic Field

The earth has a magnetic field that superficially resembles what we might expect if a bar magnet were embedded within the planet (Figure 14–4). Lines of magnetic force emerge vertically from the north and south **magnetic poles** and sweep around the planet, becoming parallel to its surface at the **magnetic equator.** At points between the magnetic poles, the magnetic lines of force intersect the earth's surface at angles that reflect the **magnetic latitude.** However, simple observations make it abundantly clear that the earth's magnetic field cannot be compared to that of a simple magnet in any precise way.

Mariners observed many centuries ago that their compasses pointed in slightly different directions every year. That observation led to the establishment of magnetic observatories, which maintain careful records of compass behavior, and to field expeditions, which were to locate the exact geographic positions of the magnetic poles. We now know that the magnetic poles wander about one degree in longitude every five years, remaining in the general vicinity of the rotational poles but not coinciding with them except perhaps briefly. Historic observation has also revealed substantial fluctuations in the strength of the earth's magnetic field both locally and for the entire planet. Furthermore, there are theoretical reasons for rejecting the idea that the earth might be permanently magnetized.

For all substances there is a critical temperature—called the **Curie point**—at which they lose their magnetic properties. That temperature varies considerably for different materials but, for all, it is lower than the temperatures that prevail within the earth. Our planet is simply too hot to maintain a permanent magnetic field.

It seems that the earth's magnetic field exists because the planet is a rotating body with an electrically conductive core. That makes it possible for the earth to act as a sort of dynamo in which currents of electricity flow along the core's equator, generating an external magnetic field with poles near the rotational axis. There is evidence that the outer part of the earth's core is liquid and turbulent movements there could explain the local variability of the earth's magnetic field. When plotted on maps prepared at intervals of several years, the earth's magnetic field does suggest patterns of eddy currents in a turbulent liquid. However, such ideas have proved difficult to test or refine because some 3,000 km of mantle separate us from the core, making direct observation of the earth's magnetic field at its presumed source quite impossible.

If the rotating earth with its liquid and electrically conductive outer core generates its magnetic field through a dynamo effect, then it is difficult to imagine how the magnetic poles could wander very far from the rotational axis. Therefore, geologists proceed with their study of rock magnetism on the assumption that the rotational and magnetic poles have always been so close that locating one is, for practical purposes, tantamount to locating the other. Data so far gathered seem consistent with that assumption.

Fossil Compasses

Most rocks acquire a permanent magnetic polarization called **remanent magnetism** as they form. They do so because the earth's magnetic field orients grains of magnetite and several other less strongly magnetic minerals within them. For example, detrital grains of magnetite accumulating in clastic sediments align themselves in the earth's magnetic field as though they were miniature compass needles (Figure 14–5) and are then locked in place as the sediment lithifies. Although igneous and high-grade metamorphic rocks crystallize at temperatures far above

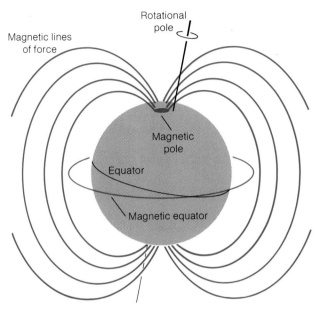

FIGURE 14–4

The earth's magnetic field can be visualized as lines of force that emerge from one magnetic pole, wrap around the planet, and plunge into the opposite pole. They are perpendicular to the earth's surface at the magnetic poles and parallel to it at the magnetic equator. Compass needles align themselves with the magnetic lines of force.

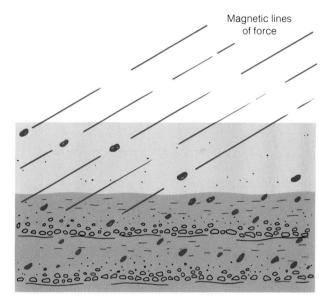

FIGURE 14–5

Clastic grains of magnetite align themselves with the earth's magnetic field as though they were miniature compass needles as they settle into a sedimentary deposit. They give the final rock a permanent magnetic orientation parallel to the direction of the earth's magnetic field at the time and place the sediment accumulated.

the Curie points of all minerals, they too acquire magnetic fields aligned with the earth's as they cool to temperatures that permit permanent magnetization.

Rock magnetism is generally weak. It is, however, easily possible to observe its existence in rocks—such as basalt or gabbro—that may contain as much as 1 percent magnetite. The simplest way is to lay a sensitive compass on a table, let its needle come to rest, and then move the rock specimen toward it. The compass needle will react to the rock's magnetic field by deflecting as it approaches to within a few centimeters and waggling back and forth as the rock moves. It is possible in principle to deduce the direction of the rock's magnetic field from a carefully planned sequence of such simple observations. However, geologists who seriously study rock magnetism use sensitive electronic instruments that can eliminate the complicating influence of the earth's magnetic field and can measure extremely weak rock magnetism.

Rocks tend to retain their original magnetic directions because the mineral grains within them are so firmly locked together that they do not easily shift. It is commonplace, for example, to find that the magnetic directions of folded rocks wrap around the structures, demonstrating that deformation had no effect on the rock magnetism. Very few natural processes with the exception of metamorphism and weathering can completely remagnetize a rock in a new direction, leaving no trace of its original magnetic orientation.

In retaining their original magnetic fields, rocks also retain some memory of where they formed (Figure 14–6). Measuring their magnetic orientation reveals both the direction to the magnetic pole and the magnetic latitude of the place where they formed. If those differ from the direction to the magnetic pole and the magnetic latitude where the rock is found today, then we must conclude that either the rock or the magnetic pole has moved.

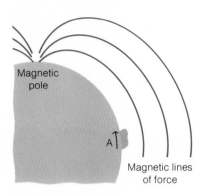
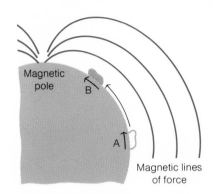

FIGURE 14-6

Rocks that formed in location A on the earth's surface will retain a memory of the magnetic latitude and direction to the magnetic pole of that place after they have moved to location B.

FIGURE 14-7

The apparent polar wandering curve for North America would very nearly fit that for Europe if it were moved east through the same angle North America would have to move to close the Atlantic Ocean. Similar relationships obtain for the polar wandering curves of other continents that reassemble to make Pangaea.

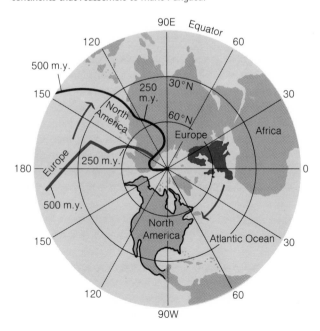

Polar Wandering Curves and Continental Movement

During the 1950s several geologists began to systematically measure the magnetic directions of rocks of known age. They may have expected to compile a record of polar wandering during geologic time but they actually produced proof that it was the continents, not the poles, that had wandered.

Figure 14-7 shows the apparent position of the north magnetic pole since the beginning of Paleozoic time as recorded in North American and European rocks. The two curves both start in the Pacific Ocean and follow roughly parallel paths northward for some distance before they rapidly converge to the present polar position in the Arctic. Similar curves compiled from rocks collected from other continents also follow their own paths across the map to the modern poles. By the late 1950s it had become obvious that there are at least as many "polar wandering" curves as continents.

It would be patently absurd to conclude that six or more pairs of magnetic poles had wandered independently during most of geologic time and then finally coalesced into one during the recent past. In any case, if multiple poles had existed, they would presumably have left multiple magnetic records in the rocks whereas in fact the magnetic record on each continent appears to refer to only one pole. The only alternative is to conclude that the continents have roamed independently over the earth's surface.

If we move the "polar wandering" curves about the map as Wegener and the other "drifters" suggested we move continents, the curves coincide just as the margins of the continents fit together. In other words, Wegener had, in effect, predicted the behavior of polar wandering curves long before the data on which they are based became available. Scientific theories can become overwhelmingly convincing when they prove to have predictive value. Thus, the polar wandering curves essentially ended the debate about whether continents move.

By 1960 the magnetic data were too numerous to ignore, too consistent to reject, and too convincing to dispute. Although there were distinguished exceptions, most of the geologic profession then admitted that continental movement is an observable phenomenon even though the mechanism remained unknown. Within a few years, another aspect of rock magnetism—the existence of magnetic reversals—brought understanding of the mechanism of continental movement. With it came the much broader theory of plate tectonics, a major revolution in scientific thought.

MAGNETIC REVERSALS

Reversed Polarity

One of the most surprising results to emerge from the study of rock magnetism was the observation that many specimens are polarized in a direction opposite to that of nearby rocks. The first reaction to that was a mixture of dismay and disbelief followed by a determined search for local explanations. However, rapidly accumulating data soon showed that reversely magnetized rocks formed in all parts of the world during certain radioactively dated intervals of geologic time. If different rocks that formed in different places during the same period of time all show reversed magnetism, we must conclude that the earth's magnetic field was actually reversed when they formed. No other reasonable explanation exists because the rocks have nothing in common beyond their ages and their reversed magnetic polarities.

No one knows why the earth's magnetic field reverses or how it happens except that it seems to occur within a geologically brief time, within no more than a few thousand years and perhaps much less. Presumably, the magnetic field fades away and then reappears oriented in the opposite direction. That seems to have happened at irregular intervals averaging every half million or so years during the Tertiary period and at frequent but less precisely dateable intervals before then. The last major reversed period ended about 800,000 years ago. No one knows when the next reversal will occur but it may be significant that the earth's magnetic field has been weakening during recent decades at a rate that will reduce its strength to zero in about 3,000 years if it were to continue. However, there is no apparent basis for assuming that the present decline will continue unchecked for the next 3,000 years so geologists hesitate to predict an imminent reversal.

The Magnetic Time Scale

Geologists are now developing a new kind of time scale (Figure 14–8) based on magnetic reversals. They expect it to enable them to correlate rocks of the same age with great precision because the magnetic periods are relatively brief and the reversals happened simultaneously all over the world. However, the new time scale will only improve the precision of correlation, not the accuracy of dating, because the magnetic periods are themselves dated radioactively.

Unfortunately, magnetic reversals distinguish rocks only according to their polarization directions, which may be the same for rocks of widely different ages, and—unlike fossils—are never uniquely characteristic. Therefore, it is necessary to know the age of the rock quite accurately before using the magnetic time scale for more precise determination. Nevertheless, magnetic reversals have already proved useful in refining the dating of many sequences of Tertiary and Pleistocene rocks. For example, they provide a convenient way to sort otherwise nondescript and completely unfossiliferous volcanic rocks (Figure 14–9). Geologists have also used magnetic reversals to refine the correlations of late Tertiary sedimentary rocks, such as those at Olduvai Gorge in East Africa, which contain the fossil remains of our earliest known ancestors.

FIGURE 14-8

The magnetic time scale consists of a series of longer "epochs" during which the earth's magnetic field remained dominant in one orientation punctuated by much shorter "events" during which it briefly reversed. The last such event happened about 30,000 years ago. It is possible to use the record of magnetic epochs and events preserved in sections of sedimentary or volcanic rocks to refine dates already closely established by other means. In this hypothetical case, an unconformity complicates matters just as they commonly do in nature.

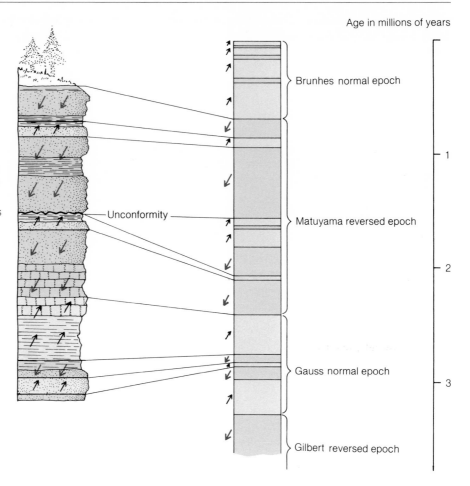

Age in millions of years

Brunhes normal epoch

Unconformity

Matuyama reversed epoch

Gauss normal epoch

Gilbert reversed epoch

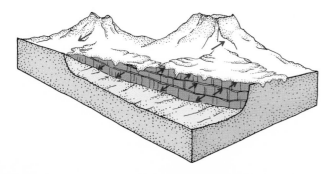

FIGURE 14-9

Magnetic reversals may clarify otherwise baffling relationships. In this hypothetical example, a geologist looking at lava flows exposed in the stream bank might be able to associate them with the two small volcanoes by matching directions of magnetic polarization.

MAGNETIC ANOMALIES AND SEAFLOOR SPREADING

While some investigators measured rock magnetism in the laboratory, others surveyed the earth's magnetic field in the oceans. The two kinds of work were quite independent and appeared for many years to be unrelated. No one could have predicted that the data they produced would eventually explain the movement of continents and lead from there to development of plate tectonic theory.

Magnetic surveying has been a basic research and exploration technique among geologists for decades. They use an instrument called a **magnetometer,** of which there are several varieties, to measure the strength of the earth's magnetic field. Geologists doing detailed work carry a magnetometer from spot to spot on the ground to prepare a magnetic-field map of a small area; those involved in broader studies tow magnetometers behind a ship or airplane to prepare a more generalized map of a larger area. In either case, the resulting map shows local areas of abnormally high or low magnetic-field strength, which geologists call **magnetic anomalies.**

For geologists working on the continents, the problem of interpreting magnetic maps is essentially one of relating the anomalies to the distributions of strongly or weakly magnetic rock types. Such maps become especially useful when they reveal the existence of bodies of rock not exposed at the surface. For example, sulfide ore bodies commonly create large magnetic anomalies (Figure 14–10) because they contain strongly magnetic minerals; large masses of rock salt create anomalies because they are

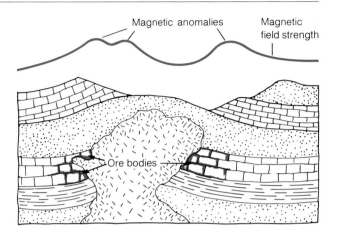

FIGURE 14–10
Many valuable ore bodies that are not exposed at the surface have been found by drilling strong magnetic anomalies.

so weakly magnetic. Geologic interpretation of magnetic maps has been the first step toward discovery of many valuable ore bodies and oil fields.

Magnetic Stripes

When magnetic surveys of the oceans finally began to appear during the 1950s, they revealed alternating belts of high and low magnetic field strength, such as those shown in Figure 14–11. No such magnetic anomaly pattern had ever been mapped on a continent. The oceanic anomalies average about 20 km in width, although that varies consid-

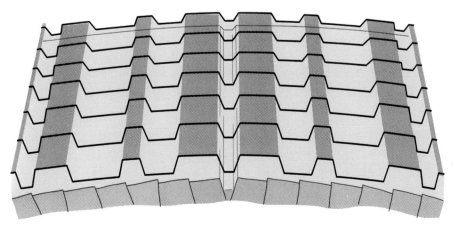

FIGURE 14–11
Magnetic maps of the oceans invariably reveal anomalies in a pattern of broad stripes that trend parallel to the crest of the oceanic ridge and are symmetrical as mirror images on its opposite sides.

FIGURE 14-12

A pattern of broad oceanic ridges winds about the earth almost as though it were a continuous seam. The mid-Atlantic ridge has been known since the last century but the existence of a connected system encircling the planet was not discovered until the 1950s.

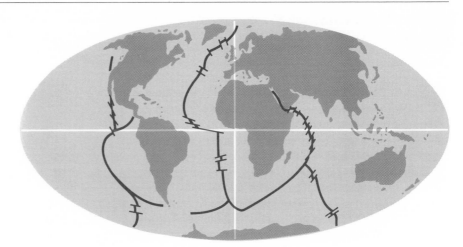

erably, and look on a map like a set of broad stripes trending generally parallel to the crest of the broad **oceanic ridge.**

The system of oceanic ridges is a major feature of the earth's surface. A ridge winds along the floor of each ocean basin and connects from one to another to make a continuous global pattern some 65,000 km in total length that suggests the stitching on a ball (Figure 14–12). The oceanic ridge averages more than 1,000 km in width and rises gently in a series of broad steps (Figure 14–13) to an average elevation some 2,000 m above the general level of the ocean floor. In a few places, most notably Iceland, the crest of the oceanic ridge rises above sea level and becomes accessible to direct observation. The narrow rift valley that follows the crest of the oceanic ridge divides both the ridge and the pattern of magnetic stripes on opposite sides into symmetrical halves as though it were a mirror.

Interpreting the pattern of magnetic stripes on the oceanic maps presented a novel problem because bottom samples of the ocean floor reveal only basalt beneath the surface veneer of muddy sediments. There is no evidence of alternating bands of different kinds of rock on the ocean floor. It therefore seemed for some time that there was no apparent reason for the pattern of regularly alternating high and low magnetic anomalies. In other words, geologists could not extend their traditional techniques of interpreting magnetic anomalies to the oceanic maps. That should have been no surprise. Wegener observed in 1924 that we then knew very little about the ocean floors but he predicted that they would prove to be utterly unlike the continents. That must rank among the more prophetic passages in the history of science.

Seafloor Spreading

In 1960 the late Harry Hess, a great visionary, suggested that the continents and ocean floors move together by drawing away from the rift in the crest of the oceanic ridge. Volcanic eruptions, Hess suggested, filled the opening gap with basalt lava flows, which became new oceanic crust (Figure 14–14). Although Hess occasionally referred to that idea as ''geopoetry'' because it ranged so far in advance of the available data, it quickly became more generally known as **seafloor spreading.** Many geologists found

FIGURE 14-13

Considered broadly, the oceanic ridge is a gentle swell in the ocean floor with a narrow rift in its crest. Considered in detail, it rises in a series of steps created by fractures in the oceanic crust that make the bottom topography rugged.

Ridge crest rift

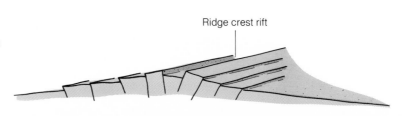

the idea immediately appealing because it explained how continents could move along with the oceanic crust instead of by plowing through it. It was inescapably clear by 1960 that continents do somehow move. Furthermore, quite a few samples of basalt had been collected from the ocean floor by that time and radioactive dates had shown that they get progressively older with increasing distance from the oceanic ridge. That is exactly the pattern one would expect to find if new seafloor were forming at the crest of the oceanic ridge and moving away in both directions. Within a few years the pattern of magnetic stripes combined with the earth's history of magnetic reversals to provide the final convincing evidence.

Proof of Seafloor Spreading

Consider Figure 14–15 and imagine the pattern of magnetic anomalies the mechanism of seafloor spreading should create. New ocean floor moving away from the

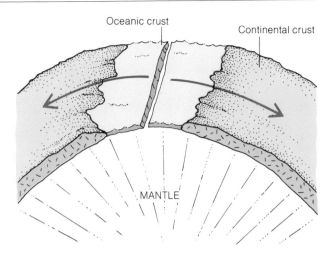

FIGURE 14–14

The concept of the seafloor spreading envisions the ocean floor moving away from the crests of the oceanic ridges and sweeping the continents along with it to create new ocean basins.

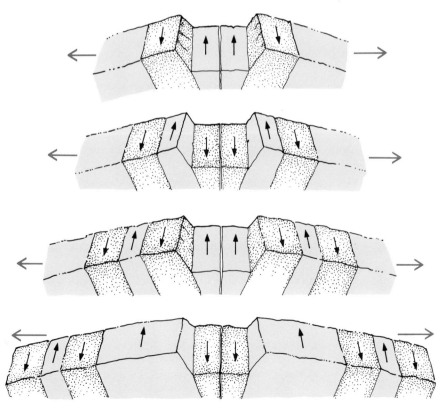

FIGURE 14–15

The mechanism of seafloor spreading as demonstrated by the oceanic pattern of rock magnetism. Basalt lava flows erupt into the ridge-crest rift to make new oceanic crust, which then splits and moves down opposite sides of the ridge, carrying the record of the earth's magnetic field reversals with it. Some authors have compared the rift in the ridge crest to a recording head and the moving oceanic crust to a magnetic tape.

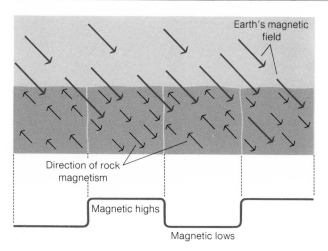

FIGURE 14–16

Magnetic highs form where the lava flows in the ocean floor are magnetized in a direction parallel to that of the earth's magnetic field so the two reinforce each other. Magnetic lows form where the magnetic polarity of the rocks in the ocean floor opposes that of the earth's magnetic field so the two tend to cancel each other.

crest of the oceanic ridge should consist of alternate bands of basalt lava flows. Such flows should be magnetically polarized in opposite directions according to whether they erupted while the earth's magnetic field was in its present or its reversed orientation. If the basalt is magnetized in a direction parallel to the earth's magnetic field, the two will add together to make a magnetic high (Figure 14–16). However, if the rock is magnetized in a direction opposite the earth's magnetic field, the two will tend to

cancel each other to make a magnetic low (Figure 14–16). Therefore, the alternating bands of oppositely polarized basalt lava flows on the ocean floor should create precisely the striped pattern of alternating high and low anomalies that appears on magnetic maps of the ocean. The mirror image relationship between the patterns of magnetic anomalies on opposite sides of the oceanic ridges adds weight to the argument because that is also precisely what one would expect if the seafloor were spreading away from the crest of the oceanic ridge. It is hard to imagine that a single theory could so elegantly fit so many different and otherwise unrelated types of data into a simple and coherent picture unless it were valid.

Furthermore, the theory of seafloor spreading successfully predicted, and explains, the existence of exactly similar patterns of magnetic anomalies following the ridges in different ocean basins. They all record the same sequence of worldwide magnetic reversals. Still more weight was added to the proof when the research vessel *Glomar Challenger* began drilling long deepsea cores. Geologists then expected to find the close correspondence between the patterns of magnetic anomalies and the sequence of reversals in the sediments that Figure 14–17 schematically illustrates. And it did indeed exist, just as expected! Since nothing in scientific theory is more persuasive than successful prediction, the mechanism of seafloor spreading is now regarded as an established fact, not just a theory. General acceptance of that concept during the mid-1960s led almost immediately to development of the much broader and more comprehensive body of plate tectonic theory, the subject of the next chapter.

FIGURE 14–17

Deepsea cores reveal exactly the same vertical sequence of magnetic reversals as appears horizontally in the pattern of magnetic anomalies in the oceanic crust.

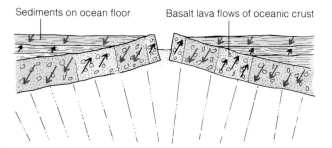

SUMMARY

The science of geology began and developed for more than a century with the basic assumption that the present geographic disposition of continents and ocean basins is a primordial feature of the earth's surface. Geologists recognized vertical movements of the earth's crust but did not seriously entertain the idea of significant horizontal displacements until 1912. Then the new theory of continental drift proposed that several continents had once been assembled into a single supercontinent that broke into fragments, which then dispersed about the globe. The direct implication that continents move long distances was generally rejected until the late 1950s primarily because no one could explain how they could move.

During the late 1950s measurements of rock magnetism provided new and almost unequivocal evidence that the continents had indeed moved as the proponents of continental drift had long contended. Most geologists then accepted the existence of continental movement as an observable phenomenon and began a serious search for the responsible mechanism.

The new theory of seafloor spreading proposed that continents move along with the ocean floor away from the rift in the crest of the oceanic ridge while volcanic eruptions fill the rift to create new ocean floor. Radioactive age dates showing that the basalts on the ocean floor get progressively older with increasing distance from the ridge supported the idea. Final confirmation came largely from comparison of the oceanic pattern of magnetic highs and lows with the record of the earth's magnetic-field reversals. The record of magnetic reversals is preserved laterally on the ocean floor just as it is vertically in sections of sedimentary rocks. Recognition of seafloor spreading as a basic geologic process led to development of the comprehensive theory of plate tectonics, which includes seafloor spreading within a much broader conceptual framework.

KEY TERMS

continental drift   magnetometer
Curie point   oceanic ridge
magnetic anomalies   remanent magnetism
magnetic equator   seafloor spreading
magnetic latitude   tectonics
magnetic poles

REVIEW QUESTIONS

1. Why is it not reasonable to regard the earth as a permanently magnetized body comparable to an ordinary bar magnet?

2. In what parts of an ocean basin would you expect to find the oldest and the youngest rocks?

3. What kinds of evidence provide clues to the age of an ocean basin?

4. What kind of magnetic anomaly would you expect to find over a body of strongly magnetic rock that formed during a time when the earth's magnetic field was reversed?

5. Why is it unreasonable to interpret ''polar wandering'' curves as evidence of polar wandering?

6. What does the mirror image pattern of magnetic stripes on opposite sides of the crest of the oceanic ridge imply about the way in which the ocean floor spreads away from the ridge crest?

SUGGESTED READINGS

Marvin, U. B. *Continental Drift, the Evolution of a Concept.* Washington, D. C.: Smithsonian Institution Press, 1973.
    A historical review of the long controversy over continental drift and the emergence of plate tectonics.

Sullivan, W. *Continents in Motion: The New Earth Debate.* New York: McGraw-Hill Book Co., 1974.
    A very readable account of plate tectonics. Written for a nontechnical audience by an unusually skilled author.

Takeuchi, H., S. Uyeda and H. Kanamori. *Debate about the Earth.* Rev. ed., San Francisco: Freeman, Cooper, Co., 1970.
    A readable and exciting account of the disputes that led to development of plate tectonics. Written by people who actually participated.

Tarling, D. and Tarling, M. *Continental Drift.* Garden City, New York: Doubleday & Company, 1971.
    Written primarily for a nontechnical audience.

Wegener, A. *The Origin of Continents and Oceans.* New York: Dover Publications, 1966.
    A translation of the book that launched the big debate.

Wilson, J. T., ed., *Continents Adrift: A Scientific American Book.* San Francisco: W. H. Freeman & Co., 1972.
    An anthology of many early *Scientific American* articles about continental drift.

CHAPTER FIFTEEN

Basalt pillows lying on the floor of a road metal quarry near Remote, Oregon. They were erupted on the floor of the Pacific Ocean during Eocene time but look exactly like modern ones photographed in the rift of the mid-Atlantic ridge. The pillow in the center of the picture is about 1 m long.

PLATE TECTONICS

If the earth creates new crust at the crest of the oceanic ridge system, then it must destroy old oceanic crust elsewhere if it is to remain the same size. Tracing the course of the oceanic crust from its creation to its destruction led to development of the theory of plate tectonics.

A VIEW OF THE EARTH

Plates

Gaining that new perspective on the earth was more a matter of assembling already well-known facts into a coherent global picture than one of gathering new information. Geologists had been aware for many years of the linear ocean deeps, or **trenches,** that rim much of the margin of the Pacific Ocean and also define the eastern margin of the Caribbean Sea. Study of earthquakes had revealed that many occur along well-defined surfaces called **Benioff zones** that slant down from the floors of the trenches like deep knife cuts into the body of the earth. Most other earthquakes were known to occur either along the crest of the oceanic ridge system or along lines that connect ridges and trenches. Earthquake waves were known to slow while passing through a region of the upper mantle called the **low velocity zone,** which forms a continuous shell around the earth between depths of about 100 and 250 km. Those observations seemed unrelated until geologists realized

that they describe the boundaries of **plates,** large segments of the earth's rigid outer rind, or **lithosphere,** which is about 100 km thick.

For reasons, or perhaps a variety of reasons, that geologists do not fully understand, lithospheric plates move. They appear to glide on the weak rocks of the low velocity zone—now called the **asthenosphere.** They move in directions that carry them away from the oceanic ridges where new lithosphere forms and into oceanic trenches where old lithosphere vanishes into the mantle. Most earthquakes occur along the boundaries of the moving plates, which tend to be zones of intense geologic activity. Figure 15–1 shows the outlines of the seven major plates that tile most of the earth's surface. The figure also shows the outlines of some of the lesser plates that fill the spaces between them, making a complete and constantly shifting mosaic that covers the entire planet.

Lithosphere and Crust

Figure 15–2 illustrates the distinction between the lower part of the lithosphere, which consists of relatively cool and rigid mantle peridotites and the upper part, or **crust,** which consists of entirely different kinds of rocks. The continental crust is a thick raft of relatively light sialic rocks, presumably granite, gneiss, and schist, that extends to an average depth of about 40 km although that varies consid-

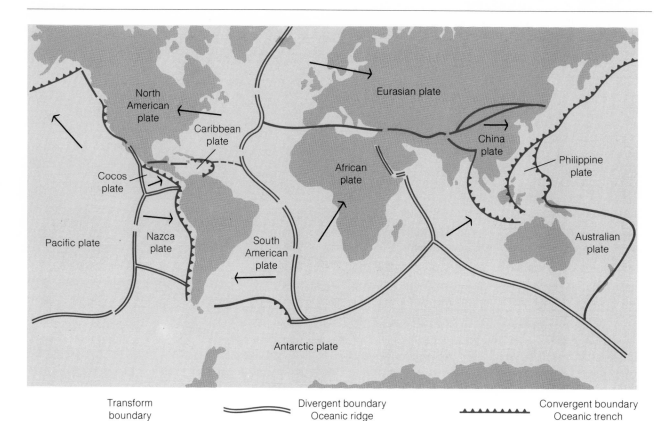

Transform boundary

Divergent boundary
Oceanic ridge

Convergent boundary
Oceanic trench

FIGURE 15–1 (above)

General outline of the earth's major plates. The small arrows indicate plate movement away from oceanic ridges, into oceanic trenches, and along major transform faults.

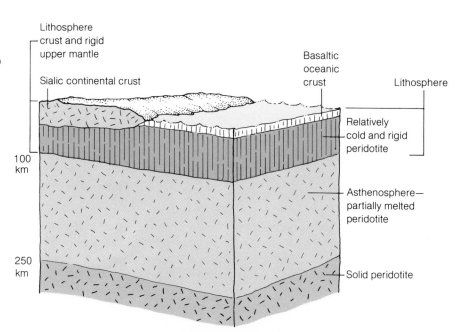

FIGURE 15–2

A schematic section through the outer part of the earth.

erably from one region to another. The oceanic crust consists of basalt lava flows and dikes underlain by gabbro that reach an average depth of about 10 km. Although the base of the crust is inaccessible to direct observation, we can nevertheless locate it rather accurately because it reflects echos of earthquake waves and the shock waves from large explosions. The lower surface of the earth's crust is called the **Moho** or, more formally, the Mohorovicic discontinuity. However, the asthenosphere is the zone of the earth most directly involved in plate movement.

The Asthenosphere

The weak rocks of the asthenosphere probably consist of partially melted peridotite. We can observe in deep wells and mines that the temperature of the upper several kilometers of the earth's crust increases downward at an average rate of almost 3°C for every 100 m. The temperature must continue to increase at greater depths, although presumably at a slower rate, and any reasonable projection suggests that the peridotites of the upper mantle must exist at a temperature that would melt them if they were at the surface. However, peridotite, like most rocks, expands as it melts so the great pressure exerted by the overlying material keeps the rocks in the lower part of the lithosphere solid despite their high temperature. Evidently, the asthenosphere is a zone of the upper mantle in which the effects of increasing temperature overcome those of high pressure and partially melt the rock. Below the asthenosphere, the effects of pressure again prevail over those of temperature, and the rocks are solid. Current estimates place the proportion of molten rock within the asthenosphere in the range between 1 and 10 percent.

Plate Movement

Figure 15–3 summarizes the simple set of rules that governs plate movement by differentiating between three different types of plate boundaries. Plates slide in both directions directly away from the crests of oceanic ridges, or **divergent boundaries,** where new lithosphere and oceanic crust form. Plates collide at **convergent boundaries** where old lithosphere and oceanic crust disappear if the boundary happens to be a trench where one plate slides beneath the other and vanishes into the mantle. Plates slide past each other along **transform boundaries,** which

begin and end in the divergent or convergent boundaries that they connect in all possible combinations. However, as Figure 15–1 shows, those rules impose no apparent order on the global pattern of plate size, shape, and motion. Plates appear to move independently of each other in a random pattern that changes with time.

Rate of Movement

Geologists measure the rate of plate motion in several ways, depending upon the time span involved. It is possible to measure short-term movement simply by using standard surveying techniques to observe displacement of points on opposite sides of a plate boundary. For example, the San Andreas fault of California, a transform plate boundary, moves at a rate of several centimeters per year. Dividing the widths of the magnetic stripes on the ocean floor by the durations in years of the corresponding magnetic periods yields average rates of movement over much longer timespans. Again, the results fall within the range of several centimeters per year. Dividing the width of an ocean basin, such as the Atlantic, by the age of the youngest matching rocks on its opposite side gives a result in the same range. Although that fairly consistent rate of plate movement is much slower than a snail's pace, it is rapid by geologic standards. It is fast enough to displace adjacent plates about the length of a long human stride during a single lifetime, fast enough to open or close an ocean basin within the geologically brief time of 100 million years.

Direction of Plate Movement

Our situation in observing plate movements resembles that of passengers in a ship who can watch other ships pass but lack fixed landmarks from which to measure their directions or speeds. They have no way of knowing whether both ships are moving or whether one is stationary and the other under way. So far as we know, all the lithospheric plates move so there is no fixed reference point on the earth's surface from which to observe the global pattern of their motion. For example, we can observe that a large slice of coastal California, which belongs to the Pacific plate, is moving northward relative to the rest of the state, which is part of the North American plate, by displacement along the San Andreas fault (Figure 15–4). However, there is no way to be sure whether the Pacific plate is moving

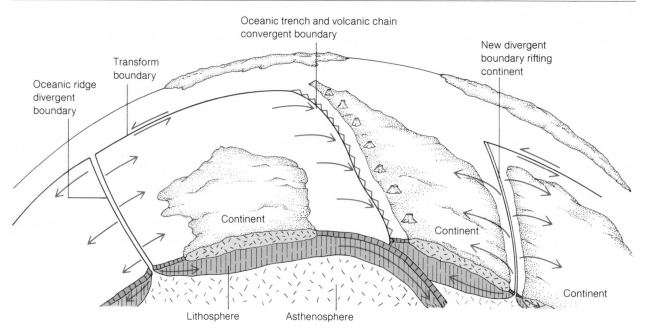

FIGURE 15-3

The general scheme of plate tectonics. Plates move away from divergent boundaries at the crests of oceanic ridges where new oceanic crust and lithosphere form. They move toward convergent boundaries at oceanic trenches where old lithosphere and oceanic crust sink into the mantle. Plates slide past each other along transform boundaries without gaining or losing area. Continents are floating rafts of sialic rock embedded in the upper parts of some plates with which they move, splitting at new divergent boundaries and colliding at convergent boundaries.

north or the North American plate south because the relative motion between them would be the same in either case. That problem poses no great obstacle to the local interpretation of plate movements but it does complicate their analysis on a global scale.

One way to cope with the lack of stationary global landmarks is to arbitrarily assume that one plate is stationary; imagine that it is firmly nailed to the mantle and then stand on it, figuratively speaking, and watch the others pass. That stratagem makes it possible to work out a global scheme of plate movement that will vary depending upon which was chosen as the fixed reference. Many geologists prefer to start with either the African or the Antarctic plate. The lithosphere appears to be thicker beneath the African plate than elsewhere so it seems reasonable to imagine that it may be more firmly fixed in position. Furthermore,

both the African and Antarctic plates are almost completely surrounded by divergent plate boundaries. That makes it easy to imagine either of them as being fixed while all other plates move away. Still another approach to unravelling the pattern of plate motion starts with observation of certain isolated volcanic centers.

Mantle Plumes

About two dozen centers of volcanic activity seem, unlike most volcanoes, to exist independently of plate boundaries. Geologists call them **hotspots.** The Hawaiian and Yellowstone park volcanoes, for example, are remote from any plate boundary. Volcanoes on the Azores, Galapagos, and Easter Islands seem not to depend upon plate movements for their existence even though they do lie near plate boundaries. Most hotspots consist of a small center of current-eruptive activity perched at the end of a long chain of extinct volcanoes that become progressively older with increasing distance from the active center. The active Hawaiian volcanoes, for example, are at the southeastern end of the chain and the other islands are extinct volcanoes that become older northwestward. The trend continues far beyond the islands in a long chain of submerged extinct volcanoes called **seamounts,** which extends for thousands of kilometers across the floor of the North Pacific Ocean.

Many geologists argue that the volcanic hotspots mark **plumes** of hot rock rising like thunderheads through the mantle while the lithospheric plates slide across them. They suggest that the hot rock rising in the mantle perforates the lithosphere to form a volcano, which remains active until plate movement carries it past the plume when a new volcano starts as the old one stops (Figure 15–5). If so, then the chains of extinct volcanoes trailing away from active hotspots record movement of a plate over a fixed point in the mantle, a landmark from which to observe plate movements.

In fact, many of the hotspots scattered around the world do seem to remain in a more or less permanent arrangement, like a constellation of stars, as the lithospheric plates move across them. Several long chains of seamounts on the floor of the Pacific Ocean, including those extending northwest from Hawaii, change direction through a similar angle at extinct volcanoes of about the same age. It is easy to imagine that pattern recording a change in the direction of movement of the Pacific plate and difficult to conceive of another equally simple and logical explanation. Similarly, hotspots on divergent plate boundaries—of which Iceland is a good example—have double chains of volcanoes trailing away from them, one on each plate, that seem to record similar histories. Unfortunately, hotspots can illuminate only the most recent pattern of plate movement because they rarely seem to last more than about 100 million years.

Poles of Rotation

Figure 15–6 shows that we can think of a plate moving across the surface of the earth as though it were rotating about an imaginary axis drawn through the center of the

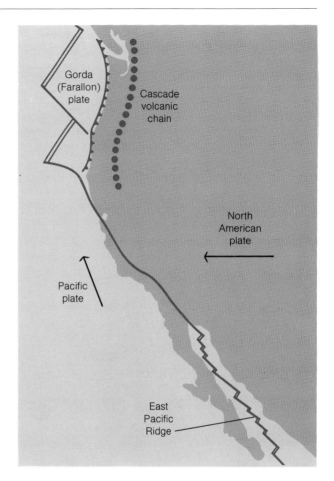

FIGURE 15–4

The San Andreas fault is a transform plate boundary that connects the north end of the east Pacific ridge in the Gulf of California to the south end of the Gorda ridge off Cape Mendocino on the northern California coast. New oceanic crust is forming in the Gulf of California as Baja California parts company with the rest of Mexico and moves north. If present trends continue, the slice of continent west of the San Andreas fault will eventually move out into the northern Pacific Ocean as a detached microcontinent.

planet. The place where that axis intersects the earth's surface is called the plate's **pole of rotation.** The axis of plate movement is imaginary so nothing actually happens at the pole of rotation. We can, however, use it to trace the motion of any place on the plate by stabbing the point of a compass into a globe at the plate's pole of rotation and

FIGURE 15–5

Volcanic hotspots appear to mark places where an ascending plume of hot rock within the mantle perforates the lithosphere. A trail of extinct volcanoes with an active one at its head marks the path of the plate across the plume.

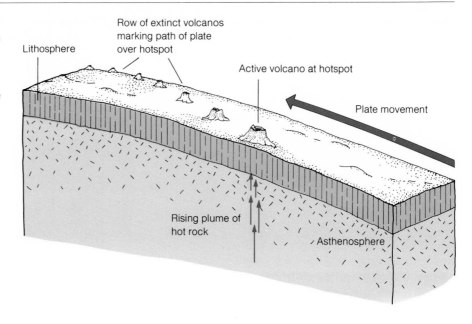

FIGURE 15–6

It is convenient to envision the motion of a plate as rotation about an imaginary axis drawn through the center of the earth and intersecting the surface at the pole of rotation for that particular plate. The rate of angular displacement increases with distance from the pole of rotation. Features that record plate movement trace lines corresponding to parallels of latitude drawn from it. The pole of rotation generally lies on another plate.

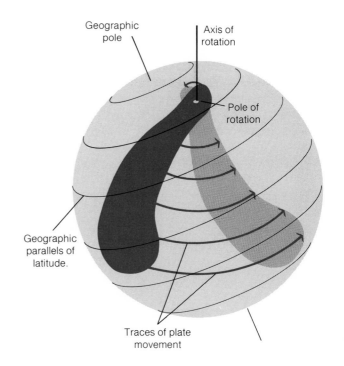

tracing a small circle in the appropriate direction. The motions of a number of points on the plate trace on the globe as though they were parallels of latitude drawn with respect to its pole of rotation, and the rate of plate motion increases with distance from the pole. Volcanic hotspots leave traces on the plate that do indeed appear on a map as though they were parallels of latitude drawn from its pole of rotation.

DIVERGENT PLATE BOUNDARIES

The Rift in the Ridge Crest

Iceland is a good place to begin discussion of divergent plate boundaries because it is on the crest of the mid-Atlantic ridge. It is one of the few places where an oceanic ridge rises above sea level. The rift in the ridge crest makes a sharply defined valley that angles across the country. It separates the western part of Iceland, which belongs to the North American plate, from the eastern part, which is on the Eurasian plate. Land surveys repeated over the years show that valley growing wider as the plates move away from each other. Occasionally, long fissures open in its floor and basalt lava wells up out of them.

The largest lava flow observed during recorded human history erupted during the summer of 1783 when approximately 12.5 km^3 of basalt lava poured out of the Laki fissure, which was about 25 km long, and covered some 565 km^2 of the valley floor. Several new fissures have opened since then and produced similar but smaller eruptions. The activity in Iceland appears to be typical of the entire oceanic ridge system.

Many authors have compared the movement of lithospheric plates away from divergent boundaries to that of ice floes drifting apart along an open crack. Imagine water welling up into the gap between the floes and freezing onto their trailing edges, enlarging them at the same rate they move away from the crack. Then compare the water to molten basalt magma rising into the rift between separating lithospheric plates and erupting there as lava flows to form new oceanic crust. Like the imaginary ice floes, the lithospheric plates appear to grow at the same rate they move away from divergent plate boundaries.

Mantle Convection

Most geologists believe that oceanic ridges form where masses of hot rock rising within the mantle reach the earth's surface. Mantle rocks evidently accumulate heat from radioactive decay until they get hot enough to rise to the surface and create the bulge we recognize as an oceanic ridge. They lose much of their heat through frequent volcanic eruptions along the crest of the ridge. Most of that activity passes unnoticed because it occurs on the floor of the ocean.

Pillow Basalt

Basalt lava erupted underwater behaves quite differently from that poured out on land. Streams of lava chill against the cold water forming rigid outer walls that briefly contain the molten basalt. Then it bursts through them to make a new stream that repeats the process. Dripping candle wax likewise repeatedly bursts through a thin outer skin of cold wax to form a new trickle that soon encases itself within a chilled outer skin. The result in the case of the underwater eruption is a lava flow consisting of a heap of more or less cylindrical masses of basalt with dimensions resembling those of barrels. Such lava flows are generally called **pillow basalts** because their appearance where sectioned in a roadcut or cliff suggests a pile of oversized sofa pillows (Figure 18–6). The upper kilometer or so of the oceanic crust consists mostly of such pillow basalts erupted in the rift at the crest of the oceanic ridge and then carried away with the moving ocean floor.

Continental Rifting

Where a new divergent boundary happens to cross a continent embedded in a lithospheric plate, it splits that continent along a rift that widens to become a new ocean basin as the fragments of the divided continent move apart. Modern examples illustrate all stages of the process.

New divergent boundaries seem to start with a more or less symmetrical three-cornered tear in a plate. Two of the arms lengthen into a long rift that becomes the new plate boundary while the third fails to grow. The Gulf of Aqaba

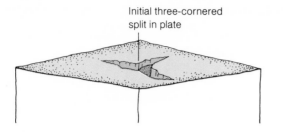

Initial three-cornered
split in plate

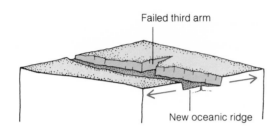

Failed third arm

New oceanic ridge

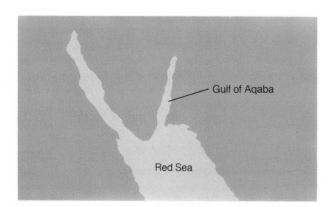

Gulf of Aqaba

Red Sea

FIGURE 15-7

New divergent plate boundaries appear to begin as three-cornered
tears that extend in two directions to split the plate along the new
oceanic ridge while the third arm fails to develop and ultimately fills
with sediment. The Gulf of Aqaba has been interpreted as a modern
example of a failed third arm branching off from the Red Sea, a new
ocean. The deep sedimentary fillings in ancient examples commonly
contain large reserves of oil and gas.

(Figure 15–7) has been interpreted as a failed third arm
branching off the northern end of the Red Sea, which is a
geologically young ocean basin, a recently opened continental rift. Ancient examples of such failed third arms are
now filled with deep accumulations of sedimentary rocks,
which commonly contain reservoirs of oil and gas. Finding
such filled rifts—called **aulacogens**—is an important economic application of plate tectonic theory.

The East African rift valleys, which contain a long row of
large lakes, appear to be a continental rift marking a new
divergent plate boundary still in the initial stages of formation. It may eventually split a large slice of East Africa from
the rest of the continent just as the rift that follows the floor
of the Red Sea has separated the Arabian peninsula. However, the East African rift valleys have existed for some millions of years and do not appear to be growing wider at an
appreciable rate. It seems that movement on that incipient
plate boundary is either extremely slow or stalled. Nevertheless, the row of towering volcanoes that follows the
margins of the rifts—Kilimanjaro, Ruwenzori, and others—
is vigorously active so the zone is not dead.

The Red Sea is a continental rift in a more advanced
stage of development. From the geologic point of view, it is
a small ocean because an oceanic ridge follows its center
line and its floor consists of basalt lava flows with the typical oceanic pattern of magnetic stripes. It is possible to see
across the southern margin of the Red Sea and therefore
to observe the continuing separation of Arabia from Africa
by ordinary surveying methods. If the present movement
continues, the Red Sea may eventually become a broad
ocean. The Atlantic was just such a narrow rift early in
Mesozoic time.

The opposing coasts of the Red Sea are the broken
edges of the continent, which stand high because they are
still on the flanks of the oceanic ridge that separated them.
They will gradually decline in elevation if continued plate
movement carries them off the oceanic ridge. Meanwhile,
erosion of the high coastal mountains will bevel the up-
turned edges of the new continental margins. When they finally move off the ridge, the erosionally thinned continental
margins will subside below sea level and become the foun-
dations for broad coastal plains as they acquire a veneer of
sediments. Figure 15–8 shows the probable sequence of
events.

Trailing Margins

Geologists refer to continental coasts moving away from a divergent plate boundary as **trailing margins.** The coasts of the Atlantic Ocean are a good modern example—they move away from the mid-Atlantic ridge that separated them early in Mesozoic time. Trailing continental margins are typically stable regions relatively free of earthquakes, volcanoes and other evidence of geologic activity; they are not plate boundaries.

Most trailing margins carry thick burdens of sedimentary rocks that form broad coastal plains and continental shelves. Figure 15-9 shows their general pattern with the oldest sedimentary formations lying at the inner edge of the coastal plain and the younger formations succeeding them seaward. The series of successively younger formations crops out as broad bands that trend generally parallel to the coast.

The oldest coastal plain sediments provide a rough index to the age of the ocean because they must have accumulated within a geologically short time after the initial rift opened. Those on the opposite sides of the Atlantic were laid down about 200 million years ago, a time that corresponds well with other evidence of the age of that ocean. Until the advent of plate tectonics, many geologists had been concerned about the absence of older sedimentary formations from the modern coastal plains. That problem no longer exists now that we recognize the oceans as relatively young and constantly changing features of the earth's surface.

CONVERGENT PLATE BOUNDARIES

Convergent boundaries swallow old lithosphere if one of the two colliding plates sinks through an oceanic trench into the mantle. The sinking plate leaves most of its veneer of oceanic sediments in the trench as a thick accumulation of material of sialic composition that eventually returns to the continental crust whence most of it came. Some of the basaltic oceanic crust melts off the sinking plate where it enters the asthenosphere at a depth of about 100 km and returns to the surface through a chain of volcanoes that parallels the trench. At still greater depth, the remainder of

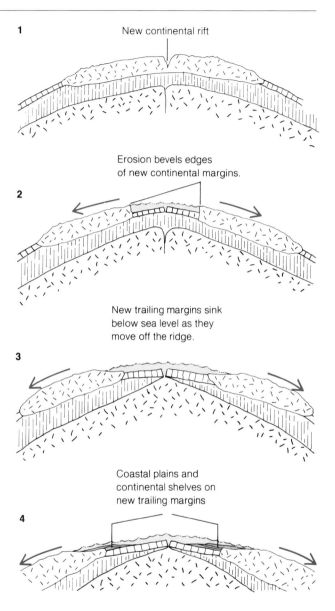

FIGURE 15-8

New oceans begin as a developing ridge system splits a continent and then sweeps the fragments aside. Erosion bevels the new trailing margins before they move far enough down the ridge to sink below sea level and start to acquire a blanket of coastal plain and continental shelf sediments.

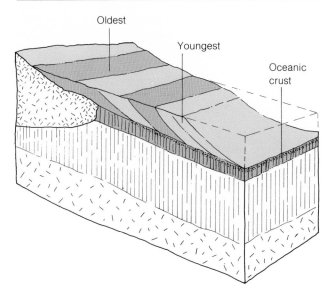

Oldest

Youngest

Oceanic
crust

FIGURE 15-9

Layers of sediment begin to accumulate along the edge of a continent shortly after it forms through rifting. Therefore, the age of the sedimentary rocks that lie on the continental crust along the inner margin of the coastal plain approximately dates the adjacent ocean.

the sinking plate finally gets hot enough to lose its identity as cold and rigid lithosphere and thus vanishes into the mantle. That combination of processes makes convergent plate boundaries the scene of varied and intense geologic activity.

The course of events at a convergent boundary depends largely upon whether the colliding plates happen to contain continents. Plates that carry only oceanic crust may vanish into the mantle, leaving virtually no trace of their passage across the face of the earth except the masses of oceanic sediment that were scraped off their surfaces as they slipped into a trench. Sedimentary rocks are too light to sink into the mantle.

Continental Sutures

Continents are also too light to sink into the mantle. Therefore, a plate must stop sinking when the disappearance of all the intervening oceanic crust brings a continent to a convergent boundary. Then, as Figure 15-10 shows, the

boundary will reverse and the other plate start sinking beneath the continent. That sequence of events probably explains why such a large proportion of the modern convergent plate boundaries involve Pacific Ocean floor sinking beneath a continent.

If both of the plates colliding at a convergent boundary contain continents, the two continents will eventually meet after all the intervening oceanic crust has been consumed at the trench. Then the two continents will suture together and the convergent boundary will soon shift to some other place where oceanic crust can sink into the mantle. Convergent plate boundaries tend to assemble larger continental masses whereas divergent boundaries dismantle them and disperse the fragments.

One clear example of an ancient continental suture is the join between Europe and Asia along the line of the Ural Mountains. That mountain range consists essentially of old oceanic sediments crushed between the colliding continents as though between the jaws of a vise. The magnetic "polar wandering" curves compiled on opposite sides of the Urals are quite different for rocks formed before about 200 million years ago but similar for those formed since then. It seems that Europe and Asia really were separate continents until about 200 million years ago.

The Tibetan Plateau marks a currently active continental join between the Indian subcontinent, which moved northeast after having separated from Africa, and the Asian continent. In that region, plate movement still grinds the two continental masses into each other along their line of suture, greatly thickening the continental crust. Many geologists have interpreted the thickening as evidence that the two continents overlap in the region of the Tibetan Plateau with Asia riding over India. However, recent research makes it seem more likely that compression along the line of the suture thickened the crust.

Ocean Deeps and Volcanic Arcs

Figure 15-11 shows the typical geographic expression of a convergent plate boundary: a long oceanic trench on the side of the sinking plate and a parallel line of volcanoes perched along the edge of the overriding plate, the one that is not sinking. The volcanoes follow the inner of a pair of concentric circular arcs and the trench the outer.

The clearest North American example of a convergent

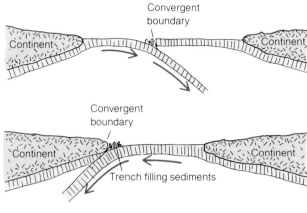

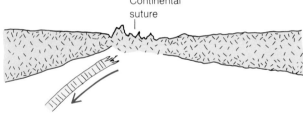

1 Convergent boundary forms in ocean basin.

2 One continent arrives at boundary; other plate begins to sink.

3 Second continent arrives at convergent boundary.

4 Sedimentary rocks that filled the trench crushed into mountains along the line of suture as convergent boundary ceases to function.

FIGURE 15-10

A hypothetical chain of events leading to collision and suture of two continents.

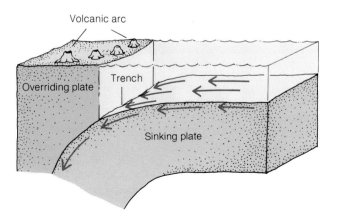

FIGURE 15-11

The typical geographic expression of a convergent plate boundary. An ocean trench on the surface of the sinking plate and a parallel volcanic arc on the overriding plate.

plate boundary is in the Aleutians. There, a long chain of active volcanoes follows the inner side of a trench along a circular arc that extends from the south coast of Alaska far out into the North Pacific. The floor of the Pacific Ocean is sinking northward through the trench and passing beneath the volcanic Aleutian Islands. The tightly curving volcanic arc of the Windward and Leeward Islands with a deep trench along its seaward side provides a much smaller but almost equally clear example along the eastern margin of the Caribbean Sea.

The Cascade volcanoes, which extend from northern California into southern British Columbia, are another typical volcanic chain. They are associated with a convergent plate boundary immediately off the West Coast where the Pacific Ocean floor is sinking beneath the North American continent as it moves westward away from the mid-Atlantic ridge. However, the trench that should exist off the coast west of the Cascades does not appear as a topographic feature, presumably because it is stuffed completely full of sediment.

Ocean Trenches and New Mountains

Oceanic trenches collect sediment both through the direct deposition from turbidity flows and as the sinking plate scrapes its burden of sediment into them when vanishing into the mantle. Trenches that lie near continents receive abundant supplies of sediment from both sources and may fill completely whereas those remote from large land areas remain nearly empty. The Aleutian trench, for example, is very deep near its outer end and becomes progressively shallower as it approaches the Alaskan mainland. Its filling of deformed oceanic sediments finally rises above sea level in the mountains of Kodiak Island and the Kenai Peninsula.

Convergent plate boundaries complete the **rock cycle** shown in Figure 15–12. Weathering, erosion, and sedimentation convert continental rocks into oceanic sediments. They are eventually swept into trenches where metamorphic and, to a lesser extent, igneous processes convert them back into the metamorphic and igneous rocks of the sialic continental crust.

Sediments packed into the depths of an oceanic trench gain heat partly through decay of radioactive elements contained within them and partly by absorption, through conduction, of heat from the rocks that surround them. As temperatures rise within the trench filling, the sediments there recrystallize to form metamorphic rocks and, in some cases, partially melt to form large masses of granitic rock. The chemical compositions of sialic continental rocks and oceanic sediments are virtually identical so we can think of those two broad categories of rocks as being essentially the same stuff in two different states. Weathering, erosion, and sedimentation convert it into one state; metamorphism and partial melting into the other.

Oceanic trenches exist as long as the plate on which they rest continues to sink. When that movement stops, the light rocks that filled the trench float up to become a linear range of mountains composed of severely deformed and more or less recrystallized oceanic sediments, a new mass of sialic continental rocks. Much of the California Coast Range, for example, consists of dark gray sandstone (Figure 15–13) and colorful beds of chert that were deposited as turbidites and siliceous oozes on the floor of the Pacific Ocean during Mesozoic time. Then they were swept into a trench that disappeared during early Tertiary time, permitting them to rise as a new range of mountains that extended the continent westward.

Finding that the moving ocean floor scrapes its load of sediments into ocean trenches solved a problem that had troubled most geologists when they still assumed that the present ocean basins had been receiving sediment throughout geologic time. They could not explain why the oceans had not filled with sediment. Plate tectonic theory resolved that problem by showing that the ocean floors are not stable platforms but moving conveyor belts that return sediments to the continents. The processes of plate tectonics constantly renew the entire surface of the planet by creating new ocean floor at the crests of the oceanic ridges and new sialic continental crust in the depths of the oceanic trenches.

TRANSFORM PLATE BOUNDARIES

Adjacent lithospheric plates slide past each other along transform boundaries without gaining or losing area. Geologists use the term **fault** to refer to any surface separating

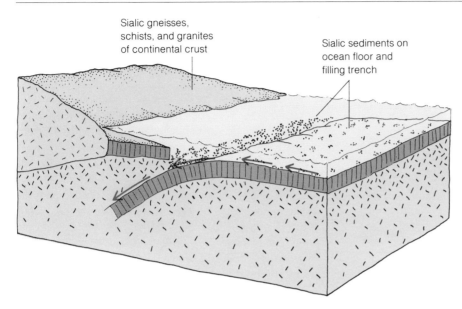

Sialic gneisses, schists, and granites of continental crust

Sialic sediments on ocean floor and filling trench

FIGURE 15-12
The sialic rocks of the continent weather into soils that are eroded, transported, and finally deposited on the ocean floor as clastic sediments of sialic composition. Then the moving ocean floor sweeps them into the trench where metamorphic and igneous processes convert them back into sialic continental rocks.

FIGURE 15-13
Layers of sandstone that were originally deposited on the Pacific Ocean floor and then swept into a trench. Now they form part of the California Coast Range.

Ridge-to-ridge

Trench-to-trench

Ridge-to-trench

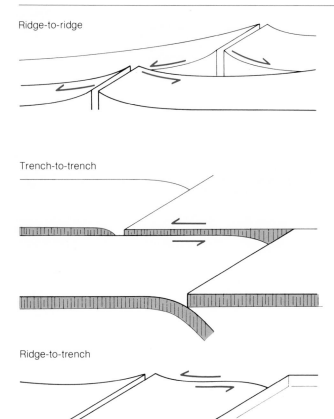

FIGURE 15–14

Transform boundaries can connect ridges to ridges, ridges to trenches, and trenches to trenches in a total of 14 possible right- and left-handed combinations, of which these are a sampling.

lithospheric plates they separate slide away from ridge crests or sink into trenches.

The most abundant type of transform boundary offsets the crests of the oceanic ridges into numerous short segments. Those ridge-to-ridge offsets make maps of the oceanic ridge system look as though they might have been chopped into numerous short lengths and then reassembled with the segments of ridge crest somewhat misaligned. No part of the oceanic ridge system lacks those transform offsets, which generally amount to less than 100 km although a few reach lengths in excess of 1,000 km.

Figure 15–14 shows that the plate segments on opposite sides of the ridge-to-ridge transforms move past each other in opposite directions as long as they remain between the offset lengths of ridge crest. After they pass the ridge crests, the plate segments on opposite sides of the transform move together, and as parts of the same plate. Although the ridge-to-ridge transforms actually function as plate boundaries only between the offset lengths of oceanic ridge crest, their traces extend far beyond the ridges as linear scarps on the ocean floor. Those scarps trace parallels of latitude drawn from the pole of plate rotation and therefore provide an excellent record of the directions of fairly recent plate movement. They also offset the pattern of magnetic stripes on the ocean floor because that forms at the ridge crest where relative movement between plate segments does occur.

No generally accepted explanation yet exists for the origin or the function, if any, of the numerous transform offsets in the oceanic ridge system. Some geologists suggest that they make minor adjustments in the trend of the ridge that serve to keep it oriented so the lithospheric plates move away in a direction exactly perpendicular to its crest. Others argue that the transform offsets in the ridge crest merely reflect sets of fractures that existed in the crustal rocks before the ridge formed (Figure 15–15).

The San Andreas Fault

The notorious San Andreas fault of California is one of a system of more or less parallel faults that together form a transform boundary between the North American and Pacific plates. Figure 15–16 shows that the San Andreas fault connects the north end of the east Pacific ridge in the Gulf

bodies of rock that slide past each other, and commonly describe transform plate boundaries as **transform faults.** Figure 15–14 illustrates several of the 14 possible types of transform boundaries that are classified according to the various combinations in which they connect ridges and trenches and the relative directions in which they move. The lateral movement along transform boundaries is absorbed in the ridges and trenches where they end. Therefore, they must move at exactly the same rate at which the

FIGURE 15-15

One theory to explain the origin of the numerous ridge-to-ridge transforms holds that they descend from old fracture systems that controlled the line along which a new divergent boundary split a plate.

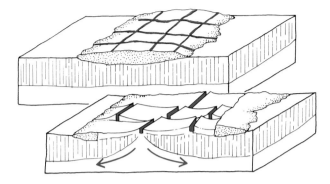

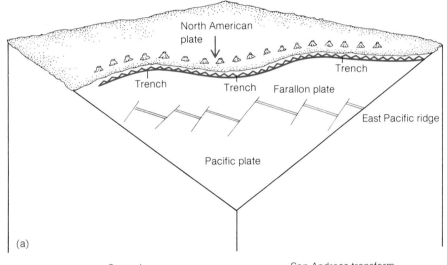

(a)

FIGURE 15-16

The San Andreas fault formed as the convergent boundary at the leading edge of the North American plate gobbled up most of the Farallon plate and the corresponding section of the east Pacific ridge. (a) During early Tertiary time, the Farallon plate was spreading eastward from the east Pacific ridge and sinking beneath the North American plate at a convergent boundary along the West Coast. (b) Now, two segments of the ridge and a small area of the Farallon plate survive seaward of the Cascade Range. When they are gone, the San Andreas and Queen Charlotte Islands faults will join to make a continuous transform boundary.

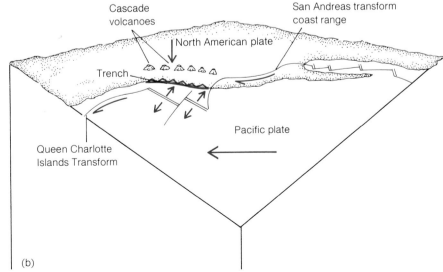

(b)

of California to the south end of the Gorda ridge off the coast of northern California. It follows a somewhat wandering course and the part between the coast of northern California and the Gorda ridge is called the Mendocino fault.

The San Andreas fault is not a typical ridge-to-ridge transform because it is much younger than the ridge segments it connects. It appears to have formed only about 30 million years ago when an oceanic trench that formerly existed off the coast of California swallowed the last of a large segment of the old Farallon plate. That brought the corresponding segment of the oceanic ridge to the trench, and the two cancelled each other as the last of the plate that had separated them disappeared. As the Farallon plate that had separated them vanished, the North American and Pacific plates came into contact along the line of the San Andreas fault, which connects the surviving segments of the oceanic ridge.

As the ridge and trench eliminated each other and the trench filling rose to become the Coast Range, a large slice of western California became attached to the Pacific plate and began moving northward along the trace of the San Andreas and other parallel faults. The plate boundary appears to be a bit "soft" and ill-defined because the movement occurs on a system of faults instead of along a single fracture. No one knows exactly how far north the part of California west of the San Andreas fault—the Salinian block—has moved during the last 30 million or so years but the distance appears to be at least 550 km and may well be greater than that. If the present movement continues, the Salinian block will eventually move out into the North Pacific as a microcontinent completely detached from North America.

The North American plate still moves west away from the mid-Atlantic ridge and the last of one of the remnants of the old Farallon plate is now disappearing into the convergent boundary off the West Coast opposite the Cascade volcanoes. If that trend continues, the two short lengths of surviving oceanic ridge will arrive at the convergent boundary a few million years hence and repeat the sequence of events that created the California Coast Range and the San Andreas fault. As the trench and ridge cancel each other, the sediments now packed into the trench will rise to make a new coast range and the San Andreas fault will propagate northward to connect with the Queen Char-

lotte Islands fault off the coast of British Columbia, which has already formed in the same way.

SUMMARY

Plate tectonic theory portrays the outer 100 km of the earth as a sort of rigid rind, the lithosphere, which consists of more than a dozen segments called *plates* that move independently about the surface of the earth. Plates move away from each other along divergent boundaries that follow the crests of the oceanic ridges, which apparently mark places where masses of hot rock rising through the mantle reach the surface. New oceanic crust forms at divergent boundaries as basalt lava flows erupt in the crest of the ridge to fill the gap that opens as the plates slide away from each other at the crest of the ridge. New lithosphere forms as the rocks of the upper mantle cool and become rigid by losing heat through the crest of the ridge.

New divergent boundaries form as masses of rock within the mantle accumulate heat released through decay of radioactive elements and then rise to form the oceanic ridges, which last as long as it takes to dissipate the heat, perhaps a few hundred million years. Where new divergent boundaries cross a continent, they split it and sweep the fragments apart, opening a new ocean basin between them.

Plates collide at convergent boundaries where the heavier of the two, invariably one bearing oceanic crust, sinks into the mantle, forming an oceanic trench. Basaltic oceanic crust melts off the sinking plate as it reaches the asthenosphere at a depth of about 100 km and returns to the surface through a chain of volcanoes on the overriding plate. At still greater depth, the sinking plate gets hot enough to blend back into the mantle and so disappears. The sinking ocean floor leaves its cover of sediments in the trench to form a thick accumulation of rock of sialic composition. When the convergent boundary ceases to function, that mass of material packed into the trench rises to become a new mountain range.

Lithospheric plates slide past each other along boundaries that transform their gain or loss of area at divergent or convergent boundaries into lateral movement across the earth's surface. Transform boundaries begin and end in divergent and convergent plate boundaries, which they may connect in any possible combination. They are capable of both dismembering and assembling continents.

By producing new oceanic crust at divergent boundaries and new continental crust at convergent boundaries, the processes of plate tectonics constantly renew the earth's surface. Without them, the ocean basins would long since have filled with sediment, which would have displaced the water they contain over the entire surface of the planet.

KEY TERMS

asthenosphere	pillow basalt
aulacogen	plate
Benioff zone	plume
convergent boundary	pole of rotation
crust	rock cycle
divergent boundary	seamount
fault	trailing margin
hotspot	transform boundary
lithosphere	transform fault
low velocity zone	trenches
Moho	

REVIEW QUESTIONS

1. What kinds of observations provide information about the locations of plate boundaries?

2. Under what circumstances can a plate vanish completely leaving little or no record of its former existence?

3. What kinds of evidence provide information about the rates and directions of plate movement?

4. Where do plates grow larger, where do they grow smaller, and where do they move without changing size?

5. What kinds of plate boundaries dismember continents, what kinds assemble them, and what kind can do either?

SUGGESTED READINGS

Bird, J. and B. Isacks, eds. *Plate Tectonics: Selected Papers from the Journal of Geophysical Research.* Washington, D. C.: American Geophysical Union, 1972.

A compilation of original research papers that played a major role in the development of plate tectonics theory.

Calder, N. *The Restless Earth.* New York: Viking Press, 1973.

An outstanding and readable discussion of plate tectonics. Addressed to a nontechnical audience.

Hallam, A. *A Revolution in the Earth Sciences.* Oxford, England: Clarendon Press, 1973.

An excellent treatment of plate tectonics and the history of its development.

Scientific American. "Continents Adrift and Continents Aground," San Francisco: W. H. Freeman & Co., 1976.

A fascinating anthology of semitechnical articles from *Scientific American* that provide a first-hand view of the development of plate tectonic theory.

Wyllie, P. J. *The Way the Earth Works.* New York: John Wiley, 1976.

A brief discussion of plate tectonics and related topics. Quite readable.

CHAPTER SIXTEEN

A tight fold in metamorphic rock. This specimen was a muddy sandstone before recrystallization at high temperature and pressure transformed it into gneiss. (Bernie Rosenblum)

FOLDS AND FAULTS

We normally think of rocks as rigid and unyielding, the epitome of strength and stability. And it is true that many rocks, indeed many large regions of the continental crust, do in fact remain almost perfectly stable for long intervals of geologic time. However, we also find severely deformed rocks wrinkled into folds and broken by faults. Such deformation generally occurs in association with plate movements and is typically most intense in regions near plate boundaries or former plate boundaries. Those who specialize in the study of rock deformation call themselves **structural geologists.**

TYPES OF DEFORMATION

Rocks bend into **folds** where stresses in the earth's crust deform them without breaking them as though they were ductile solids, such as modelling clay or beeswax. **Brittle deformation** occurs if rocks break to form a fracture, which is called a **joint** if the opposite sides do not move significantly, and a **fault** if they do. Whether rocks actually break or change shape without breaking depends partly upon the rock type, partly upon the rate at which the deforming force is applied, and partly upon the circumstances of pressure, temperature, and water content. Ge-

ologists generally find evidence that ductile and brittle deformation occur together. However, brittle deformation tends to dominate near the surface; ductile deformation, on the other hand, becomes an increasingly important factor with depth.

Structural Style

Folds and fractures permanently record the directions of the crustal forces that create them in a distinctive regional pattern that geologists refer to as **structural style**. Therefore, systematic study of the deformation structures in a region generally enables geologists to reconstruct ancient stress patterns and perhaps to infer something of their causes. In many cases, analysis of deformed rocks leads ultimately to reconstruction of ancient plate movements. However, such work is more often of immediate economic value in the search for new mineral resources.

DUCTILE DEFORMATION

Figure 16–1 shows one of the most astonishing sights in nature—an outcrop exposing layers of rigid rock crumpled into tight folds. The layers appear to have wrinkled like so

FIGURE 16-1

Tight folds in Cretaceous limestone. They formed as the rocks moved along the fault that truncates the fold on the right. The sagebrush in the foreground is about one-half meter high.

much canvas, yet they are now as rigid as any rock. How is it possible for solid rocks to behave as though they had so little strength?

Geologists address that question in several ways. Many observe folded rocks in the field and prepare maps and cross sections showing folds and faults. Others deform rocks experimentally to determine under what conditions various kinds of rock behave as brittle or as ductile solids. Most geologists have examined thin sections of deformed rocks under the microscope and have witnessed what happened to individual mineral grains. A few geologists engage in almost purely theoretical consideration of what occurs within a solid mass of rock as it changes shape.

Bending Rocks

One of the simplest experiments in rock deformation was conducted quite by accident during the last century when many European cities adorned their parks with benches that consist basically of a long slab of marble supported at both ends by upright slabs. Many of those old benches

now droop slightly in the middle because the long marble slab is slowly sagging under its own weight. If a slab of marble can bend visibly under its own weight in the course of a century, then imagine what it might do in a million years.

More deliberate experiments in rock deformation typically involve squeezing a specimen in a press until it either bends or breaks. In general, such experiments show that rocks tend to break if subjected to a rapidly applied force and are more likely to bend if the force is applied slowly. Earth movements are typically slow so most forces within the earth's crust must be applied slowly. Experiments also show that any kind of rock can be made to behave as a ductile solid if it is slowly squeezed while kept under a sufficiently high confining pressure. We can be sure that pressures deep within the earth's crust are extremely high. Other experiments have shown that rocks deform more easily if they are hot and wet than if they are dry and cold. Both high temperatures and water certainly exist deep within the earth's crust. As the laboratory experiments would lead us to expect, field observation shows that rocks deformed at depth tend to behave as ductile solids whereas those deformed near the surface are more likely to break.

Experimental work and field observation show that different kinds of rock vary greatly in their response to deforming forces. Shale, for example, flows easily under experimental conditions that break sandstone. Thus, field geologists often find folded beds of shale next to faulted beds of sandstone. Geologists use the term **competent** to describe the behavior of rocks that remain rigid and deform by breaking. They describe the more ductile rocks that change shape without breaking as **incompetent**. Figure 16–2 shows one type of structure that often develops when rocks with different degrees of ductility deform together. However, the distinction between competent and incompetent behavior is relative and must be applied to rocks that deformed together under similar conditions of temperature and pressure. Virtually all rocks are brittle under surface conditions and become increasingly ductile as pressure and temperature rise with increasing depth.

Rocks become weak under metamorphic conditions and develop complex structures, such as those shown in Figure 16–2.

Microscopic comparison of deformed and undeformed specimens of otherwise similar rocks sheds some light on what happens as rocks change shape. The boundaries of individual crystal grains in many deformed rocks are crushed, which suggests that the grains may have slipped past each other during deformation like kernels of corn in a sack. Many deformed rocks contain elongated mineral grains that appear to have stretched without breaking through internal slippage along weak planes in their crystal structures. Many deformed rocks contain broken mineral grains offset along small fractures that functioned as miniature faults within the rock. Chemical processes also contribute to rock deformation as mineral grains dissolve at points of greatest pressure and grow into areas where the pressure is lower. In general, the processes involved in deforming rocks resemble those that contribute to flow of glacial ice.

FIGURE 16–2

When these rocks were squeezed vertically, the competent layers broke into pieces that separated to form the structures that look like rows of sausages while the surrounding incompetent rock flowed in around them.

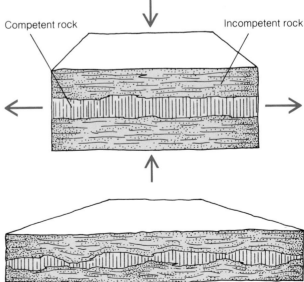

FIGURE 16-3

An anticline flanked by two synclines in tightly crumpled "ribbon chert." It was originally siliceous ooze deposited far offshore on the floor of the Pacific Ocean but now forms a roadcut near New Almaden, California. The area shown is about 4 m across. (D. W. Hyndman)

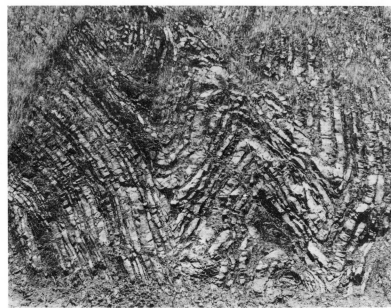

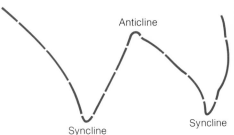

The Strength of Rocks

One interesting theoretical approach to rock deformation derives from attempts to build models that will simulate the behavior of rocks within the earth's crust. The basic problem is to find materials that will behave like rocks if deformed on a much smaller scale during a relatively brief time. One method involves scaling the physical properties of rocks down mathematically to fit the space and time dimensions of the model, and then seeking real materials with properties resembling those obtained by calculation.

Finding materials weak enough to fit the calculated requirements of a geologic scale model is difficult. One of the few familiar substances with the appropriate mechanical properties is a gelatin dessert left too long in a warm room. Another is dry portland cement powder. Both materials have been used in many geologic scale model experiments with convincing results. Thus, theoretical calculations and experimental observation both show that on the scale of the earth's crust and in the vastness of geologic time, rocks do indeed behave as though they had virtually no strength.

Anticlines and Synclines

Field geologists see folded rocks in all degrees of complexity, ranging from some with simple bends to others with forms so complex they suggest two liquids swirled together. However, thoughtful analysis shows that virtually all folds consist basically of sets of arches, called **anticlines**, and troughs, called **synclines**, that follow one after the other like the crests and troughs of waves on the ocean (Figure 16–3). Analysis of complex folds typically reveals that they consist of several trains of anticlines and synclines superimposed on each other. In some cases, successive periods of deformation involving forces applied from different directions create structures consisting of folds that are themselves folded.

Figure 16–4 illustrates the basic anatomy of a simple fold in which the opposite flanks tilt away from the crest at similar angles. Such a fold is designated symmetrical because a surface called the **axial plane,** which contains the hinge lines of all the individual layers of rock, divides it into mirror image halves. More tightly deformed rocks commonly bend into asymmetric folds, which have one flank tilted more steeply than the other. Asymmetric folds do not form mirror images about the axial plane, which may itself be folded. In some cases, one flank of the fold may tilt beyond the vertical. Some folds become so far overturned that the axial plane becomes horizontal and half the rock layers are upside down (Figure 16–5).

Fold crests generally tilt so that one end of the structure plunges into the ground to create a hairpin-shaped outcrop pattern, such as that shown in Figure 16–6. In fact, geologists generally find that fold crests plunge in both directions. Therefore, we can imagine folds as having shapes resembling those of boats, a capsized boat if the fold is an anticline and an upright one if it is a syncline. In some regions the folds are so short that the anticlines and synclines become radially symmetrical domes and basins shaped like bowls instead of like boats.

Strikingly photogenic folds such as those typically illustrated in geology books are hard to find. If that were not so, many geologists would be without work. In most regions

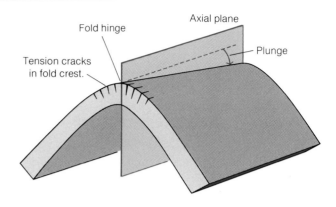

FIGURE 16–4

A diagram of a simple fold showing the axial plane that passes through the hinge lines of all the rock layers and divides the fold into halves. Most folds are at least slightly asymmetrical with one flank more steeply tilted than the other. In more complex cases, the axial plane itself is folded. Tension cracks develop along the crests of anticlines, making them more vulnerable to erosion.

FIGURE 16–5

Large recumbent folds exposed in the south slope of Mount Crandell overlooking Waterton, Alberta near the eastern margin of the Canadian Rockies. The axial plane lies nearly horizontal.

FIGURE 16-6
One fold slid over another along a thrust fault in this area on the south side of McCartney's Mountain, Montana. (Bureau of Land Management)

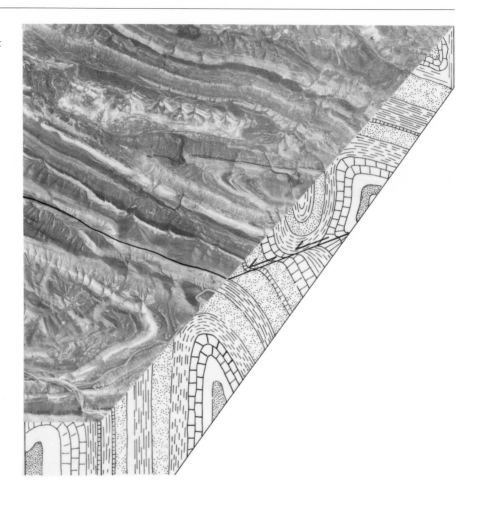

folds manifest themselves as tilted layers of rock exposed in isolated outcrops. Geologists find those outcrops, measure the **strike** of the layers—which is the direction of a horizontal line traced on their surfaces—and their **dip**—the angle at which they tilt (Figure 16–7). The geologists plot the strike and dip data on a map along with the areas in which the various rock formations are exposed. Only then does it generally become possible to assemble the isolated items of information into a coherent geologic interpretation.

Figure 16–8 shows how a layer of impermeable rock, such as shale wrapped around the crest of an anticline, may trap oil and gas by blocking their upward migration

through the water, filling pore spaces in the more permeable rock beneath. For many years the quest for new reserves of oil and gas consisted largely of a search for anticlines. By sometime during the 1920s, most of the large anticlines exposed at the surface in North America had been found and tested by drilling—they all seem to have at least one old well somewhere along their crests. Then geologists turned their attention to a search for anticlines that lack surface expression. Finding those requires assembling data from wells, gravity and magnetic surveys, and seismic surveys that involve the use of artificial earthquake shock waves. Most of the subsurface anticlines had been found and drilled by the end of the 1950s and petroleum

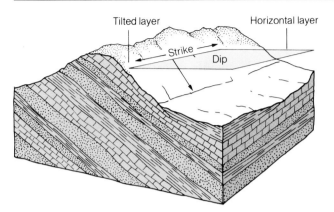

FIGURE 16–7

The strike of a tilted layer is the direction of the line of intersection of an imaginary horizontal plane with its surface; the actual direction the shore of a lake would trend if it were to lap onto the surface. The dip is the angle at which the layer tilts, measured in the direction a marble would go if it were rolled down the surface. Geologists generally measure both dip and strike with a combination instrument called a **Brunton compass.**

FIGURE 16–8

Petroleum geologists look for structures, such as folds or faults, that may trap oil or gas in places where an impermeable cap rock, such as shale, blocks upward migration of oil and gas through porous and permeable reservoir rocks, such as sandstone. In many regions those structures are not exposed at the surface and must be sought entirely by studying well records and other subsurface methods.

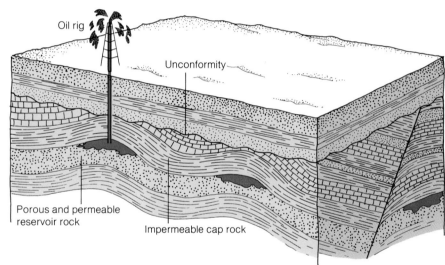

exploration has since focussed on much more subtle kinds of oil traps. It is now more true than ever that the difference between success and failure in petroleum exploration—the difference between profit and loss—hinges upon the quality of geologic reasoning used in selecting drill sites.

BRITTLE DEFORMATION

Joints

Rocks break as they deform under stresses applied either too rapidly or under too little confining pressure to permit them to behave as a ductile solid. All rocks, whether other-

wise deformed or not, contain joints. In most rocks the joints occur in several sets of parallel fractures that intersect to form simple geometric patterns, such as those shown in Figure 16–9. Joints tend to make rocks break naturally into more or less regularly shaped blocks. Most deformed bodies of rock also contain faults that may have moved distances, ranging from a few meters to hundreds of kilometers.

Faults

Like folds, faults are rarely so obvious in the field as in the photographs and diagrams commonly used to illustrate textbooks. Structural geologists devote considerable effort to the task of accurately locating faults, plotting their traces on maps, and determining the amount and direction of relative offset on opposite sides. It is rarely possible to see the fracture surface itself. Fault movement generally grinds the rocks near the fracture surface into rubble (Figure 16–10), which weathers rapidly and does not make re-

sistant outcrops. Streams tend to erode their valleys along those zones of broken rock so straight valleys suggest the existence of a fault, especially if several straight valleys align along a trend that crosses several drainage basins. The best evidence of faulting is the discovery of an outcrop pattern, such as that shown in Figure 16–11, that cannot

FIGURE 16–9

An outcrop of granite broken, as always, by a geometrically regular pattern of joints.

FIGURE 16–10

A small normal fault slicing through an outcrop of gneiss splays into two branches separated by closely fractured rock. A zone of finely crushed rock several centimeters wide follows the main fault. The area shown is about 2 m across.

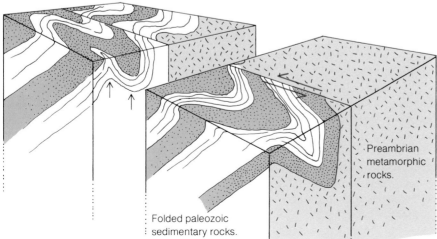

Folded paleozoic
sedimentary rocks.

Preambrian
metamorphic
rocks.

FIGURE 16–11

A vertical air view of a strike-slip fault in
the Tobacco Root Mountains of Montana
that slid Precambrian metamorphic rocks
several kilometers to bring them adjacent
to much younger Paleozoic sedimentary
rocks. (U.S. Forest Service)

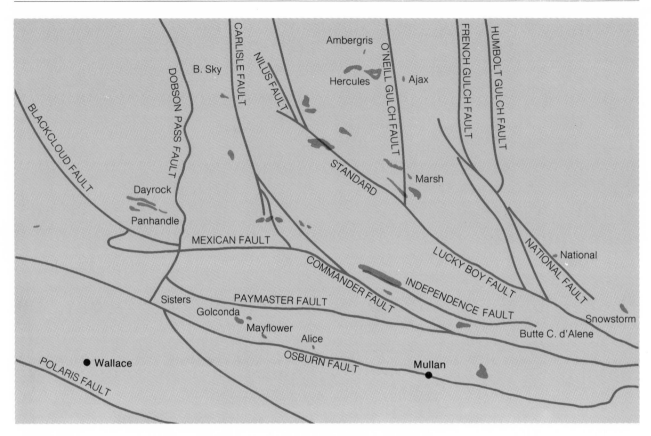

FIGURE 16-12
A map showing some of the larger faults that slice the Coeur d'Alene mining district of northern Idaho into a complex pattern of slivers that has provided employment for many geologists. These faults are part of the Lewis and Clark fault system, which extends from the northern Cascades of Washington into western Montana.

be explained without assuming that the rocks have been offset along a fracture.

The difficulty of interpreting faults is further complicated because the larger ones tend to move along swarms of more or less parallel fractures instead of along a single fault surface. Many mines that work ore bodies in such complexly faulted rocks depend for their continued operation, in some cases almost on a day-to-day basis, upon successful geologic interpretation of the faults. Figure 16–12 shows a greatly simplified map of the fault pattern in such a district. The geologists and engineers who supervise operations there devote much of their time to inter-

preting faults. Many such districts contain mines that resumed production years after having been abandoned because they were "worked out." Production began again when a new geologic interpretation located part of an ore body that had been displaced along a fault. Mining exploration geologists often examine and drill old and long-abandoned properties, hoping to develop a better interpretation of the folding and faulting that will enable them to find ore their predecessors missed.

Geologists also investigate faulting in the laboratory by breaking rocks experimentally and through studying the behavior of models. For example, Figure 16–13 shows the

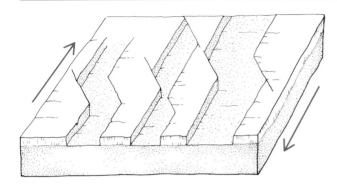

FIGURE 16-13

Southeastern Oregon, shown below, and nearby parts of adjacent states have a distinctive pattern of active faults: Normal faults that trend generally north to south and break the region into blocks that rise and sink to form ranges and basins, and right-lateral strike-slip faults that trend generally northwest to southeast. Similar patterns can be simulated experimentally by subjecting a model consisting of a layer of clay on a flexible base to shear by moving its west side north and its east side south.

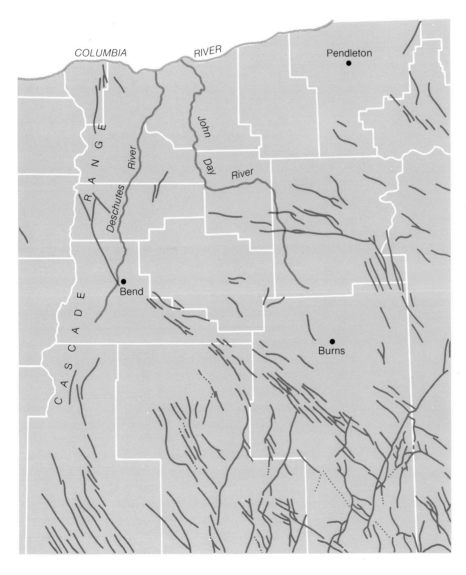

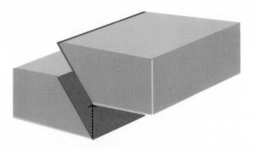

FIGURE 16–14

Geologists describe faults according to the dip and strike attitude of the fault plane and the direction the opposite blocks moved with respect to each other. Any direction of relative movement is possible.

pattern of actively moving faults that exists in a large area of southeastern Oregon and northeastern California. Vertical movements on one set of faults that trends almost north to south creates mountain ranges and intervening basins while horizontal movement on another set of faults that trends nearly northwest offsets them laterally. An almost exactly similar set of fractures forms when a simple model of the earth's crust is sheared in such a way that its west side moves north relative to its east side. In this case, the model may actually simulate the natural situation. The Pacific plate is moving north relative to the North American

plate and it may well be dragging the western part of the continent along. If so, that motion would create the same pattern of forces and fractures that exists in the model.

Fault Anatomy

Figure 16–14 shows a somewhat idealized view of the basic elements of a fault: A fracture surface called the **fault plane** and the displaced blocks of rock on either side. In most cases, the fault plane is a somewhat idealized feature because faults tend to move along numerous more or less parallel fractures distributed through a **fault zone**. Many fault zones are at least several meters in width and some are more than a kilometer wide. Fault zones may contain enough fracture porosity and permeability to make good aquifers in rocks that would not otherwise produce water.

Geologists describe faults in terms of the strike and dip orientation of the fault plane or fault zone and the direction of relative movement along it. If the fault plane is inclined, the rocks above it are called the **hanging wall**; those below it the **foot wall**. Those are old miner's terms that refer to the position of the rocks as perceived from a mine opening driven along the fault plane (Figure 16–15).

Normal Faults

Geologists use the term **normal fault** to refer both to vertical faults that moved vertically and to inclined faults in which the hanging wall moved down relative to the foot wall. As Figure 16–16 shows, displacement along an inclined normal fault lengthens and thins the earth's crust. Geologists generally interpret such movement as evidence of crustal tension.

The Basin and Range Province of Nevada and neighboring states contains numerous mountain ranges separated by broad valleys. Figure 16–17 shows schematically how each of those ranges and valleys is a block of the earth's crust bounded by normal faults, which in that region trend generally north to south. Movement along the faults since late Tertiary time has raised the ranges relative to the valleys and greatly extended the earth's crust in an east-west direction. The westernmost of those Basin and Range normal faults has raised the Sierra Nevada several thousand meters during the past few million years—that is considered a large displacement for a normal fault. The vertical

FIGURE 16–15

The hanging wall of a fault is the block that is overhead in a mine opening driven along the fault surface; the foot wall is underfoot. Many mine openings actually do follow faults because they commonly control the occurrence of ore.

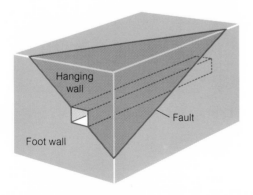

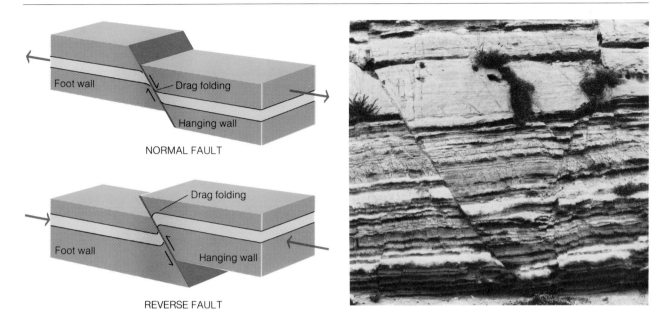

NORMAL FAULT

REVERSE FAULT

FIGURE 16–16

Movement on a normal fault extends the earth's crust whereas displacement on a reverse fault shortens it. Notice that it is possible to miss some layers of rock entirely in drilling through a normal fault and to intersect them twice in drilling through a reverse fault.

movement along the fault created the steep eastern face of the Sierra Nevada, a **fault scarp** that rises abruptly like a long, straight wall from the lower desert to the east.

Reverse Faults

Inclined faults in which the hanging wall moves up relative to the foot wall are called **reverse faults** or **thrust faults**. Reverse fault movement, which is opposite to that of normal faults, shortens and thickens the earth's crust. In most cases, geologists interpret such movement as evidence of crustal compression, and reverse faults are indeed a conspicuous feature of regions near convergent plate boundaries. The Himalaya region of northern India and neighboring countries where the Indian subcontinent is now jamming into the southern margin of Asia is an area of active reverse faulting on an enormous scale. Reverse faults also dominate much of the structure along the eastern margin of the Rocky Mountains in the United States

FIGURE 16–17

The Basin and Range Province, shown schematically here, consists of a number of mountain ranges and intervening basins that trend generally north to south and consist of blocks of the earth's crust bounded by normal faults. Vertical displacement of the blocks must also have extended the crust in an east-to-west direction. Valley fill sediments floor the basins because the region has an arid climate.

Rocks deformed
before faulting

Valley fill
sediments

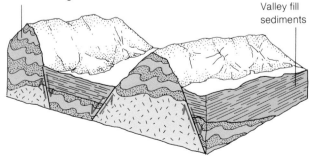

and Canada. Geologists have only recently begun to learn how to interpret the extremely complex geologic structures in that belt of intensely deformed rocks. They are finally finding petroleum there after many decades of frustration.

Overthrust Faults

Of all geologic structures few are more baffling than **overthrust faults**, which move enormous slabs of rock along fault planes that generally incline no more than a few degrees. Many overthrust faults have large displacements amounting to tens of kilometers and slide older rocks onto younger ones thus reversing the normal stratigraphic sequence. One of the best-known examples is in Waterton-Glacier National Park of Alberta and Montana. There, a slab of Precambrian sedimentary rocks at least 3 km thick moved more than 50 km east on the Lewis overthrust fault before finally coming to rest on top of Cretaceous sedimentary rocks some billion years younger (Figure 16–18).

Large overthrust faults are an important element in many structural provinces. Many in the Rocky Mountains and elsewhere control the occurrence of large deposits of petroleum and other mineral resources. Therefore, their interpretation is a matter of great economic as well as academic interest. To a large extent the problem revolves around the question of how they move.

For many years geologists generally assumed that overthrust slabs had been shoved over the rocks beneath, that some force had pushed them from behind. However, calculations show that the friction along the surface of an overthrust fault must be great and that rocks are too weak to withstand a force strong enough to overcome that friction without buckling into folds. The problem resembles that of pushing a wet noodle. The rocks above many overthrust faults, including those above the Lewis overthrust, are almost undeformed except near the leading edge of the slab. They certainly show no evidence of the kind of buckling we might expect to find if they were pushed from behind. Furthermore, there is nothing behind many overthrust slabs that might have pushed them. The area behind the Lewis overthrust, for example, is a large valley that could not have pushed anything. The best available evidence is adverse to the idea that overthrust slabs were pushed from behind.

An alternative theory proposes that overthrust slabs move downslope under the influence of gravity, which pulls every particle of rock and therefore eliminates the need for a push from behind. However, the problem of overcoming friction along the fault surface still remains. One theory proposes that high fluid pressure within the pores of the rock may tend to separate the mineral grains thus reducing internal cohesion and minimizing friction. Experimentally, high internal fluid pressures have been shown to reduce the strength of rocks. Furthermore, high fluid pressures have been encountered in many deep oil wells. The high fluid pressure mechanism probably does explain movement along some overthrust faults. However, it cannot explain all such faults. Some, including the Lewis overthrust, involve rocks that contain virtually no open pore space and therefore virtually no fluids.

Still another theory proposes that overthrust slabs may move by gliding on a layer of extremely ductile rock, such as shale. Field geologists quite commonly find a thick section of rock that has moved on a slippery bed of shale, leaving the rocks beneath essentially undisturbed. And it is also true that the base of many overthrust slabs consists of a formation that normally lies on top of a thick bed of shale. However, many overthrust faults are not associated with shale formations.

In most cases, overthrust faults are part of a complex assemblage of structures that also includes tight folds and numerous reverse faults. The problem is not just a simple matter of sliding one slab of rock over another. Geologists are now beginning to view overthrust faults in the context of the other structures with which they are associated. That makes the problem simpler in some ways and more complex in others. We need no longer explain how the entire thrust sheet could move at once over the fault surface because it might instead move in short hitches, perhaps nudged along from beneath by movements of other faults in the complex. Seen in that light, overthrust faulting may indeed involve compressional forces instead of simple downhill sliding. However, it is not necessary to assume that the compressional force is applied at the back of the moving slab.

Strike-Slip Faults

Figure 16–19 shows how the opposite sides of **strike-slip faults** move past each other horizontally. The movement is in a direction parallel to the strike of the fault plane. Despite their horizontal movement, strike-slip faults can lo-

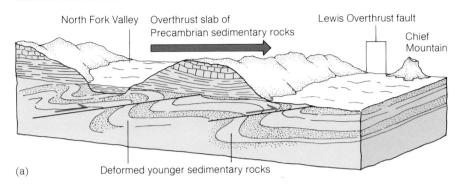

North Fork Valley   Overthrust slab of   Lewis Overthrust fault
                    Precambrian sedimentary rocks   Chief
                                                               Mountain

(a)   Deformed younger sedimentary rocks

(b)

FIGURE 16-18

(a) A thin slab of Precambrian sedimentary rocks slid at least 50 km eastward over the much younger sedimentary rocks of the high plains on the Lewis overthrust fault. Undeformed early Tertiary sedimentary deposits in the North Fork Valley show that movement was complete by Eocene time. Chief Mountain is an outlier of the overthrust slab isolated by erosion. (b) The steep upper slopes of Chief Mountain are in Precambrian sedimentary rock, and the gentler lower slopes are in the younger Cretaceous sedimentary rock. The break in slope marks the Lewis overthrust fault.

cally create vertical fault scarps resembling those that mark normal faults if they happen, for example, to offset a hill across a valley. They also offset streams and ridges, making them jog in the direction of fault movement. Because the fault plane invariably stands vertically, strike-slip faults follow a straight course across the countryside, hardly wavering as they cross hills and valleys. They make conspicuous linear elements in the landscape.

Geologists distinguish two kinds of strike-slip movement, which they recognize by imagining themselves standing on the fault and looking in either direction along the line of its trace on the ground surface (Figure 16–19). The fault is *left-handed* or **left-lateral** if its movement brings the rocks on the left side of the trace toward the observer and *right-handed* or **right-lateral** if the reverse is true. The San Andreas fault of California is right-lateral because the rocks

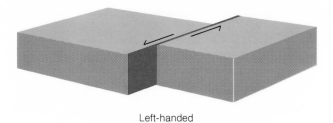

Left-handed

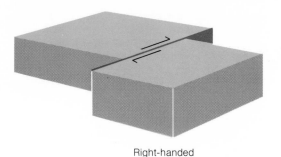

Right-handed

FIGURE 16–19

Strike-slip faults move horizontally on a vertical fracture surface in directions that bring either the right or left side of the fault toward an observer looking in either direction along its length.

on the west side of its trace move north. Imagine yourself standing on that fault and looking either north or south along its trace. No matter which way you look, the rocks on your right will be coming toward you relative to those on your left. We observe the relative, not the absolute, movement on faults.

A few strike-slip faults have displacements measured in hundreds of kilometers although most move much shorter distances. Those with large displacements are generally associated with transform plate boundaries. The San Andreas fault and the similar faults that trend parallel to it are an example. Many smaller strike-slip faults are associated with overthrusting. The edges of overthrust slabs that trend parallel to the movement direction must be strike-slip faults and many overthrust slabs have tears within them that show strike-slip displacement.

SUMMARY

Rocks may deform either as brittle solids, in which case they break; or as ductile solids, in which case they change shape permanently without breaking. Most rocks are brittle under surface conditions but become increasingly ductile as temperature and pressure increase with depth, especially if they contain water. Under sufficiently high temperatures and pressures, all rocks become ductile and the earth's crust in general responds to long-term stresses as though it had little strength. Ductile deformation of layered rocks generally creates folds. Folds vary greatly in complexity according to (1) the number and severity of the episodes of deformation that caused them and (2) the prevailing temperature and pressure conditions.

Brittle deformation breaks rocks into joints, which do not move significantly, and into folds, which do. All rocks contain joints, which typically occur in patterns composed of several sets of intersecting fractures. Faults may move in any direction; some move distances measurable in kilometers or, in a few cases, hundreds of kilometers.

Plastic and brittle deformation often occur simultaneously, as there is folding in some parts of a body of rocks and fracturing in others. The resulting patterns of folding and faulting generally reflect the directions of the stresses that created them and ultimately the plate movements responsible for those stresses.

KEY TERMS

anticline	fault	fold	joint	reverse fault	structural geologist
axial plane	fault plane	foot wall	left-lateral	right-lateral	structural style
brittle deformation	fault scarp	hanging wall	normal fault	strike	syncline
competent	fault zone	incompetent	overthrust fault	strike-slip fault	thrust fault
dip					

REVIEW QUESTIONS

1. What factors determine whether rocks bend or break when subjected to deforming forces?

2. Why is normal faulting generally regarded as evidence of crustal extension whereas thrust faulting appears to be evidence of compression?

3. If you were to investigate an eroded fold and find that the oldest rocks are exposed in its core, would you conclude that it is an anticline or a syncline?

4. Would you be more inclined to expect to find thrust faults or normal faults associated with tight folding of sedimentary rocks? Explain.

5. If you found that a fault had moved older rocks on top of younger ones, would you conclude that it had normal or reverse movement?

SUGGESTED READINGS

Billings, M. P. *Structural Geology.* 3d ed. Englewood Cliffs, New Jersey: Prentice-Hall, 1972.

A standard introductory text in descriptive structural geology and general reference for more years than most geologists can remember.

Compton, R. R. *Manual of Field Geology.* New York: John Wiley, 1962.

Describes the field techniques involved in the observations, mapping, and interpretation of deformed rocks.

Hills, E. S. *Elements of Structural Geology.* New York: John Wiley, 1963.

Another standard introductory text in the structures of folded and faulted rocks.

CHAPTER SEVENTEEN

The masonry house in the foreground was almost totally destroyed by the earthquake that struck Charleston, South Carolina in 1886.
(J. K. Hillers, U.S. Geological Survey)

EARTHQUAKES

Few experiences are as unsettling as feeling the solid ground quiver as though it had turned to jelly. This experience is likely to prove as expensive as it is dangerous because earthquakes—of all natural disasters—cause the most extensive property damage and the most casualties. Earthquakes strike suddenly, usually without warning, and in some cases violently enough to destroy cities and decimate their populations.

Earthquakes are closely associated with movement along plate boundaries. In fact, a map showing their world-wide distribution reveals almost exactly the same pattern as one outlining the lithospheric plates. However, there are exceptions that include some extraordinarily violent events: Charleston, South Carolina and Lisbon, Portugal, neither of which is near a plate boundary, have both experienced devastating earthquakes. The most violent tremors in recorded North American history struck in 1811 and 1812 in southeastern Missouri, another region remote from plate boundaries. If similar shocks were to occur there now, they would devastate Memphis and seriously damage St. Louis. The area extending north from Yellowstone Park to Helena, Montana and south to Salt Lake City is likewise remote from a plate boundary and it also experiences frequent earthquakes, including some that have been severe.

Destructive as they may be, earthquakes inform us greatly about the interior of the earth. We could not obtain the data they provide from any other source. Our deepest drill holes barely prick the surface of the earth's crust. Only earthquake shock waves actually penetrate deep into the earth and return to the surface with information about the interior. Thus, **seismology,** the branch of geology that deals with earthquakes, is the chief source of knowledge about our planet's interior. Exploration seismologists, who generate their own small earthquakes instead of waiting for the natural ones, apply the methods of earthquake seismology to the practical search for structures within the earth's crust that may contain mineral resources.

EARTHQUAKES AND FAULTS

In many cases, the ground is visibly offset along a known fault trace after an earthquake. Small fault scarps appear during an earthquake, little cliffs exist where none had been before, roads and fences suddenly acquire unplanned jogs, and surveyors find that property lines no longer conform to the descriptions on their plats. The great San Francisco earthquake of 1906, for example, accompanied sudden right-handed, strike-slip displacement (Figure 17–1) of about 5 m along the part of the San Andreas fault that passes close to the city, as much as about 7 m in Tomales Bay a few kilometers to the north. Comparable vertical movements have occurred along normal faults during great earthquakes, such as the Good Friday earthquake that devastated much of Anchorage in 1964.

Imagine a model of the earth's lithosphere consisting of two thick slabs of foam rubber abutted against each other along their cut edges. Then suppose that the cut edge is a fault and the slabs of foam rubber are segments of the lithosphere moving along it. Slabs of foam rubber will not move past each other without sticking as though they were blocks of polished wood. Instead, the cut edges will tend to catch in places and then bend and stretch until the stuck section finally pulls loose with a sudden jerk—an earthquake. That is how the rocks move along a fault and that is why fault movement produces earthquakes.

Close observation of active faults shows that they may move smoothly and continuously—a phenomenon called **tectonic creep**—and then stick and remain apparently stationary for a long time. It is those stationary faults or sections of faults that are likely to suddenly snap and produce an earthquake. For example, some lengths of the San Andreas fault creep steadily while others, including those that pass through the San Francisco and Los Angeles areas, show no apparent movement and therefore appear to be stuck. Nevertheless, ordinary surveying methods can be used to measure the slow bending of the rocks adjacent to the stationary sections of the fault, and it can be observed increasing steadily from year to year.

ELASTIC DEFORMATION

Rocks, like foam rubber slabs, steel springs, and most other solids, are capable of a certain amount of elastic deformation: They change form in response to a force and then recover their shape and release the stored **strain energy** when the deforming force is removed. However, beyond a certain point, which depends upon the characteristics of the material, elastic deformation passes into brittle or ductile deformation. Both of these are permanent and neither disappear with the deforming force nor release strain energy when it is removed. We can think of seis-

mology as being the study of elastic deformation in rocks as opposed to structural geology, which deals with brittle and ductile deformation.

Thinking of rocks as elastic solids may seem strange because they generally seem brittle. However, anyone who has watched a large sheet of plate glass waver as it is eased into position has observed that glass, a silicate substance chemically comparable to most rocks, bends quite easily. Furthermore, the elastic properties of fiberglass are widely exploited in bows, vaulting poles, and fishing rods.

As rocks slowly flex near the stuck section of a fault (Figure 17–2), they store energy as though they were a drawn bow. We can measure that slow bending and the amount of stored energy with instruments called **strain gauges.** One type consists essentially of a fine wire tightly stretched between two rigid pins set in a rock outcrop. As the rock changes shape, it stretches the wire, drawing it thinner and changing its electrical resistance, a change that can be measured with precision. Some of that energy stored in the rock will eventually be used in creating fractures or folds. However, a large proportion of the stored energy will be released as earthquake shock waves when the fault finally slips into the position it would have reached had it been moving steadily through tectonic creep.

SEISMOGRAPHS

The sudden jolt of a fault breaking loose and slipping strikes the earth a sharp blow, starting a complex series of earthquake shock wave vibrations called **seismic waves**. Such waves spread across the earth's surface like ripples on a pond and echo through its interior, sometimes for hours after the event. Many have compared the echoing to the ringing of a gong. However, if a tape recording of earthquake shock waves is played at a speed that brings them within the audible frequency range, they sound more nearly like a dull thud. The earth rings like a lead gong.

Detecting and recording those seismic waves poses an interesting problem. Our normal procedure of planting ourselves firmly on the solid earth and then watching the moving object pass fails if the earth is itself the moving object. The general solution consists basically of a heavy weight suspended from a weak spring (Figure 17–3) that couples the weight so loosely to the earth that it hangs relatively

FIGURE 17-1

Fence offset by right-lateral movement on the San Andreas fault during the 1906 San Francisco earthquake. The man is standing on the trace of the fault facing along its length. The side on his right moved toward him and vertically upwards. (G. K. Gilbert, U.S. Geological Survey)

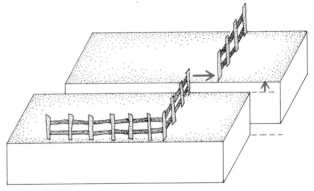

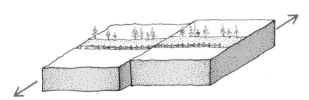

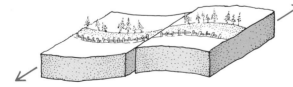

FIGURE 17-2 (left)

When a section of a fault sticks, the rocks around it slowly bend until they accumulate enough strain energy to break the stuck section loose. Then the fault slips to where it would have been had it not stuck, with a sudden jolt that releases the stored energy as earthquake waves.

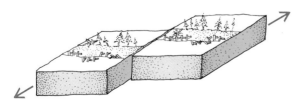

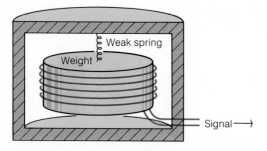

FIGURE 17-3

One simple way to build a seismograph: If the weight suspended from the spring is a magnet, its motion within the surrounding coil will generate a signal in the form of an electrical current that can be amplified and recorded. This is only one of many possible designs.

FIGURE 17-4

Part of a seismogram recorded at the earthquake observatory of the University of Montana showing a rhythmic pattern of small vibrations. Their origin is unknown but they may record storm surf pounding against the Pacific coast. Observatory seismographs are sensitive enough to detect ground displacements so small that a line of the same length could not be seen under a powerful microscope. The regularly spaced dots above the trace of the instrument pen mark minutes.

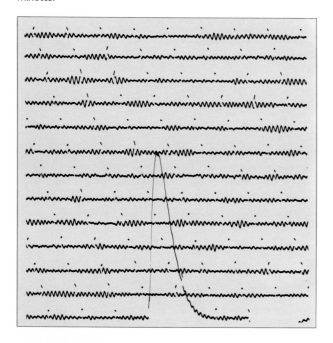

motionless while the ground shakes beneath. Thoughtful design and electronic amplification enable such instruments, called **seismographs,** to record extremely small ground motions, such as those in the seismogram shown in Figure 17–4. Most seismographs are designed to sense ground movement in only one direction and within a restricted range of frequencies. Thus, a complete seismic observatory contains a number of instruments.

SEISMIC WAVE MOTION

Compressional Waves

Seismograms, such as that in Figure 17–8, show that earthquakes consist of three different types of wave motion. The simplest are the compressional waves, often called **P waves,** which travel like sound waves in air as a series of alternating compressions and expansions. P waves pass through the body of the earth and through any kind of matter, whether it is solid, liquid, or gas. Figure 17–5 shows that as a P wave passes, the particles of rock move back and forth in a direction parallel to that in which the wave is traveling.

Shear Waves

Another kind of seismic waves called **shear waves,** or **S waves,** move laterally in a twisting motion transverse to the direction of wave advance. Figure 17–6 shows the relationship between the movement of individual particles of rock and the direction of wave advance schematically. Sometimes that style of movement becomes vividly apparent to farmers who happen to be working their fields during an earthquake and see their straight rows wriggling before the tractor as the shear waves pass. It is a most disconcerting experience. Like P waves, shear waves also pass through the body of the earth but only through its solid portions; they do not penetrate liquids.

Surface Waves

Figure 17–7 depicts **surface waves,** which somewhat resemble broad swells on the open ocean and travel along the surface of the earth without passing through the interior.

The Earthquake Experience

Sudden movement on a fault starts all three kinds of wave motion at once, and people near their source feel them as a confused jumble of ground motions. However, each type of seismic wave travels at its own speed so they soon separate. People or seismographs some distance from the source of an earthquake detect its shock waves in sequence with the fastest coming first and the slowest last. The compressional waves strike first as a series of sharp jolts. If you are inside, you may feel as if a truck has just collided with the house. Then a few seconds of quiet follow while plaster dust sifts out of the ceiling and the cat scurries for cover. Next come the shear waves with their wriggling motion that waltzes dishes off the shelf and snaps chimneys, dropping their debris through the roof. Surface waves travel slowest and follow the longest path around the earth so they arrive last. Surface waves feel like a long series of ground swells that rock everything slowly back and forth, toppling walls already weakened by the compressional and shear waves. Some people feel seasick while the surface waves pass.

Earthquakes, even big earthquakes, never last more than a few minutes at most. However, lesser earthquakes called **aftershocks** may continue for several days after the main event.

Locating Earthquakes

Figure 17–8 shows a seismogram with a quiet interval between arrival of the compressional and shear waves. The length of that interval increases as the shock waves move farther from their source. Seismologists use its length as a means of calculating the distance to the earthquake. However, the interval between arrival of the P waves and S waves does not indicate the direction from which they came. Therefore, it is possible only to draw a circle, such as one of those shown in Figure 17–9, with the seismograph station at its center and the distance to the earthquake as its radius. The earthquake must have occurred somewhere along the circumference of such a circle.

Another circle drawn in the same way from another seismograph station will intersect the first at two points, one of which must lie near the site of the earthquake. A third circle drawn from still another station will intersect the

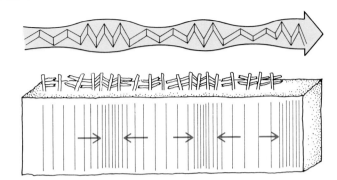

FIGURE 17–5
P waves travel as a series of compressions and expansions that move the individual particles of rock back and forth in the direction the wave is traveling.

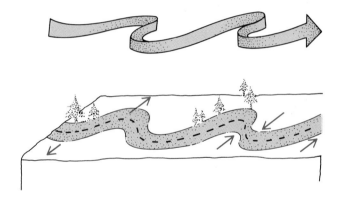

FIGURE 17–6
Shear waves pass with a wriggling motion transverse to the direction of wave advance.

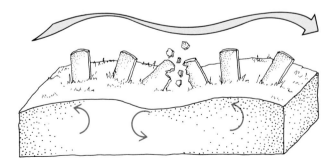

FIGURE 17–7
Surface waves pass with a rolling motion similar to that of swells on the open beach.

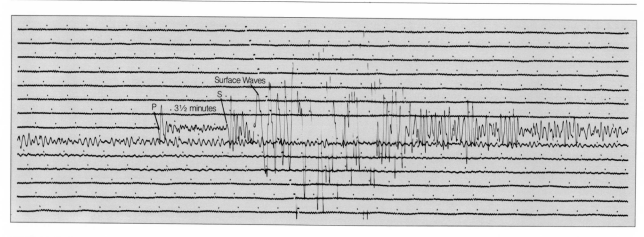

FIGURE 17–8

Seismic waves from an earthquake in the Gulf of California west of Hermosillo, Mexico as recorded at the earthquake observatory of the University of Montana on November 20, 1977. The time interval between the P- and S-wave arrivals shows that the event occurred at an angular distance of about 19.5°. Comparison of that interval with the amplitude of the ground motion shows that it had a magnitude of 5.6. The vibrations on the line below that in which the earthquake appears are continued ground motion from the same event recorded during the next hour. (A. I. Qamar)

first two at only one point. That definitely locates the **epicenter** of the earthquake, the point on the earth's surface directly above the internal site of actual movement, which is called the **focus**. Figure 17–10 illustrates the distinction between the epicenter and the focus.

A central clearing house collects reports on the time interval between arrival of the compressional and shear waves from a number of seismograph observatories immediately after every major earthquake. With that information it is possible to figure out not only the location but, as we shall see in a following section, also the size of the earthquake in a short time. That explains why newspapers are often able to report the location and size of a large earthquake before any information arrives from the actual scene.

Focal Depth

Of course, the time interval between the arrivals of the first compressional and shear waves actually reflects the distance from the focus of the earthquake—not from the epi-

center—to the seismograph station. If the earthquake was fairly shallow and fairly remote, the difference between the distance from the seismograph to the focus and to the epicenter is too small to matter. However, that difference may become significant if the earthquake was deep and relatively close to the seismograph station. In that case, distance circles drawn from the seismograph stations in a vertical plane instead of on the map will intersect within the earth at the focus of the earthquake and thus reveal its depth. Figure 17–11 illustrates the point. Lithospheric plates sinking along convergent boundaries typically generate deep-focus earthquakes, a few from depths as great as 700 km.

Direction of Fault Movement

Try kicking a rug to see how that pushes it in one direction while pulling it in the other. Fault movements likewise push the rocks in some directions, sending a compression as the first P-wave motion, and pull them in other directions, sending an expansion as the first motion. It is possible to

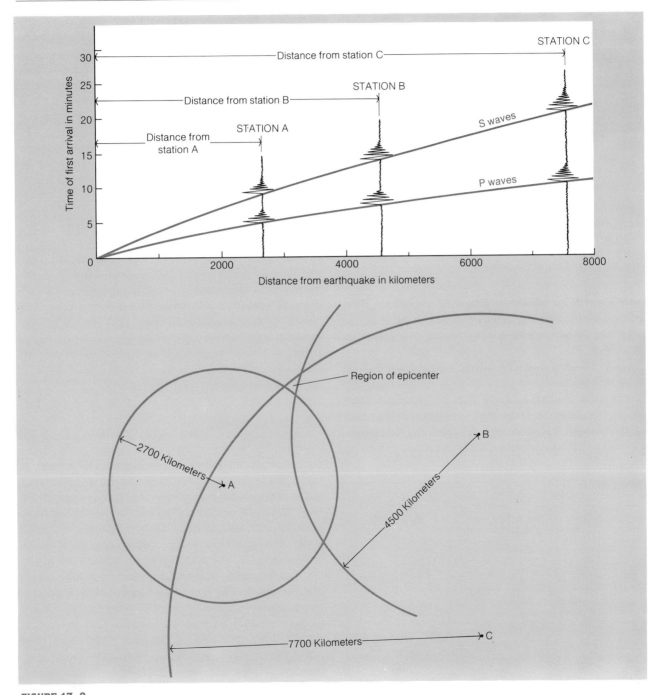

FIGURE 17-9

The time interval between the P- and S-wave arrivals increases with distance from the earthquake, making it possible to draw a circle of appropriate radius from any seismograph station that received the vibrations knowing that the event must have occurred somewhere on its perimeter. The area where three such circles intersect uniquely locates the epicenter. There is commonly, as shown here, some uncertainty in that location, which further analysis of the data generally reduces.

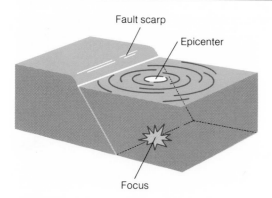

FIGURE 17-10

The **focus** of an earthquake is the point of actual fault movement within the earth. The **epicenter** is the point on the earth's surface directly above the focus. There will be surface displacement of the fault scarp only if the focus is at the surface.

distinguish those movement directions on seismograms and to infer from them both the orientation of the fault that produced the earthquake and the relative direction of its movement.

Figure 17–12 shows a hypothetical distribution of first-wave motions as recorded by an array of seismographs distributed around the epicenter of an earthquake. In such a case, the pattern could be interpreted either as the result of right-handed movement on a northwest trending strike-slip fault or left-handed movement on a northeast trending

FIGURE 17-11

If the distance circles to an earthquake are drawn on a vertical section through the earth instead of on map as in Figure 17-9, they will intersect at the focal point within the earth.

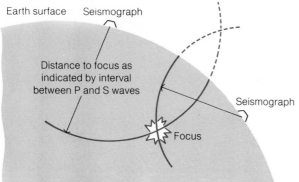

fault. Such patterns are always ambiguous in that way; the dilemma is inherent in the nature of the data. In many cases, it is possible to resolve such ambiguities if the earthquake originated along a known fault. Even if the fault is not shown on the geologic maps, it may be possible to resolve the problem through knowledge of the structural style of the region, its distinctive pattern of faulting and folding. For example, if the hypothetical data shown in Figure 17–12 had been collected in southeastern Oregon where northwest trending strike-slip faults are common and northeast trending structures unknown, the most probable choice would be clearly apparent.

Detecting Nuclear Tests

First-motion studies of compressional waves also help detect nuclear explosions, which cause moderate earthquakes easily detectable by sensitive seismographs a long distance from the event. However, explosions differ from natural earthquakes because they push the rocks outwards in all directions. Therefore, seismographs everywhere record a compression as the first-wave motion. Blasts are also distinctive in producing relatively little surface-wave motion.

Seismic Sea Waves

If slippage on a fault shifts a large section of the ocean floor vertically, it suddenly displaces the water above that part of the bottom thus starting a wave. Large underwater landslides or volcanic eruptions may have a similar effect. Such waves race across the open ocean at speeds as great as 800 km per hour, depending upon the water depth. As long as they remain in deep water, those waves remain very low and ships generally notice nothing unusual as they pass. However, the same wave may pile up to great height when it gets into shallow water, especially along coasts that have a gently shelving bottom that extends far offshore. Most people call these phenomena tidal waves, which is misleading because such waves have nothing to do with tides. Many authors prefer to call them **tsunami,** a Japanese word that means ''harbor wave.'' The expression **seismic sea wave** is at least aptly expressive even if a bit cumbersome. Seismic sea waves may cause far more damage than the earthquakes that started

  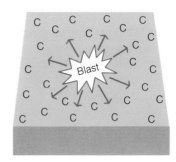

FIGURE 17-12

Seismographs arrayed around an earthquake epicenter will register compressions (C) and expansions (E) as the first P-wave motion in the same pattern for the two kinds of strike-slip displacement shown above. Resolution of the ambiguity depends upon observation of surface displacement or knowledge of the local pattern of fault movements. Other types of faults likewise create distinctive patterns of first P-wave motion, which also aid in determining their trends and movement directions. Blasts send compressions as the first P-wave motion in all directions.

them as they suddenly flood coastlines as far as thousands of kilometers from their source. They are especially common in the Pacific but have been known to occur in all oceans.

If the fault displacement dropped a section of the ocean floor, the first wave to arrive will be a trough that drops the water level far below the lowest tide. When that happened after the Lisbon earthquake of 1755, large numbers of people seized the sudden opportunity for clam digging and were then washed away when the correspondingly high wave crest arrived less than half an hour later. If the fault movement raised a section of the ocean floor, the first seismic sea wave arrival will be a crest. In most cases, a series of successively higher wave crests follow the first one during a period of several hours so people who flee for high ground are well advised to keep moving.

An international warning system issues a seismic sea wave alert every time a major earthquake shakes the floor of the Pacific Ocean. Many of those warnings turn out to be false alarms because it is impossible to determine in the short time available whether the fault displaced the ocean floor vertically, in which case it would generate a seismic sea wave, or horizontally, in which case it would not. Nevertheless, people who live along vulnerable coasts should ensure that they observe seismic sea waves from high ground.

EARTHQUAKE INTENSITY AND MAGNITUDE

To most people—especially to those in the affected area—the most important aspect of an earthquake is its **intensity,** which is a measure of how violently the ground shook, what collapsed, and what remained standing. However, it is of greater scientific importance to measure an earthquake's **magnitude,** which is the amount of energy it released.

Earthquake Intensity

Seismologists estimate the violence of local ground movement by using the **Modified Mercalli Intensity Scale** to grade the severity of damage in degrees. The least-intense earthquakes can be detected only by instruments and cause no damage; earthquakes then range upwards to those that cause total demolition of all structures.

As we might expect, earthquake intensity maps generally show progressively less damage with increasing distance from the epicenter. In most cases, the pattern of intensity distribution also reflects the trend of the fault and the stability of the underlying bedrock. For example, intensity maps of the 1906 San Francisco earthquake show areas of most-severe damage along the trend of the San

Andreas fault, and also in areas along the shore that are underlain by soft sediments or unconsolidated fill.

Visualize the effect of differing ground conditions on earthquake intensity by imagining yourself rapping the rim of a large bowl filled with soft custard and then watching the pudding quiver while the rigid bowl hardly moves. Then compare the custard to loose sediments or uncompacted fill and the bowl to the solid bedrock. The point is nicely illustrated by a number of occasions in which miners have emerged from their shift to find great surface devastation from an earthquake that they hardly felt while working within the solid rock underground. Developers and land use planners operating in areas of great earthquake hazard should keep the soft custard in mind and avoid building on sites underlain with loose sediments or fill.

Knowing an earthquake's intensity does not solve the problem of figuring out how big it was because a large earthquake some distance away may produce the same local effects as a much smaller one nearby. Neither does the intensity scale contribute to the investigation of earthquakes coming from epicenters underwater or in uninhabited regions where there is no structural damage to register the violence of ground motion. Seismologists also need an absolute scale of measurement to grade earthquakes in terms of the amount of energy they release regardless of how much local ground motion they may cause. The problem of determining the absolute magnitudes of earthquakes resembles that of interpreting ripples on a lake.

The Richter Magnitude Scale

A small ground motion may come from a minor earthquake nearby or from a major one at a distance just as a small ripple on an otherwise calm water surface may tell of a minnow splashing nearby or of a large fish jumping some distance away. There is no way to determine what the ground motion or the ripple may mean without knowing how far they have traveled. Earthquake magnitude scales relate local ground motion to the distance from the epicenter and therefore indicate the amount of energy the earthquake released.

Seismologists generally use the **Richter Magnitude Scale** in which each whole number jump reflects a 10-fold

increase in the amplitude of the local ground motion, which in turn reflects an approximately 30-fold increase in the amount of energy released at the earthquake focus. The actual relationship between magnitude and energy release depends to a large extent upon the type of earthquake, upon such factors as the length of fault surface that slipped.

In principle, records from all seismograph observatories should give the same magnitude figure regardless of their location because the amount of ground motion should decrease predictably with increasing distance from the focus. However, magnitude estimates do vary somewhat from one observatory to another because local variations in the type of bedrock affect the behavior of seismic waves. Magnitude estimates also vary according to which kind of seismic wave motion is used in the measurement. Nevertheless, it is possible to obtain a reasonably accurate measurement of the energy released by a distant earthquake, or by a nuclear test, from any distance at which seismographs can detect the ground motion.

Any earthquake with a Richter magnitude greater than about 5.5 is likely to damage buildings; those with Richter magnitudes much greater than 6 are generally quite destructive if they strike in populated regions. Earthquakes of Richter magnitude 8 release approximately as much energy as 1 billion tons of TNT and strike somewhere in the world once in every several years. The largest earthquakes so far recorded have Richter magnitudes of 8.9. Many seismologists suspect that may be the greatest possible earthquake because they believe the lithosphere is too weak to withstand any greater stress without moving.

ECHOES FROM WITHIN

Seismographs record a large earthquake as a long series of vibrations only a few of which come directly from the focus of an earthquake to the instrument. The others are secondary waves reflected as echoes from the surface of the earth or from surfaces within the earth, contacts between different kinds of rocks that carry seismic waves at different speeds. Figure 17–13 shows how seismologists interpret complex seismograms to trace the many paths seismic waves may follow in passing from their source to the seismograph. Such interpretations are a means of investigating the internal structure and composition of the earth.

The Moho

The simplest and most persistent echo comes from a surface generally called the *Moho* after the Yugoslav seismologist named Mohorovicic who discovered it many years ago while studying seismograms of shock waves from nearby quarry blasts. In most continental regions the Moho lies at a depth of about 40 to 50 km and many geologists interpret it as a reflection from the base of the sialic continental crust (Figure 17–14). Others disagree and contend that the Moho is at the base of a layer of basaltic material that may form the lowest part of the continental crust. A similar echo returns from a surface at a depth of about 10 km beneath the ocean floor. Most geologists interpret that echo as the base of the oceanic crust although there is some disagreement about what change in rock types it may represent.

The Low Velocity Zone

Figure 17–15 illustrates the general tendency of seismic waves to accelerate as they penetrate more deeply into the mantle except in the so-called **low velocity zone,** which

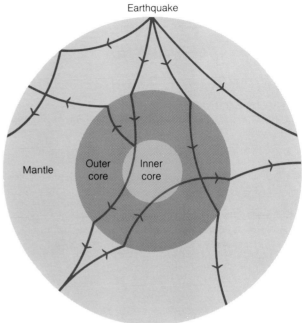

FIGURE 17–13

A diagram showing some of the paths seismic waves follow as they reflect off the surface of the earth, the core, and the inner core in various combinations. S waves do not pass through the core, apparently because its outer part is liquid. All the paths are curved because seismic wave velocity increases as the rocks become denser at depth.

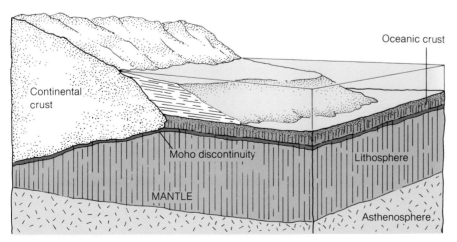

FIGURE 17–14

The Moho discontinuity marks the base of the rather thick continental and much thinner oceanic crusts on the upper surface of the lithosphere. It strongly reflects seismic waves because the peridotites in the mantle propagate them at much higher velocities than do the crustal rocks above.

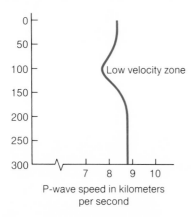

FIGURE 17-15
Seismologists discovered the low velocity zone in the upper mantle many years before it was equated with the existence of the asthenosphere.

FIGURE 17-16
The shadow zone is a ring-shaped area of the earth where seismographs detect no vibrations because they lie beyond the range of seismic waves that graze the core as they pass through the mantle. However, some seismic waves pass through the core, which focuses them into the area directly opposite the earthquake. Therefore, instruments within the ring of the shadow zone detect strong vibrations.

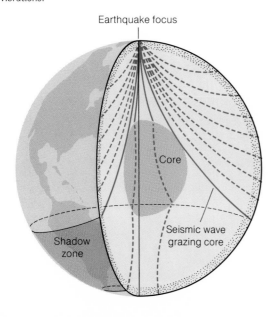

lies at a depth between about 100 and 250 km. Seismic waves that follow paths through long stretches of the low velocity zone not only slow down considerably but also lose a large part of their shear wave component. However, they do not reflect as echoes. Both the slow speed and the poor propagation of shear waves suggest that the low velocity zone consists of weak rocks that are probably at least partly melted. Of course, the low velocity zone is now recognized as the asthenosphere on which lithospheric plates glide about the surface of the earth.

The Core

Seismographs receiving vibrations from large earthquakes generally record an echo that comes from a depth of about 2,900 km and apparently reflects from the surface of the earth's core. The core also creates a **shadow zone** (Figure 17-16) on the side of the earth opposite the earthquake where seismographs lie beyond the range of shock waves that barely graze the core and therefore detect nothing. However, seismographs located near the middle of the shadow zone in the area directly opposite the earthquake commonly record strong vibrations focused into that region by the core, which acts as though it were a lens embedded within the planet. However, only compressional waves pass through the core so it seems that at least the outer part of it is liquid and therefore incapable of propagating shear waves. There are reasons to suspect that the inner part of the core is solid but these reasons lie considerably beyond the scope of this discussion.

The Composition of the Earth

Seismic waves travel at characteristic speeds through different kinds of rocks. Thus, the time they spend passing from their source to a seismograph provides an excellent clue to the types of rocks along the path. Travel times through the continental crust are what one would expect of seismic waves passing through sialic rocks and thus confirm the view of continents that we gain through direct observation at the surface. The considerably faster passage of seismic waves through the mantle suggests that peridotite does indeed compose most of that part of the earth. Wherever we find surface exposures of rocks that appear to have come from the mantle, they do consist of peridotite. Seismic wave velocities in the earth's core are consistent with the assumption that it is made up of an alloy of iron and nickel similar to that found in many meteorites.

If we assume that the earth's core does indeed consist of an alloy of iron and nickel and that the mantle is composed of peridotite, then we can calculate the mass and density that a planet so constituted should have. The calculated mass corresponds almost exactly to the known mass of the earth as determined through measuring its gravitational field. Furthermore, the calculated distribution of mass with depth corresponds to that determined from observation of the earth's rotation about its axis. We see that completely independent lines of evidence bearing on the earth's internal composition converge from different directions to suggest that the planet does indeed have an iron-nickel core embedded within a peridotite mantle. Although the evidence is almost entirely circumstantial, it nevertheless assembles into a convincing picture of the earth's interior.

Seismic Exploration

Geologists solve many routine problems by extending the principles of earthquake seismology to the analysis of artificial shock waves. They locate crustal structures and identify concealed masses of rock in basically the same way that seismologists locate the boundary between the mantle and core and identify the low velocity zone.

One of the simplest applications of seismic techniques to geologic problems is in finding the depth to bedrock, an important problem for those concerned with designing large excavations or heavy buildings. The usual technique involves generating a small shock wave at the surface—a sledgehammer blow suffices in many cases—and then using a portable seismograph to detect the echo returning from the bedrock surface. The instrument measures the time it takes for the echo to return and that indicates the depth of the soil cover accurately enough for most purposes. The cost is a small fraction of that of drilling a test hole.

Figure 17–17 schematically illustrates a different technique called **seismic refraction,** which has many applications, one of the simplest being to determine the depth of the water table. Geologists probing for the water table often use a sledgehammer or large firecracker to generate

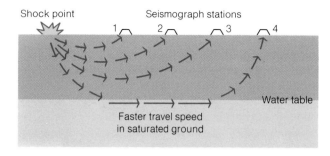

FIGURE 17–17

Using the seismic refraction technique to probe for the water table. As the seismograph steps farther from the shock point, the P-wave travel time increases in direct proportion to the distance until the instrument reaches station 4. There, and at points beyond, it receives P waves that travel part of the distance through the higher velocity material below the water table. Therefore, the P waves arrive sooner than they would had they traveled all the way through dry ground.

their seismic waves. They move the shock point farther from the instrument with each successive blow. As the shock point moves farther away, the seismograph receives wave motion that passed through deeper levels of the ground. The travel velocity of the shock waves increases sharply when the instrument begins to receive those that passed beneath the water table because seismic wave velocity is much greater in water-saturated ground than in dry ground.

The principal use for both reflection and refraction techniques is in the search for buried structures that may contain trapped oil or gas. The equipment consists basically of a number—as many as several dozen—portable seismographs arranged in geometric array and connected to a central recording station mounted in a truck or on a ship if the work is being done at sea. Shock waves powerful enough to penetrate deep into the earth's crust return echoes from numerous reflecting horizons or change velocity with depth if refraction techniques are used. In most projects the array of seismographs is moved along a line to create a more or less continuous profile of the structures at depth. Exploration seismic work is expensive but nevertheless profitable because drilling holes to obtain similar information would cost far more.

PREDICTING AND MANAGING EARTHQUAKES

Until recently, earthquakes seemed the most capricious and uncontrollable of all natural events. Now, after many years of research, the goal of predicting at least some kinds of earthquakes seems almost within our grasp. Geological engineers talk seriously of preventing or controlling at least some of them. As so often happens in scientific research, some of those results came from unexpected sources.

People who live in California expect to experience earthquakes and, if they are wise, design their buildings with that expectation in mind. People who live in New York merely expect to read in their newspapers about earthquakes that struck elsewhere. In a sense, that general sense of the occurrence of earthquakes is a rudimentary form of earthquake prediction in the same way that people who live in North Dakota can predict cold weather during the winter. But that sort of general expectation does not solve the more specific problem of forecasting exactly when, where, and how violently an earthquake or a blizzard may strike.

It is possible to anticipate events a bit more specifically in areas where the active faults are well known and under close observation. For example, geologists have known the trace of the San Andreas fault for more than 80 years and have watched most of it closely ever since the great 1906 earthquake devastated San Francisco. The only segments of the fault that seem to pose an imminent threat of releasing a large earthquake are those that have been locked and therefore accumulating strain for many years; those that are showing no apparent movement. It is impossible to predict theoretically when they may break loose and slip because there is no way to tell how tightly they are stuck. Nevertheless, knowing how long the fault has been stuck does at least enable seismologists to estimate how great the earthquake will be when it does finally strike. Suppose we know how fast a fault moves and when a segment of its length stuck. We can then easily predict how far it will slip and how great an earthquake it will cause if it slips at any given future date even though we cannot yet predict that date.

Earthquake Forecasts

For many years, close surveillance of active faults by arrays of seismographs seemed the avenue of research most likely to lead to a method of forecasting earthquakes. The idea is to learn to detect and recognize whatever distinctive kinds of minor earthquakes the rocks might release as they begin to break during the days or weeks before the fault finally slips and releases a major earthquake. However, that approach has not produced satisfying results and the prospects of its doing so now seem bleak. Even if it did work, the method would suffer from the serious disadvantage of being too expensive to use except locally in areas of greatest risk. However, other approaches have begun to emerge from observations that seemed irrelevant until a few years ago.

Many survivors recalled having heard unusual numbers of dogs howling mournfully in the streets of San Francisco during the night before the disastrous earthquake of 1906. In fact, reports of animals behaving erratically during the hours before a major earthquake are so numerous that it seems possible they may actually detect something that gives them a strong sense of foreboding. The Chinese have made a systematic effort to develop a system of earthquake prediction based on observation of animals and did successfully alert the population of Haicheng several hours before a major earthquake flattened 90 percent of the city on January 28, 1975. The casualty toll was remarkably small for so large and devastating an earthquake; there is little doubt that the forecast saved many thousands of lives.

Within recent years, an entirely different method of earthquake prediction based on fracturing of the rocks near a fault has been developed in the Soviet Union and widely pursued elsewhere. It seems that formation of new fractures in the heavily strained rocks near a locked fault expands and weakens them. The expansion causes an increase in surface elevation that may amount to several centimeters, and the weakening slows passage of seismic waves through the area. Those first warnings may come months or perhaps years in advance of the earthquake.

In the days immediately preceding the earthquake the rate of fracturing near the fault increases rapidly. Ground water seeps into the newly formed fractures thus lowering the static water level in wells and increasing the rate at which seismic waves pass through the area. Meanwhile, the rapidly fracturing rocks lose large quantities of radon gas, one of the radioactive intermediate steps in the long decay chain from uranium to lead. The radioactive gas raises the level of background radiation in well water. Therefore, a sharp drop in the water level in wells accompanied by a rapid rise in the velocities of seismic waves passing through the area and by a rapid increase in background radiation appear to indicate that an earthquake will strike within a few days. The size of the affected area seems to at least roughly indicate the magnitude of the impending event.

Unfortunately, no method of forecasting earthquakes can promise more than limited success for a variety of reasons. Most regions of active tectonic movement contain numerous faults at least some of which are likely to escape observation until after they have already produced an earthquake. Furthermore, many destructive earthquakes originate at depths far beyond the reach of surface observation and many involve movement on faults that do not break the surface. Neither is there much hope of predicting the occasional destructive earthquakes that devastate areas remote from plate boundaries, such as those that have struck Boston, Charleston, and southeastern Missouri. Perhaps it is just as well that the prospects for successfully predicting earthquakes are so limited because many thoughtful people fear that a successful forecasting method might create serious problems.

The prospect of millions of people abruptly fleeing a large city, leaving it to looters and descending as refugees on a countryside unprepared to receive them is almost as harrowing as that of an earthquake. Long-range forecasts of earthquakes might cause severe economic damage by disrupting normal business operations and depleting inventories. Unscrupulous real estate and insurance operators could pillage worried people during the months between issuance of a long-range forecast and the actual earthquake—if there were an earthquake; an earthquake forecast, like a weather forecast, might be wrong. That happened in China after the authorities ordered the partial evacuation of Peking because seismologists predicted that devastating aftershocks would follow the catastrophic Tangshan earthquake of July 28, 1976. Millions of people camped out in the streets and parks until mid-August, waiting for an earthquake that never came. The political problems involved in handling earthquake forecasts will be extremely delicate and difficult to handle.

Precautions against Earthquakes

Sensible land use planning and rigorous enforcement of stringent building design codes provide the best protection against earthquakes. Figure 17–18 shows an example of disastrously poor land use planning, a large tract of new homes built directly across the trace of the San Andreas fault south of San Francisco. Much of that ground should have been reserved for use as parks that would have been virtually immune to earthquake damage. Furthermore, the area around San Francisco Bay contains many examples of extensive development on soft and water-saturated sediments that will shake with exceptional violence during earthquakes. Building codes in earthquake prone areas should forbid construction of load-supporting masonry walls because they collapse into heaps of rubble during earthquakes while wood- or steel-framed and reinforced concrete structures generally survive. The big earthquake casualty tolls typically come from regions where people live in stone, brick, or mud houses that bury their occupants as they fall. Many earthquakes also start large fires that burn unchecked because the fire department has difficulty negotiating rubble-filled streets and finding water mains that work.

Managing Earthquakes

The idea of managing earthquakes seemed preposterous until the U.S. Army inadvertently experimented with the problem in 1962 when it began to dispose of waste liquids, mostly nerve gas, by pumping them down a deep disposal well at the Rocky Mountain Arsenal near Denver. Swarms of minor earthquakes began almost immediately and later analysis of the records by members of the U.S. Geological Survey showed that the earthquake swarms occurred almost synchronously with waste liquid disposal. Evidently,

FIGURE 17-18

View along the line of the San Andreas fault in the San Francisco peninsula. The fault runs almost exactly down the middle of the picture, beneath the lake in the background and the new housing developments in the foreground. (R. E. Wallace, U.S. Geological Survey)

the liquids were lubricating a fault zone known to exist near the bottom of the well, permitting it to slip. The Geological Survey later conducted more deliberate experiments with known faults in oil fields, using water instead of nerve gas as a lubricant, and succeeded in moving stuck faults, causing more minor earthquakes. We can reasonably assume that the method might succeed in any fault zone locked within the reach of injection wells, to depths as great as approximately 8,000 m.

Some geologists envision a well-planned series of water-injection wells gently unlocking the stuck sections of the San Andreas fault in the Los Angeles or San Francisco areas and releasing its stored energy in a series of small tremors rather than in one great earthquake. Success would certainly save thousands of lives and untold costs in property damage. Unfortunately, no one can offer absolute assurance that such a project would indeed release the fault gently. The whole thing might rip loose at once, exposing the people managing the project to blame for causing a great earthquake. It would probably do them little good to argue that the earthquake would have come eventually regardless of the injection wells; that the lubrication project minimized losses by triggering the earthquake before it could develop even more potential. Perhaps we can wait until after the earthquake and then lubricate the fault, hoping to prevent its sticking again.

SUMMARY

Earthquakes occur when a fault that had been stuck suddenly slips. The movement releases strain energy that had been stored in elastically deformed rocks near the stuck section of the fault as a series of shock waves.

The shock waves consist of: (1) compressional waves, which move as a series of alternating compressions and expansions of the rock; (2) shear waves, which move in a direction transverse to that in which the particles of rock are moving; and (3) surface waves, which resemble swells on the open ocean. Compressional and shear waves travel through the body of the earth whereas surface waves move across its surface. Shear waves cannot pass through liquids and their disappearance reveals the existence of melted rock somewhere along their paths.

Compressional waves move fastest so they arrive at a seismograph first, followed by the shear waves and finally by the surface waves, which not only move more slowly but also follow a longer path around the earth's surface. Seismologists use the time interval between arrival of the compressional and shear waves as a measure of the distance to an earthquake. Such measurements made at three different seismograph observatories make it possible to locate the source of an earthquake by triangulation.

An array of seismograph stations surrounding an earthquake permits observation of the regional pattern of first compressional wave motion. Such data can be used to infer the orientation of the fault and the direction of its movement and can also distinguish between natural earthquakes and explosion blasts.

Echoes of seismic waves from reflecting surfaces within the earth reveal the inner structure of the planet. The speeds at which seismic waves move provide information about the identities of the rocks they pass through. Such data suggest that the earth consists of an iron-nickel core surrounded by a peridotite mantle. That interpretation is thoroughly consistent with other data obtained independently through other types of observation.

Study of earthquake wave velocities and propagation of shear waves reveals the existence of a low velocity zone in the mantle, which consists of partially melted rock. The low velocity zone corresponds to the asthenosphere.

Earthquake forecasting may soon become possible on at least those faults that are locked closely enough to the surface to permit observation. It is at least conceivable that the behavior of some faults may be controlled by lubricating them with water injected through wells. However, the best precaution against injury and damage from earthquakes is sensible land use planning to restrict development of the most hazardous areas and building design codes that require steel, wood frame, or reinforced concrete construction.

KEY TERMS

aftershock	seismic sea wave
epicenter	seismic wave
focus	seismograph
intensity	seismology
low velocity zone	shadow zone
magnitude	shear wave
Modified Mercalli Intensity Scale	strain energy
P wave	strain gauge
Richter Magnitude Scale	surface wave
S wave (shear wave)	tectonic creep
seismic refraction	tsunami

REVIEW QUESTIONS

1. What kind of fault movement generates earthquakes and what kind does not?

2. What does the absence of S waves from the seismograph record of an earthquake imply?

3. Why are equally large earthquakes far more destructive to human life and property in some countries than in others?

4. What kind of observational evidence makes it possible to distinguish those parts of a large fault that are capable of generating a large earthquake in the near future from those parts that are not?

5. If you were concerned about earthquake hazards, how might you select a homesite that would minimize the in-tensity of ground motion? Could such a selection influence the magnitude of any earthquake that might strike?

6. What kinds of seismograph data would you need to ascertain the location and magnitude of a distant earthquake? Could that data also provide information about the intensity of the event?

7. Why do some large earthquakes on the ocean floor generate seismic sea waves whereas others do not?

8. Explain why some earthquakes are preceded by a drop in the ground-water level accompanied by an increase in the level of radioactivity.

SUGGESTED READINGS

Clark, S. P. *Structure of the Earth.* Englewood Cliffs, N.J.: Prentice-Hall, 1970.

An excellent treatise on the internal composition and layering of the earth.

Hodgson, J. H. *Earthquakes and Earth Structure.* Englewood Cliffs, N.J.: Prentice-Hall, 1964.

An excellent nontechnical book.

Iacopi, R. *Earthquake Country.* Menlo Park, Ca.: Lane Book Company, 1964.

A detailed discussion of the San Andreas fault in particular and earthquakes in general. Written for a non-technical audience.

Richter, C. F. *Elementary Seismology.* San Francisco: W. H. Freeman & Co., 1958.

A standard text on seismology.

CHAPTER EIGHTEEN

The most abundant rock type at the earth's surface: Pillow basalt erupted at a divergent plate boundary in the rift at the crest of an oceanic ridge. This outcrop is in the San Juan Islands, Washington. (D. W. Hyndman)

TRENCHES AND MOUNTAINS

Shortly after the middle of the last century, geologists discovered that many mountain ranges contain enormous thicknesses of sedimentary rocks that appear to have been deposited in seawater. Because mountain ranges tend to be linear and tend to follow the margins of continents, those early geologists concluded that the first step in mountain-building is deposition of sediments in a deep linear trough trending roughly parallel to the edge of a continent.

GEOSYNCLINES

Early geologists called those hypothetical troughs **geosynclines** and later generations elaborated the concept, distinguishing a number of different types. The most important were: **eugeosynclines,** which filled with oceanic sediments associated with basalt lava flows and other volcanic rocks; and **miogeosynclines,** which filled with sediments that accumulated in much shallower water close to shore and lack volcanic rocks. We now recognize that eugeosynclinal assemblages actually form mostly on the deep ocean floor and that miogeosynclinal sediments accumulate on the continental shelf. But those are relatively recent insights.

Over the course of more than a century, geologists developed increasingly complex theories explaining the formation of eugeosynclinal troughs: How such troughs might be created along the edges of continents, fill with sediments, and then rise to convert the sedimentary filling into a mountain range, which in many cases contains a core of metamorphosed sediments and granite. Most of the various theories viewed the earth's crust as an "elastic skin" that might rise or sink vertically but did not break or move horizontally. Those were the days of elevator tectonics. Some geologists theorized that the geosynclinal trough sank beneath the weight of the sediments accumulating in it—a limited mechanism at best. Others suggested that the sediments filled the trough as it sank, for vaguely specified reasons.

If geosynclines existed in the past as ancient mountain ranges formed, then presumably they must exist today where future ranges are forming. The discovery of deep ocean trenches late in the last century appeared to many geologists to provide eligible candidates for modern geosynclines. They are linear troughs of great depth, many of them next to continents, where deep sections of sedimentary rock might well accumulate. One of the earlier and more revealing efforts to investigate ocean trenches began during the 1930s with surveys of the earth's gravity field.

TRENCHES AND GRAVITY ANOMALIES

Geologists have routinely used gravity surveys for more than 50 years as a means of locating large masses of uncommonly light or heavy rock underground, an important aspect of the search for new mineral deposits. They use an instrument called a **gravimeter,** which is sensitive to local variations that amount to only a few millionths of the total

FIGURE 18–1

An example of a local gravity anomaly. Large masses of salt as much as several kilometers in diameter that intrude coastal plain sediments in Louisiana and Texas can be located by finding negative gravity anomalies even though they may have no surface expression. The salt intrusions are important economically because they commonly trap oil and gas around their margins and the rocks capping them contain sulfur. Some also support salt mines.

Earth's gravity field.

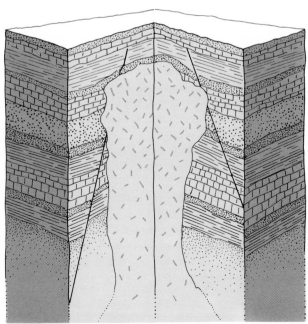

gravitational field strength. It is always impressive to set a gravimeter on the floor, read it, and then move it onto a low table and read it again. The instrument will clearly show that it has been moved one-half meter farther from the earth's center of mass.

Geologists doing a gravity survey read the instrument at a number of stations distributed through the study area and then correct the readings for the effects of elevation, the tidal pull of the sun and moon, latitude, and the gravitational attraction of nearby mountains. Any discrepancies between the corrected readings and the theoretical value of the earth's gravitational field strength are called **anomalies.** The discrepancies are termed *positive anomalies* if the observed field strength is greater than it theoretically should be and *negative anomalies* if it is less. Local gravity anomalies generally reflect variations in the density of the underlying bedrock (Figure 18–1) whereas regional anomalies usually indicate that the earth's crust is not in perfect floating equilibrium.

Regional gravity anomalies generally suggest that the earth's crust should be rising or sinking to restore itself to isostatic balance. The Hawaiian Islands, for example, coincide with large, positive gravity anomalies that reflect the weight of the great volcanoes resting on the oceanic crust, and the islands are indeed sinking at a measurable rate. Regional negative gravity anomalies exist in areas where the great glaciers melted at the end of the last Ice Age and those regions are indeed rising. However, the correlation between gravity anomalies and vertical crustal movements breaks down around oceanic trenches.

Gravity surveys over ocean trenches reveal the strongest negative anomalies ever measured. But the trenches are clearly not rising and therefore appear to be operating in flagrant defiance of the laws of gravity. Discovery of strongly negative gravity anomalies over deep ocean trenches caused the same consternation that might be experienced if one witnessed a ship sinking lower as it was unloaded instead of rising higher. As in the case of the ship, we must conclude that the ocean trenches are not floating properly. Since nothing pushes ocean trenches down from above, it seems clear that something must be pulling them down from below.

One theory that intrigued a number of geologists for many years held that the ocean trenches are, in effect, large dimples marking places where descending arms of convection currents in the mantle pull the oceanic crust down. Similar dimples appear in the surface of a slowly simmering pan of thick soup. Such a trench would last and accumulate sediments until the convection current stopped moving. Then it would float up, as the negative gravity anomaly so clearly demands, and raise its filling of sediments above sea level to become a long chain of folded mountains. That theory ingeniously explained many of the observed facts without invoking horizontal movement of the ocean floor, which had not yet been observed.

The advent of plate tectonics led immediately to a new interpretation of the thick sequences of severely deformed rocks that compose many mountain ranges. Geologists now regard such sequences of rocks as oceanic sediments swept together at convergent plate boundaries, as tectonic rather than as depositional assemblages. Although the traditional geosynclinal terminology is still widely used, many geologists now prefer to call those thick sedimentary accumulations **subduction complexes.** That term emphasizes their association with convergent boundaries where one plate sinks into the mantle, a process often referred to as **subduction.**

SUBDUCTION COMPLEXES

Tectonic Assemblage

Trailing continental margins, such as the east coast of North America, quietly accumulate large aprons of sedimentary rocks that form broad coastal plains and continental shelves. Those rocks consist mostly of mudstones, sandstones, and limestones, all of which contain fossils and sedimentary structures that reflect deposition near a continental coast. Such a trailing margin eventually becomes a convergent plate boundary. Then, the ocean floor sliding beneath the continent jams the sedimentary rocks of the old coastal plain and continental shelf back against the continent as though they were a carpet shoved against a wall. The crumpled coastal plain and continental shelf become a range of folded mountains composed of sedimentary rock types that geologists broadly describe as miogeosynclinal.

Much of the Sierra Nevada consists of miogeosynclinal sedimentary rocks as does part of the Klamath Mountains along the border between California and Oregon. The same trend continues north through the Blue and Wallowa Mountains of central and eastern Oregon; some geologists believe they can trace it still farther north in the Kootenay Mountains of eastern British Columbia. All of those mountain ranges contain rocks that appear to have been deposited along a trailing continental margin during late Paleozoic time. Then that old trailing continental margin became a convergent plate boundary early in Mesozoic time. That change happened as the North American plate began to move west away from the newly developing mid-Atlantic ridge, creating a convergent boundary where the western edge of the continent began to ride over the floor of the Pacific Ocean.

Under such circumstances, an oceanic trench develops offshore where the sinking lithospheric plate begins its long slide into the mantle. Most of the sediments that cover the oceanic crust of the sinking plate and some of the pillow basalts that compose that crust are scraped into the trench. The sinking plate stuffs slice after slice of sedimentary and volcanic rocks into the trench for as long as the convergent boundary lasts, perhaps for as long as 100 million years. However, the line of plate collision will eventually shift and then the sinking plate will no longer maintain the trench. When that happens, the crumpled mass of oceanic sediments and basalt that filled the trench will rise to become a range of mountains composed of rocks that geologists broadly describe as eugeosynclinal.

The so-called *Franciscan rocks* in the coast range of California and southwestern Oregon are an excellent example of a eugeosynclinal assemblage. They consist for the most part of turbidite sediments that originally accumulated over a vast expanse of the Pacific Ocean floor. Then those sediments were swept into a trench along with considerable amounts of siliceous ooze, which is now chert, and much smaller amounts of calcareous ooze, which is now limestone. There are generous amounts of pillow basalt jumbled in with the sedimentary rocks, and the entire assemblage can only be described as disorderly and confusing.

Melanges

Coastal mountain ranges formed at collision boundaries possess only the crudest kind of internal order (Figure 18–2). In general, the old continental shelf, if there was

1 Trailing continental margin with thick accumulation of coastal plain and continental shelf sediments

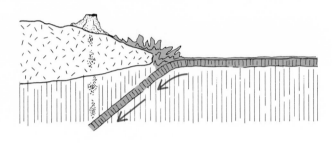

2 Convergent plate boundary develops jamming coastal plain and continental shelf sediments into folds. Volcanos start erupting above the sinking plate.

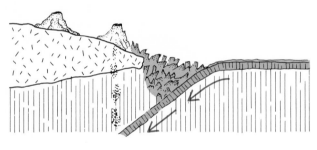

3 The sinking plate stuffs oceanic sediments into the trench packing most of it in as successive slices thrust under the seaward margin of the accumulating pile. The volcanic chain shifts position from time to time.

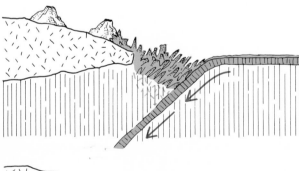

4 Metamorphism and partial melting transform the sediments deep in the trench into sialic schists, gneisses, and granite.

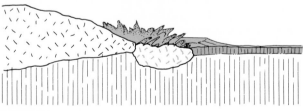

5 The convergent boundary ceases to function and the material that had been packed into the trench floats upward to form a subduction complex mountain range, a new accretion of sialic crust to the continent.

FIGURE 18-2
The steps in development of a subduction complex mountain range and addition of new sialic crust to a continent.

one, now lies closest to the old continental margin and the rocks scraped into the trench lie near the new coastline. That explains why the miogeosynclinal sedimentary assemblage generally lies landward of the eugeosynclinal material; the Sierra Nevada, for example, is landward of the Coast Range.

The eugeosynclinal assemblages are especially disorderly. The first rocks stuffed into the trench are generally the most severely deformed because all the others were jammed in after them. The sinking oceanic crust tends to drag each successive slice of sedimentary rock and pillow basalt in under the last, a phenomenon called **underthrusting.** That creates a crudely reversed age sequence with the oldest rocks, which were first into the trench, generally above the younger ones, which were dragged in beneath them. Where the rocks in such subduction complexes seem hopelessly jumbled, as they quite commonly do, geologists refer to them as a **melange.** That is one technical term that could probably be described as a cry of despair.

Continental Accretion

Because oceanic sediments consist essentially of material eroded from the continents, they have nearly the same sialic composition as continental rocks. Metamorphism and local partial melting in the depths of many trench fillings transform the rocks there into schists, gneisses, and granites exactly like those that may have been weathered to produce the sediments. Therefore, the addition of a subduction complex enlarges a continent and that is why many geologists prefer to call them **accretionary prisms.** However, the process of continental accretion goes beyond mere recycling of eroded material back into the continents and actually increases the total area of continental crust. Consider what happens as basalt, which is not a sialic rock, weathers to soil that eventually becomes sediment.

Soils derived from basalt typically consist of clays that become clastic sediments upon erosion and eventually accumulate on the continental shelves and deep ocean floors where they absorb potassium from seawater. Meanwhile, the calcium and magnesium liberated from the weathering basalt move to the ocean as part of the dissolved load of streams. Those elements tend to deposit from seawater as

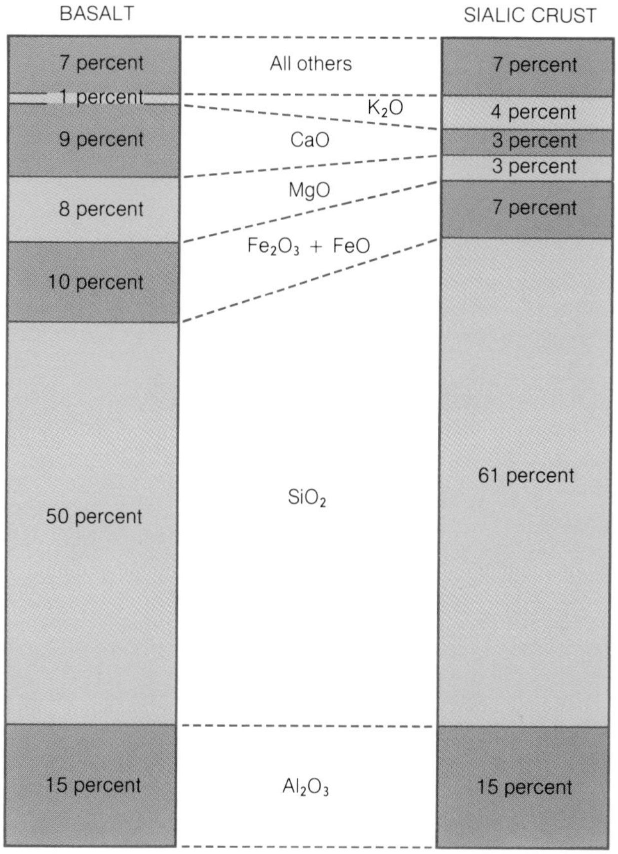

FIGURE 18-3

The estimated chemical composition of the sialic continental crust compared to that of a typical basalt. What the sialic crust lacks in iron, magnesium, and calcium, it makes up in silica and potassium.

limestone and dolomite, which form more abundantly in shallow seas flooding the continents than in the ocean basins.

Figure 18-3 shows the principal compositional differences between sialic rocks and basalt. The most striking distinctions are that sialic rocks contain much more silicon and potassium than basalt, and much less magnesium and calcium. Those distinctions correspond nicely to the differentiation that accompanies deposition of sediments derived from weathered basalt: The more sialic fraction enters the ocean basins while the fraction enriched in calcium and magnesium accumulates on top of the continents. Most of the sediments deposited in ocean basins eventually enter subduction complexes whereas those laid down on top of the continents generally escape that fate.

We know of no opposing process that converts sialic material into rocks resembling basalt in composition except on a local scale. Therefore, it seems reasonable to conclude that the earth's inventory of sialic material is increasing and radioactive age dates do indeed show that the continents have grown. However, most of the continental crust formed rather early in geologic time so few geologists would attribute its origin entirely to the continuing processes of sedimentary differentiation.

Ophiolites: Oceanic Crust on Land

During the 1950s when the composition of the mantle was still a subject of vigorous debate, a group of eminent geologists conceived the idea of drilling a hole through the oceanic crust and into the upper mantle. They called it the ''mohole'' because it would penetrate the Moho-discontinuity. Although it never did drill into the mantle, the project did develop equipment and techniques capable of drilling deep cores from the ocean floor, which have yielded a wealth of information. However, the original goal of the ''mohole'' project no longer matters because we now realize that there are many places where we can see sections of the oceanic crust exposed on land. Now a simple field trip can accomplish the goal envisioned for the ''mohole.''

Geologists have known for many years that the mass of severely deformed oceanic sedimentary rocks that comprises subduction complex mountain ranges also includes equally deformed slabs of basalt, gabbro, and peridotite. Similar masses of dark rocks likewise occur within the generally sialic complexes that comprise the continental crust. Geologists called those dark slabs **ophiolite complexes** and speculated endlessly about their origin and significance until the late 1960s. Then they were recognized as slices of oceanic crust and upper mantle sheared off the tops of lithospheric plates sinking at convergent boundaries.

Ophiolite complexes are numerous in the subduction complex ranges near the west coast of North America, especially in the Klamath Mountains of southwestern Oregon and northwestern California. They also occur in a number of other North American localities of which the Bay of Islands complex in Newfoundland is one of the best known. It is a fantastic experience to walk across a sunny hillside on outcrops of rocks that once belonged to the oceanic crust and upper mantle. However, no single ophiolite complex exposes a complete section so the diagram in Figure 18-4 represents a somewhat idealized composite of many exposures.

FIGURE 18-4

A generalized section through the oceanic crust as it is exposed in ophiolite slabs incorporated in subduction complexes.

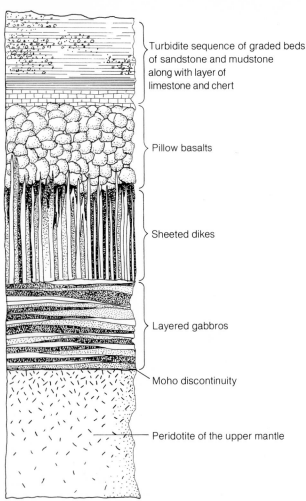

Turbidite sequence of graded beds of sandstone and mudstone along with layer of limestone and chert

Pillow basalts

Sheeted dikes

Layered gabbros

Moho discontinuity

Peridotite of the upper mantle

As one might expect, oceanic sediments, such as those shown in Figure 18–5, lie on top of most ophiolite complexes. They consist mostly of turbidites but may also include cherts and limestones derived from lithification of siliceous and calcareous oozes. Figure 18–6 shows a fairly typical exposure of the pillow basalt lava flows that lie immediately beneath the sedimentary rocks to a depth of some hundreds of meters. Much of the basalt is green instead of black. The reason for this is that hot water circulating through the flows while they were still near the crest of the oceanic ridge altered the black pyroxene in the original rock to green chlorite.

The pillow basalts generally contain scattered basalt dikes and pass downward into a section at least several hundred meters thick composed of a solid mass of vertical basalt dikes packed together like the pages of a book. They evidently formed as fissures opened in the crest of the oceanic ridge and then filled with basalt, welling to the surface where it would erupt to make the pillow lavas above. Each dike measures in its thickness of a meter or less an incremental increase in the width of an ocean. It seems that the sea floor spreads dike-by-dike.

The lower part of the oceanic crust below the dike complex consists predominantly of horizontal sheets of gabbro, the coarse-grained plutonic rock equivalent in composition to basalt. Figure 18–7 shows an exposure of that gabbro. Evidently, a large part of the rising basaltic magma squirts horizontally into the newly forming oceanic crust to form sills instead of rising higher to form vertical dikes or still higher to erupt as lava flows. The gabbro passes downward into heavy, black peridotites composed of pyroxene and olivine, clearly a slice off the top of the mantle.

Ophiolite Ore Deposits

Understanding the origin of ophiolite complexes greatly simplifies the problem of exploring for new reserves of the several ores they are likely to contain. Perhaps the most conspicuous are the massive sedimentary sulfide deposits, which form through deposition of sulfide minerals around submarine hot springs and from pools of hot brine at the crest of the oceanic ridge. They occur within the sediments on top of the ophiolite complex. Those circulating hot waters that form the submarine hot springs and brine pools are also responsible for altering the pyroxene in the pillow

FIGURE 18–5
A fold and a fault in deformed beds of sandstone and shale originally deposited on the floor of the Pacific Ocean and now exposed in the California Coast Range, a subduction complex. The area shown is about 2 m across.

FIGURE 18–6
Pillow basalt exposed in a quarry face in the Oregon Coast Range; part of an old slab of oceanic crust. The larger pillows are about 1 m long.

FIGURE 18-7
Layered gabbro in a section of oceanic crust exposed in the San Juan Islands, Washington. (D. W. Hyndman)

basalts to chlorite. In some cases, they leave veins of valuable ore minerals, most commonly copper minerals, in the basalt.

The large gabbro sills in the lower parts of ophiolite complexes provide most of the world's supply of chromium. It occurs as round globs of chromite, a chromium oxide mineral that appears most abundantly near the bases of the sills. Chromite appears here because it separates from the magma and then settles through it while both are still melted. Geologists systematically follow the bases of the sills, looking for commercially attractive concentrations of chromite.

The peridotites in the lowest parts of ophiolite complexes generally contain widely disseminated flecks of metallic platinum, which is rarely concentrated enough to be mineable from the solid rock. However, streams draining from large areas of peridotite commonly contain commercially workable placer deposits of platinum.

Both peridotites and gabbros contain small amounts of nickel, which may concentrate in the lower parts of lateritic soils to form large deposits of excellent and easily mineable ore. Many of the world's large nickel mines, including those in Cuba, produce from laterites weathered on the lower parts of ophiolite complexes. A relatively small mine of that type operates near Riddle in southwestern Oregon.

Serpentinite

Serpentine minerals form as water reacts with such mafic minerals as pyroxenes or olivines. The gabbros and peridotites in ophiolite complexes are typically serpentinized to some extent, and in some cases the peridotite is completely altered to **serpentinite,** a greenish rock composed largely of serpentine. Most samples of serpentinite are soft enough to carve with a knife. It is most familiar in the form of small, greenish figurines widely sold in souvenir shops. Outcrops of serpentinite, such as that shown in Figure 18–8, typically reveal a rock broken into numerous small chunks with surfaces polished by internal movement within the mass. The general absence of large, unfractured pieces explains why serpentinite figurines are small.

To judge from what we see in ophiolite sections, it seems that substantial amounts of serpentinite must exist near the base of the oceanic crust. Most of it slides down into the mantle at convergent plate boundaries where it must lose water as it gets hot at depth and presumably reverts back to peridotite. Some of the steam emitted from volcanoes along convergent plate boundaries may come from dewatering serpentinite.

However, serpentinite is as slippery within the earth's crust as it is in the hand and some squeezes back toward the surface. Large masses of serpentinite penetrate the mangled rocks accumulating in the subduction complex as though they were globs of grease squeezing through a bale of compacting scrap metal. On informal occasions geologists often refer to that kind of intrusion as "watermelon seed tectonics." We find masses of intrusive serpentinite woven all through subduction complexes, especially along fault zones, which evidently provide easy passageways for the slippery rock.

Eclogite and Blueschist

Many masses of intrusive serpentinite contain chunks of **eclogite,** a dark and uncommonly dense rock of red garnet and green pyroxene. Eclogite is closely related to the gemstone *jade,* which consists entirely of the green pyroxene mineral *jadeite* and seems to occur along shear zones.

The chemical composition of eclogite is identical to that of basalt. Experimental work showed many years ago that the augite and plagioclase that compose basalt turn into

(a)

FIGURE 18-8
An outcrop of serpentinite. (a) General
view of an area about 2 m across showing
fragmentation caused by internal
movement during deformation. (b) Close
view of the surface of a fragment polished
by slippage. Shown about natural size.

(b)

the garnet and green pyroxene of eclogite under extremely high pressures, such as those that prevail in the upper mantle. That came as no surprise because eclogite is much denser than basalt and should therefore form under higher pressures. However, it was quite surprising when experimental results showed that the minerals in eclogite form under fairly moderate temperatures much lower than those that must prevail in the upper mantle.

For many years geologists could not imagine any place within the earth where the appropriate combination of pressure and temperature to convert basalt into eclogite might exist. It seemed that it should be impossible to penetrate deeply enough into the earth to find pressures high enough to permit the ecologite minerals to form without also finding temperatures much too high to permit their formation. Like many other geologic problems, the eclogite dilemma disappeared with the advent of plate tectonics.

We now interpret eclogite as evidence that the sinking oceanic crust descends to depths where the pressure is high enough to convert basalt into eclogite before enough heat can soak into the rock to make that metamorphism impossible. Evidently, the sinking slab continues to regurgitate slippery masses of serpentinite after it has descended well into the mantle. Some of those masses rip pieces off the basaltic oceanic crust—at that depth transformed into eclogite—and carry them back up into the higher levels of the subduction complex. In some cases in which the chunks of eclogite are uncommonly large, it is possible to recognize in them the remains of basalt pillow structures.

Some serpentinite bodies also contain chunks of **blueschist,** a dark rock formed by metamorphism of oceanic sediments. Blueschist is so-called because it contains an intensely blue variety of amphibole. Like eclogites, blueschists contain minerals that form under conditions of extremely high pressure and only moderately high temperature. The chunks that appear along with eclogites in masses of intrusive serpentinite evidently derive from stray masses of oceanic sediment that went down with the descending plate instead of remaining behind in the trench.

Blueschists also occur on a regional scale in large parts of some subduction complexes. The Franciscan rocks of the Coast Range in northern California and southwestern Oregon, for example, contain large volumes of blueschist. Evidently, they formed deep in the trench where the pres-

sure was very high and the temperature moderate because the sediments in the accumulating subduction complex had not yet absorbed much heat from the surrounding lithosphere. Blueschists survive to reach the surface if the plate collision stops and the subduction complex rises to become a mountain range before it becomes hot enough to metamorphose into more ordinary schists and gneisses.

SUMMARY

Many mountain ranges, especially those fringing continental margins, consist largely of thick accumulations of intensely deformed sedimentary rocks. They apparently form at convergent plate boundaries where sinking oceanic crust telescopes continental shelf sediments into folded mountains and sweeps sediment off the ocean floor into trenches. After the convergent plate boundary ceases to operate, the trench disappears and its filling of deformed sediments floats up to become a mountain range.

Metamorphic recrystallization and local melting convert many trench fillings into new continental crust composed of schist, gneiss, and granite. Such new crust consists largely of material eroded from the continents, deposited in the ocean basin, and then recycled back to the continent by way of the trench. However, the processes of weathering, erosion, and deposition also differentiate rocks, creating a sedimentary fraction of sialic composition that likewise enters the trench and becomes new continental crust. The inventory of sialic material increases with time.

Subduction complexes generally contain slices of oceanic crust and upper mantle sheared off the top of the descending lithospheric plate. They provide a cross-sectional view of the oceanic crust as well as insights into the processes that operate at divergent plate boundaries where that crust forms.

Part of the lower oceanic crust consists of serpentinite, a slippery rock that tends to squeeze back toward the surface from sinking lithospheric plates and to intrude the material accumulating in the trench. Many serpentinite masses contain inclusions of metamorphic rocks formed under conditions of very high pressure and moderately high temperature. They indicate that the descending slab heats up rather slowly as it sinks into the mantle and that at least some of the serpentinite masses return from mantle depths.

KEY TERMS

accretionary prism	melange
anomaly	miogeosyncline
blueschist	ophiolite complex
eclogite	serpentinite
eugeosyncline	subduction
geosyncline	subduction complex
gravimeter	underthrusting
jadeite	

REVIEW QUESTIONS

1. Why do subduction complexes contain rocks that originally formed over a large area and during a long period of time?

2. Briefly describe the important kinds of rocks that occur in subduction complexes, how they get there, and the general way in which they are arranged.

3. Why do we believe that eclogite nodules and the serpentinites that contain them must have been carried to great depth with a sinking plate and then returned to shallow levels?

4. What are the major layers in the oceanic crust and how do they form?

5. Why did observation of large negative gravity anomalies over oceanic trenches show that they cannot possibly be sinking under the weight of the sediments they contain?

SUGGESTED READINGS

Bascom, W. *A Hole in the Bottom of the Sea.* New York: Doubleday & Co., 1961.

Engagingly written account of the mohole project by one of the leading participants. For a nontechnical audience.

Coleman, R. G. *Ophiolites: Ancient Oceanic Lithosphere?* Berlin: Springer-Verlag, 1977.

An outstanding comprehensive review of ophiolites in all their aspects by a leading authority. Although intended for a professional audience, the book is remarkably readable.

Dickinson, W. R., ed. *Tectonics and Sedimentation.* Society of Economic Paleontologists and Mineralogists. Special Publication 22, 1974.

A collection of articles, most rather technical, on the association between sediments, trenches, and mountains.

Le Pichon, X., J. Francheteau, and J. Bonnin. *Plate Tectonics.* New York: Elsevier, 1973.

A general text on plate tectonics. Includes a good discussion of the operation of convergent plate boundaries.

Talwani, M. and W. C. Pitman. *Island Arcs, Deep Sea Trenches, and Back-Arc Basins.* Washington, D.C.: American Geophysical Union, 1977.

A collection of fairly technical articles on various aspects of the processes that operate along convergent plate boundaries.

CHAPTER NINETEEN

Steam and ash blowing from Mount St. Helens during the explosive eruption of May 18, 1980. The dark color of the eruption cloud reflects its heavy ash content. (U.S. Geological Survey)

VOLCANOES AND BATHOLITHS

MOUNT ST. HELENS, 1980

On March 20, 1980 a series of small and moderate earthquakes began to shake Mount St. Helens, a Cascade volcano in southwestern Washington. A week later, small clouds of steam and ash (Figure 19-1) began to puff out of a new crater in the summit. Then, in early April, the clouds of steam and ash began to diminish in number and size, although the earthquakes continued, and many people suspected that the mountain would soon become quiet. It had last erupted in 1857.

Meanwhile, precise surveying instruments detected a slight swelling on the north flank of the volcano. The swelling continued to grow at an accelerating rate and by early May had developed into an easily visible and ominous-looking tumor. It seemed clear that a large mass of molten rock was intruding that part of the volcano. Geologists wondered whether it would crystallize beneath the surface to make a plutonic igneous intrusion or erupt through the flank of the volcano. There was also some concern that the large and growing bulge might cause massive landslides on the north flank of the mountain. In the end, all three possibilities occurred: An enormous landslide did move, the volcano did erupt, and part of the mass of magma did remain to become an igneous intrusion.

On the morning of May 18, an enormous landslide started down the flank of the volcano. Its movement re-lieved the pressure on the mass of magma within, permitting it to explode into the enormous cloud of steam and ash shown in Figure 19-2. This cloud rose to an elevation of at least 18 km within a few minutes. Meanwhile, the landslide, combined with enormous volumes of newly erupted volcanic ash, poured down the Toutle River (Figure 19-3) and then into the Columbia, blocking the navigation channels into Portland.

As Figure 19-4 shows, the explosion opened a gaping crater in the north flank of Mount St. Helens, completely changing its skyline profile. During the months following the big eruption, lava extruded from the floor of the new crater and slowly bulged into a dome that briefly interrupted its growth in each of a series of much smaller eruptions. That dome-shaped extrusion is undoubtedly the top of an intrusive mass of magma that still extends to great depth after having blown off its upper part, which was heavily charged with steam. If Mount St. Helens follows the example of Bezymianny, a closely similar volcano in Kamchatka, U.S.S.R. that erupted in exactly the same way in 1956, the lava dome will finally grow to fill most of the new crater.

The ash cloud moved directly east at an average rate of about 60 km per hour. People in its path saw it rise like a dark wall on the western horizon and then abruptly turn what had been a brilliant early summer day into gritty darkness. In Spokane, the street lights switched on in mid-

FIGURE 19-1

Mount St. Helens blowing a small cloud of steam during April, 1980. At this early stage of the eruption the rising magma was still far below the surface and the steam was emerging at a temperature only slightly above the boiling point of water. The small quantity of ash in the cloud is old material mixed with the steam, not new lava reaching the surface. (Johnnie Moore)

FIGURE 19-2
A close view of the May 18, 1980 eruption of Mount St. Helens. The snow-covered peak in the distance is Mount Hood, another Cascade volcano across the Columbia River in Oregon. (U.S. Geological Survey)

FIGURE 19-3

The enormous mud flow that poured off the flanks of Mount St. Helens and down the Toutle River during the eruption of May 18, 1980. The river will recover; similar mudflows have filled it on a number of past occasions, which may include the eruption of 1842. The picture has low contrast because medium-gray ash covers everything in the scene. (U.S. Geological Survey)

FIGURE 19-4
The area on the north flank of Mount St. Helens that had been bulging for weeks exploded on May 18,
opening an enormous explosion crater that took at least 400 m off the top of the mountain.
(U.S. Geological Survey)

afternoon. By early evening the dark pall had spread across the mountains of northern Idaho and into western Montana still turning day into night. Volcanic ash fell like snow through a windless night and people awoke the next morning to see a landscape drained of all color, covered beneath a blanket of medium gray ash, which rose in choking clouds behind everything that moved.

The 1980 eruption of Mount St. Helens was merely the most recent event, and a relatively minor one, in a region that has been vigorously active in one way or another for at least 30 million years. However, the story more logically

begins at the rift in the crest of the oceanic ridge system where volcanic activity occurs most abundantly and in its simplest form.

THE ORIGINS OF MAGMAS

Most of the earth's volcanic eruptions occur unwitnessed along divergent plate boundaries. There, in the crest of the oceanic ridge, basalt lava wells up through fissures that open between separating plates and pours out to make

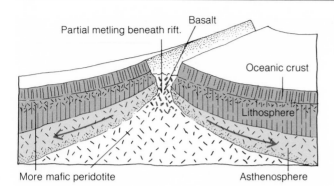

Partial metling beneath rift.

Basalt

Oceanic crust

Lithosphere

More mafic peridotite

Asthenosphere

FIGURE 19-5

Movement of lithospheric plates away from the crest of an oceanic ridge relieves pressure on the hot peridotite below, permitting it to partially melt. The results are basalt lava erupted in the crest of the ridge to form new oceanic crust and more mafic peridotite depleted in basaltic constituents that moves away as part of the growing lithospheric plate.

new oceanic crust composed of pillow lava. The oceanic ridges are regions of high, crustal heat flow, which means that the rocks beneath them are abnormally hot and therefore more likely to melt than rocks elsewhere. Nevertheless, plate movement away from the ridge crest probably plays a more direct role in generating magma by relieving pressure on the rocks beneath the ridge crest (Figure 19-5). Those rocks are already so hot that they would melt if they were at the surface so relief of pressure permits them to begin melting with no increase in temperature. However, the magma that appears is basalt, not peridotite. How can melting of peridotite rock produce basalt magma?

Partial Melting

Partial melting of a substance of one composition to yield a liquid of another is a familiar phenomenon: We all learn at an early age to suck the flavored syrup out of a warming popsicle, leaving only white ice on the stick. That is possible because the syrup—a mixture of sugar and water—melts at a lower temperature than pure ice or pure sugar. Similarly, basalt magma has the lowest melting point of any

mixture within the compositional confines of peridotite so it is the liquid that appears when the rocks of the upper mantle partially melt. Basalt is less mafic than peridotite so the solid rock left behind after it melts must be more mafic than the original peridotite. Many of the peridotites exposed in the lowest parts of ophiolite complexes are indeed enriched in mafic constituents. This is exactly what one would expect if a fraction of basaltic composition had been extracted from them. They are the equivalent of the ice left on the popsicle stick.

Volcanic Chains

The basalt lava flows of the oceanic crust return to the mantle at convergent plate boundaries and evidently melt again as they reach a depth of about 100 km, the level of the asthenosphere. Magma rising from the surface of the sinking plate penetrates the overriding plate and erupts to form a chain of volcanoes, such as the Cascades. Volcanoes of that type erupt the entire spectrum of lava compositions, ranging from basalt to rhyolite although the dominant rock type is dark andesite only slightly lighter than basalt. Where do all those different kinds of magma form?

Maps of volcanic chains show that the peaks generally align along a smooth curve at nearly uniform distance from the trench. Evidently, the magmas must melt at nearly uniform depth and therefore at nearly uniform temperature. Figure 19-6 illustrates the point. If several different kinds of magmas melted off the descending slab, we would expect them to do so at different depths and distances from the trench. The depths and distances would correspond to the magmas' melting temperatures, which range from a low of about 750°C for rhyolite to a high of about 1,200°C for basalt. That would create a staggered chain of volcanoes erupting different kinds of lavas at characteristic distances from the trench, an arrangement that does not generally occur in nature. However, we should note that Mount St. Helens may be an exception to that generalization. It lies about 50 km west of the main trend of the Cascades and erupts more sialic lavas than its neighbors.

It seems likely that the magma melting off the descending slab must be basalt or dark andesite very similar to basalt. The prevalence of those rock types in volcanic chains and the results of laboratory experiments both point to that conclusion. The more sialic lavas that erupt from volcanic chains probably form through melting of crustal rocks. We can imagine that basalt and mafic andesite magmas, which have high melting points, might melt large volumes of sialic rocks in the lower part of the crust where the rocks are almost hot enough to melt in any case. Seismologists commonly observe that earthquake shear waves weaken or disappear at shallow depth beneath some parts of volcanic chains. Shear waves do not pass through liquids so their failure to pass beneath parts of some volcanic chains suggests that they may have encountered large reservoirs of magma.

Basalt-Rhyolite Volcanic Fields

Quite another kind of volcanic province develops in areas of the continents where extension of the crust relieves pressure on the rocks at depth. In such regions, partial melting of the mantle peridotite produces basalt magma just as it does at the crest of the oceanic ridge. However, partial melting of the rocks in the lower part of the continental crust produces rhyolite magma, the sialic composition with the lowest melting point. Volcanic provinces associated with extension and thinning of the crust tend to erupt large volumes of basalt and rhyolite accompanied by relatively little lava of intermediate composition. Volcanoes of that type have been intermittently active in various large areas of the western United States since early Tertiary time. The Columbia volcanic plateau of the Pacific Northwest is the largest example.

COMPOSITION, VISCOSITY, AND STEAM

Structure and Behavior of Mafic Magmas

Figure 19–7 shows a schematic representation of the internal structure of a mafic magma, such as basalt. The relatively low silicon content makes it possible for many of the silicate tetrahedra to exist as isolated individuals or single chains only loosely linked by doubly charged ions of iron, calcium, or magnesium. The isolated silicate tetrahedra and small groups of linked tetrahedra can move fairly freely within the melt and their internal mobility gives the magma

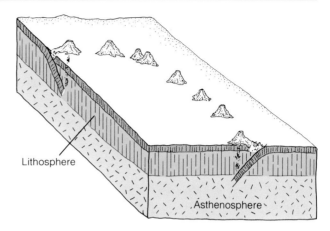

FIGURE 19-6

The active peaks in volcanic chains tend to align along very smooth curves, suggesting that they erupt magma melting at constant depth beneath them. However, there is evidence that the line of eruptions shifts back and forth in time thus distributing the igneous activity over a broad zone.

as a whole considerable fluidity. Basalt and mafic andesite lavas flow about as easily as honey.

Basalt and mafic andesite tend to erupt in large volumes of lava that spread over large areas to make thin lava flows, which may be no more than a few meters thick. They reflect the considerable fluidity of the magma in many of their surface features, such as those shown in Figure 19–8. In many cases, the surfaces of such flows freeze, leaving molten lava beneath. It drains out from under the hard crust to open caverns that may wind within the flow for distances as great as several kilometers. Those caverns, which are generally called *lava tubes*, later collapse to open sinkholes, such as that shown in Figure 19–9. They also swallow surface water, making the development of new drainage systems on basalt flows a long and erratic process.

A long sequence of basalt or mafic andesite eruptions from the same or closely spaced vents typically builds a broad volcano with gently sloping flanks (Figure 19–10) which make a low and undramatic profile. Such structures are called **shield volcanoes** because their form suppos-

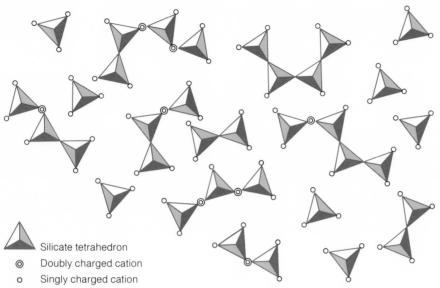

Silicate tetrahedron
Doubly charged cation
Singly charged cation

FIGURE 19-7

A schematic diagram of the structure of a mafic magma. The relatively low silica content makes extensive oxygen sharing between tetrahedra unnecessary so they tend to exist as isolated individuals or short chains of linked tetrahedra flexibly connected by doubly charged cations. Singly charged cations do not connect tetrahedra. The result is a magma with high internal mobility and therefore considerable fluidity.

FIGURE 19-8

Ropy surface of a basalt flow that formed as fluid magma continued to move beneath a thin skin of solidifying rock. The polygonal pattern in parts of the picture outlines shrinkage fractures that formed as the flow cooled—the upper end of a columnar joint set such as that shown in Figure 19-19.
(D. W. Hyndman)

FIGURE 19-9

A gaping hole yawns where part of the roof of a lava tube collapsed. Lava Beds National Monument, California. (D. W. Hyndman)

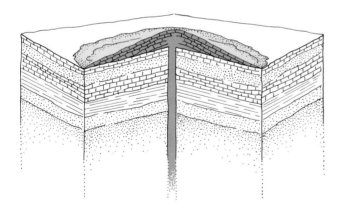

FIGURE 19-10

The profile of Mauna Loa, a large basalt shield volcano, as seen from the slopes of Kilauea, Hawaii. (D. W. Hyndman)

FIGURE 19-11

Cinder cones consist of a conical pile of basaltic fragments and generally become extinct after producing a lava flow that bursts out through the base of the cone. These cinder cones are inside Haleakala Crater at the summit of a large shield volcano on Maui, Hawaii. (D. W. Hyndman)

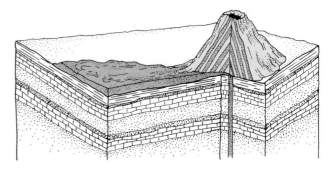

edly resembles that of a classical Roman shield. The largest active volcanoes on earth, those in the Hawaiian Islands, are shield volcanoes as is the largest now known to exist in the solar system, Olympus Mons on Mars.

If basalt or mafic andesite magmas encounter aquifers on their way to the surface, they may absorb small quantities of water and thus generate enough steam to add considerably more emphasis to the eruption. Escaping steam coughs shreds of lava as much as several hundred meters into the air where they chill as they fall. The larger pieces land near the vent as **volcanic bombs,** which typically have aerodynamically streamlined shapes and **volcanic cinders,** which are smaller and more angular. The cinders, being small, cool too quickly to acquire a streamlined form. Both volcanic bombs and cinders are full of gas bubbles at least several millimeters across, which give them a texture somewhat resembling that of coarse bread. Such bubbly basalt is called **scoria.** The finest particles drift downwind as a cloud of dark **volcanic ash** that may travel many kilometers before it finally settles.

Eruptions of steam-charged basalt make spectacular fireworks displays at night. The chunks of incandescent lava trace glowing arcs across the dark sky and then roll, still glowing red, down the slopes of the growing pile of bombs and cinders. Within the few days or weeks, occasionally longer, that the eruption lasts, the pile of cinders and bombs grows to make a small volcano called a **cinder cone** (Figure 19–11). Most cinder cones are less than a few hundred meters high and they erupt only once; the next eruption in the vicinity will produce a new cinder cone. As they near the end of their eruptive careers, after the rising lava has blown off most of its steam, cinder cones generally produce one or two lava flows. In most cases, the lava flows burst out through the base of the pile of loose cinders instead of overflowing from the crater at its top.

Structure and Behavior of Sialic Magmas

As magmas become more sialic in composition, they also become more viscous. Consider, for example, the internal structure of a rhyolite magma, which has a high silicon content and correspondingly low concentrations of calcium, magnesium, and iron. Figure 19–12 shows how such magmas develop into extensive networks of linked silicate tetrahedra because the abundance of silicon atoms forces them to share oxygens. Furthermore, the chemical bonds that link silicate tetrahedra through shared oxygens are strongly directional because they involve electron sharing instead of simple attraction between oppositely charged ions. That directionality makes the network of shared silicate tetrahedra rigid thus further restricting the mobility of individual ions within the magma. Because the ions within them lack internal mobility, sialic magmas are viscous. Rhyolitic magmas are so extremely viscous that their consistency resembles that of modeling clay.

If sialic magmas arrive at the surface without having absorbed much water, they erupt quietly by extruding as masses of viscous magma that slowly grow over a period of months into hills called **plug domes** or **lava domes.** The lava dome that began forming in the crater of Mount St. Helens during the summer and fall of 1980 is a good example. Figure 19–13 shows a typical plug dome, a craterless volcano that stands as an isolated hill, its sides cloaked in talus composed of angular blocks that cracked off the cooling surface of the plug.

Water in Magma

Many silica-rich sialic magmas do absorb water, much of which seems to serve as an extra source of oxygen atoms to surround the abundant atoms of silicon. Figure 19–14 shows how a water molecule entering a sialic magma may attach itself to one of two linked silicate tetrahedra thus breaking the oxygen-sharing bond that had connected them. Such reactions certainly increase the mobility of in-

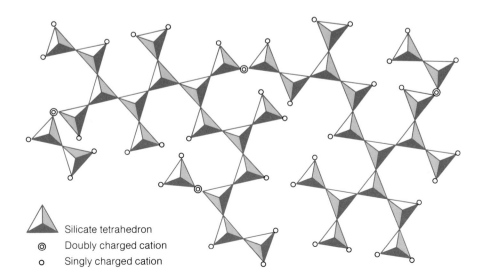

FIGURE 19-12
A schematic diagram of the structure of a sialic magma. The high silica content requires extensive oxygen sharing, which creates large networks of rigidly linked tetrahedra that inhibit internal mobility and therefore make the magma extremely viscous.

Silicate tetrahedron
Doubly charged cation
Singly charged cation

FIGURE 19-13

Plug domes are the extruded tops of masses of sialic magma that probably continue in most cases into granitic plutons at depth. Their sides are generally covered with talus so the only bedrock outcrops are at the summit, which has no crater. Shown here is Black Butte, a multiple plug dome near Mount Shasta in California.

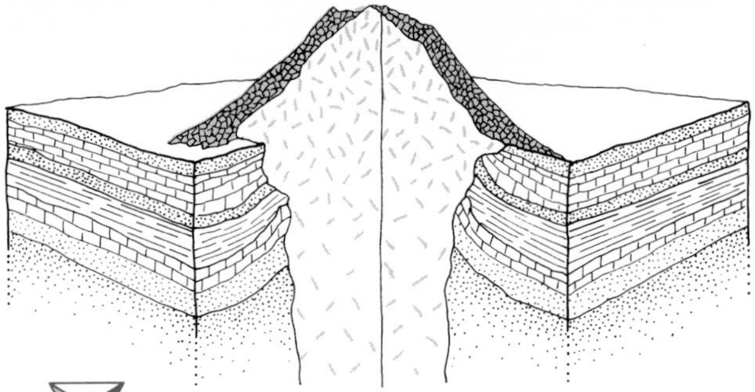

FIGURE 19-14

A molecule of water entering a magma adds another atom of oxygen, which enables two silicate tetrahedra that had been joined through a shared oxygen to separate. That increases mobility within the magma and therefore also increases its fluidity. The same reaction proceeds in the opposite direction as the magma crystallizes thus releasing steam, which increases the internal pressure and may finally cause a volcanic eruption.

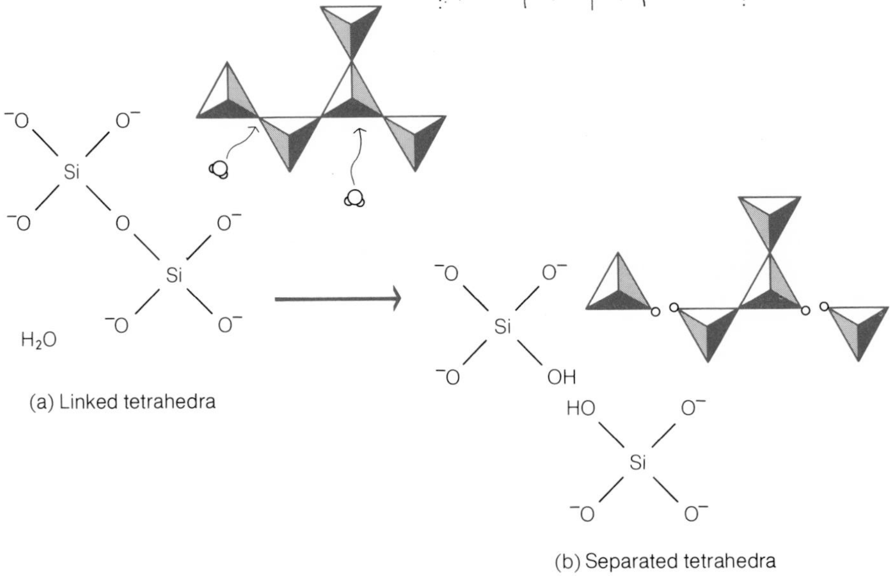

(a) Linked tetrahedra

(b) Separated tetrahedra

dividual ions within wet sialic magmas and therefore increase the fluidity of the entire mass. Nevertheless, sialic magmas remain impressively viscous even when loaded with large amounts of water.

The reaction shown in Figure 19–14 must reverse and release the water as steam when the magma crystallizes because silicate minerals contain little or no water. Therefore, the growing crystals crowd the steam into a diminishing volume of magma. That increases steam pressure within the magma for the same reason that the air pressure inside a rubber balloon increases if you squeeze it into a smaller volume. If the steam pressure within the magma becomes greater than the pressure exerted by the weight of the overlying rock, the magma will explode as though it were an overheated steam boiler. The ash erupted during the May 18 explosion of Mount St. Helens contains a large proportion of broken crystals, which must have contributed to the occasion by increasing the steam pressure within

the mountain as they grew. The post-eruption lava dome that formed in the crater of Mount St. Helens has blown off steam on a number of occasions, probably because pressure within the dome increased as crystals grew in the lava.

The combination of high viscosity and a heavy charge of steam makes wet sialic magmas extremely dangerous because the steam cannot escape quietly from the magma but instead puffs it up as though it were bread dough (Figure 19–15). Wet sialic magmas often form a rock called **pumice,** which is essentially a glass foam; many specimens are so light they will float on water. Exploding masses of steaming sialic magma commonly shoot mushroom clouds of steam and ash many kilometers into the air. Meanwhile, more ash and pumice boil out of the volcano like oatmeal overflowing a pan and spread over the surrounding countryside as an **ash flow.** Large parts of many ash flows remain hot enough to fuse into a solid rock

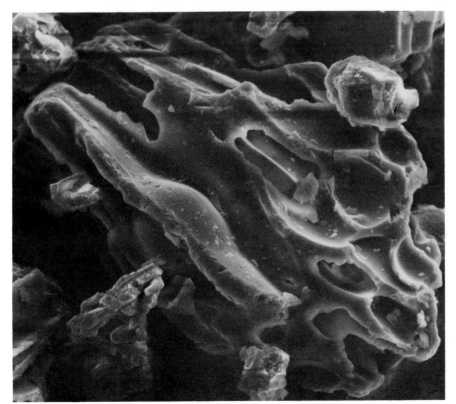

FIGURE 19–15

One piece of ash and several much smaller crystals of feldspar erupted from Mount St. Helens on May 18, 1980 as seen magnified about 1,200 times by a scanning electron microscope. The ash is volcanic glass inflated into a foam of minute bubbles by steam expanding within it at the moment of eruption. Its numerous sharp edges make it very abrasive and presumably harmful to the lungs. It is also an effective insecticide. The ash fall from the May 18 eruption effectively cleared the air of most flying insects as far as 500 km from the volcano. They did not reappear for several weeks. (Johnnie Moore)

FIGURE 19-16

Composite volcanoes produce a long series of andesite eruptions from
a central vent to build a cone composed of a mixture of lava flows and
fragmental material. In many cases, they erupt from large reservoirs of
magma at shallow depth, many of which remain as granitic plutons
after the volcano has been eroded away. Mount St. Helens, a typical
composite volcano, is shown here blowing a small cloud of steam
during April, 1980. The sinuous pattern on its flanks marks the tracks
of mud and lava flows that poured down the slopes during previous
eruptions. Composite volcanoes may erupt from subsidiary vents on
their flanks as well as from the central crater in the peak.
(U.S. Geological Survey)

called **welded ash** or **welded tuff.** For many years, geologists misinterpreted welded ash flows as rhyolite lava flows and puzzled over the question of how a lava as viscous as rhyolite could run out into thin flows.

Composite Volcanoes

Most of the lava that erupts from the volcanic chains overlooking convergent plate boundaries is andesite, which is neither as mafic as basalt nor as rich in silica as rhyolite; neither absolutely dry nor saturated with steam. Most of those andesites are somewhere intermediate between basalt and rhyolite not only in composition but also in viscosity, steam content, and eruptive behavior. On some occasions those intermediate lavas erupt as clouds of ash and cinders and on others as lava flows. A long series of such andesitic eruptions from a central vent builds a towering **composite volcano** (Figure 19–16), so-called because it consists of a mixture of fragmental material and lava flows. Composite volcanoes are the imposing peaks most people envision when they think of volcanoes, the kind that appear on travel posters.

Caldera Formation

About 6800 years ago ancient Mount Mazama, an imposing composite volcano about the size of modern Mount Shasta, disgorged some 17 km³ of rhyolite ash and pumice during a single eruption. Compare that to the May, 1980 eruption of Mount St. Helens, which produced only about 1.3 km³ of lava. As the lava poured out from Mount Mazama, the volcano sank into the space it had occupied. All that remains of the original mountain is its outer rim, which forms a circular ridge enclosing a vast subsidence crater called a **caldera.** Crater Lake, Oregon now fills the caldera that formed when Mount Mazama disemboweled itself.

Calderas are fairly common in volcanic regions and geologists originally assumed that they formed when explosive eruptions demolished large volcanoes, leaving a crater where they had stood. However, careful study of the debris blankets around calderas has shown that they typically consist almost entirely of freshly erupted lava and contain very little old rock as one might expect if a volcano had been blown to bits. Therefore, it seems clear that calderas are subsidence rather than explosion craters that probably form through a sequence of events, such as that shown in Figure 19–17. Many caldera-forming eruptions are much larger than the one that disposed of Mount Mazama.

For example, a violent rhyolite eruption of the volcano Santorini on the eastern Mediterranean island of Thera about 36 centuries ago transformed a large part of the island into a collapse caldera. Thera had been one of the major centers of the Minoan civilization and the eruption of Santorini almost certainly contributed to the collapse of that culture while possibly inspiring the ancient legend of a lost continent of Atlantis. In the greatest single volcanic eruption within historic time, the Indonesian island of Krakatoa vanished on August 26, 1883 beneath a column of ash that rose some 30 km and left a large caldera on the sea floor in which a new island has since appeared. About 30,000 people died during that eruption, many of them victims of seismic sea waves. The most recent North American caldera formed in 1912 when the Alaskan volcano Katmai ejected some 13 km³ of rhyolite, most of it in a large ash flow, and opened a caldera about 5 km in diameter as it subsided.

Glowing Cloud Eruptions

Occasionally, a strong blast of steam from a composite volcano blows a cloud of finely shredded lava as though it were a mist from a spray can. The most notorious eruption of that kind occurred on the morning of May 8, 1902 when Mount Pelee, a composite volcano on the Caribbean island of Martinique, blew such a cloud after having been intermittently active for about a month.

The mixture of steam and hot lava swept down the flank of the mountain at hurricane speed and across the small city of St. Pierre, instantly killing all but 2 of its 28,000 people and setting fire to the ruins as well as to several ships in the harbor. Surviving eyewitnesses who watched the catastrophe from ships standing in to port reported that the eruption cloud was so hot it glowed hence the name **glowing cloud eruption.** Such eruptions are also referred to as *nuees ardentes,* a French expression that means glowing cloud.

A number of people had suggested that St. Pierre be evacuated during the month before it was destroyed but the local authorities refused, basically because they were

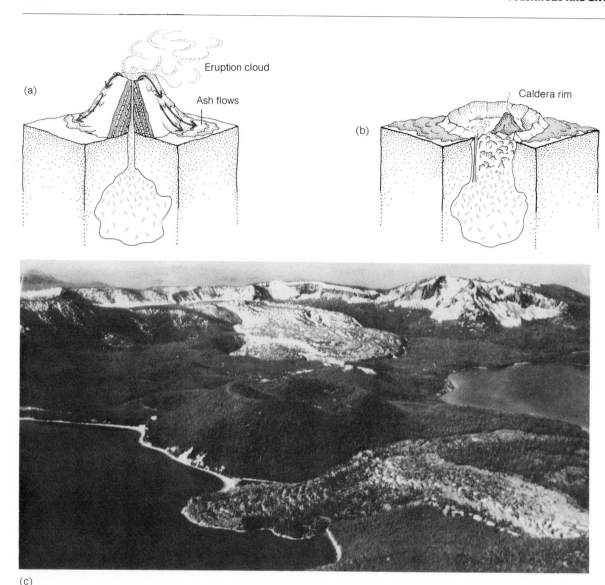

(c)

FIGURE 19–17

(a) A large reservoir of water-rich sialic magma erupts through a composite volcano as a vast cloud of ash that blows high into the air and as enormous ash flows that pour down its flanks. (b) Calderas form as volcanoes subside into emptying magma chambers during large eruptions. If more magma remains beneath the caldera, it may build a new volcano through further eruptions or it may crystallize underground to form a pluton roofed by the remains of the old volcano. The volcano settles into the emptying magma chamber, creating a broad subsidence crater rimmed by the remnants of the original cone and surrounded by ash flow deposits. (c) A view across Newberry caldera near Bend, Oregon. The pumice cone between the lakes and the large obsidian flows in the foreground and background erupted after the caldera subsided. (Oregon Department of Geology and Mineral Industries)

anxious to get out the vote for an impending election. The city had been widely notorious as a center of vice and generalized corruption and there were people who suggested that its fiery annihilation came as just retribution. However, the identity of one of the survivors, a convicted murderer doing time in the local jail, probably limited the number of sermons.

THE CASCADES AND THE COLUMBIA PLATEAU

The Western Cascades

The first Cascade volcanoes began erupting along a line somewhat west of the modern volcanic chain during Oligocene time. They started their careers by erupting basalt to build a long row of large shield volcanoes and then, as more silica-rich andesite began to appear, erected a chain of composite volcanoes on the basalt foundation. The volcanoes in the southern third of the chain produced large quantities of extremely silica-rich lavas, including rhyolite, during the later stages of their careers. Some of those rhyolite eruptions spread sheets of welded ash well into central Oregon, more than 100 km from the source. No historic eruption has spread welded ash more than a few kilometers from the vent.

After approximately 10 million years of vigorous activity, the western Cascade volcanoes snuffed out during early Miocene time. The centers of volcanic activity then shifted to central and eastern Oregon and Washington and to northeastern California to build the **Columbia volcanic plateau,** which is composed almost entirely of basalt and rhyolite. No one knows what caused the shift.

The Columbia Plateau

Figure 19–18 shows the extent of the Columbia volcanic plateau as well as the contiguous and similar Snake River plain of southern Idaho, which together form one of the world's largest volcanic provinces. Most of the eruptions that created the Columbia plateau occurred during the latter half of Miocene time and that part of the province has been relatively inactive since then. Most of the Snake River

plain formed during Pliocene time but its eastern end, which includes the Yellowstone Park volcano, is still active.

Detailed geologic field work has shown that the Columbia plateau consists of a mosaic of quite large volcanoes each of which produced basalt distinctive enough to be recognized in the field and plotted on a map. The volcanic centers in northern Oregon and eastern Washington erupted only basalt, presumably because no sialic continental crust exists beneath that region. The volcanic centers farther south where sialic continental crust does exist produced both rhyolite and basalt as do those in the Snake River plain. Evidently, the magmas originated through partial melting of the upper mantle and lower continental crust, presumably in response to relief of pressure caused by extension of the earth's crust.

FIGURE 19–18

The Columbia volcanic plateau and Snake River plain are both composed of basalt and rhyolite, as are the less overwhelmingly abundant volcanic rocks in the Basin and Range province to the south. Volcanic activity in those regions appears to be associated with crustal tension. The Cascade volcanic chain is composed dominantly of andesites and is associated with the convergent plate boundary immediately off the Pacific Coast.

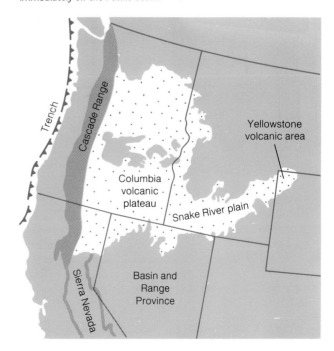

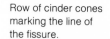

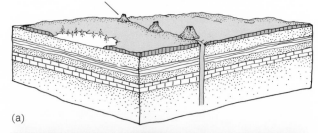

(a)

(b)

FIGURE 19-19

(a) Flood basalts well up out of a fissure and spread across vast areas. After the flow has been eroded away, the basalt-filled fissure remains as a dike. (b) A section eroded through one of the flood basalt flows of the Columbia volcanic plateau near the Oxbow of the Snake River, Oregon. The columnar jointing is typical of lava flows and also of many dikes.

Flood Basalts

Many of the eruptions in the Columbia plateau and Snake River plain produced overwhelming volumes of basalt lava. Individual lava flows have been traced all the way from eastern Oregon and Washington to the Pacific Ocean. A number of individual flows have been shown to have covered areas measured in many thousands of square kilometers with several hundred cubic kilometers of basalt, volumes several times greater than those of large composite volcanoes, such as Mount Rainier. Geologists refer to such enormous flows as **flood basalts** (Figure 19–19). Although we have eyewitness accounts of small flood basalt eruptions in Iceland, no large eruption of that type has been seen during historic time.

A small flood basalt eruption occurred on the northern border of the Snake River plain about 2000 years ago in an area now partly within Craters of the Moon National Monument. An estimated 37.5 km^3 of basalt lava welled out of a long fissure marked by a row of neatly aligned cinder cones and flooded an area of 1,550 km^2. Part of the fissure still survives as a long series of open cracks that split the surface of the lava plateau south of the flow. They are several meters wide and as much as several hundred meters deep. Those open fissures leave no doubt that the earth's crust is indeed stretching in that area, that pressure-relief partial melting should be occurring at depth.

Yellowstone Park

The Yellowstone Park volcanic area lies at the northeastern tip of the Snake River plain. Most geologists regard it as the youngest part and leading edge of that volcanic province. Past eruptions in the area have produced enormous volumes of rhyolite and modest quantities of basalt. There is evidence that the Snake River plain consists basically of a thick accumulation of rhyolite capped by a crust of basalt. Thus, we can envision it as a series of volcanoes that erupted rhyolite early in their careers and then covered themselves with basalt. If so, then the Yellowstone volcanic area can be interpreted as a part of the Snake River plain still in an early stage of development.

The Yellowstone volcanic area has a long history consisting basically of several episodes of violent activity during which enormous calderas formed and then filled with rhyolite. The calderas appear on geologic maps showing distribution of rock types but do not form topographic basins. Radioactive age dates show that the large eruptions happened at intervals of approximately 600,000 years and that the most recent occurred about that long ago. The numerous hot springs and geysers in the park clearly show that intensely hot rock still exists at shallow depth. Furthermore, seismic waves passing beneath Yellowstone Park lose much of their shear wave motion so it appears that at least some of that hot rock is still melted. No geologist would be surprised by a large or even catastrophic volcanic eruption in Yellowstone Park.

The High Cascades

About 10 million years ago eruptions ceased in most of the Columbia volcanic plateau and resumed in the Cascades where the modern volcanic chain has grown along a line east of that followed by the earlier chain of Cascade volcanoes. It is interesting that the new volcanoes lie east of the older ones because it might seem that the offshore trench should be shifting to the west as the sinking lithospheric plate stuffs more sediment into it. Perhaps the slab of lithosphere is now sinking at a flatter angle than it formerly did. If so, it would penetrate farther beneath the continent before it reaches the depth at which its cover of oceanic crust begins to melt (Figure 19–20).

Like their extinct predecessors in the western Cascades, the modern High Cascade volcanoes began by laying down a broad foundation of basaltic shield volcanoes some of which have erupted within recent centuries. Then andesite lava began to erupt and built the row of imposing composite volcanoes that appear on so many picture postcards.

Early settlers in the northwestern United States reported, with varying degrees of reliability, numerous eruptions in the High Cascades. That activity stopped in 1857 and little has happened since then except the 1916 eruption of Mount Lassen and the 1980 eruption of Mount St. Helens. The region has also been remarkably free of the deep

FIGURE 19–20

(a) It may be possible to explain the eastward shift from the older Western Cascades to the modern High Cascades as the result of a change in the angle at which the sinking plate descends. The lithosphere now sinking off the coast formed at a ridge close offshore and is therefore relatively warm and relatively light. It does not sink steeply. (b) When the Western Cascades were active during middle Tertiary time, the East Pacific ridge lay far offshore. Therefore, the lithospheric plate spreading from it was cold and dense by the time it reached the convergent boundary. That made it sink steeply thus bringing the Western Cascades close to the trench.

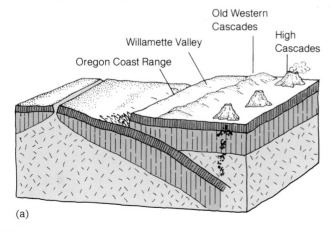

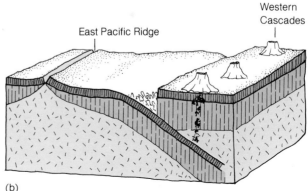

focus earthquakes typical of an area underlain by a sinking lithospheric plate. Nevertheless, many of the modern Cascade volcanoes are very young, show evidence of having been vigorously active within recent centuries, and seem likely to erupt again.

ANTICIPATING ERUPTIONS

Large volcanic eruptions commonly cause heavy property damage and numerous casualties because the fertile soils typical of volcanic regions generally support dense human populations. Reliable methods of forecasting the time and type of eruptive activity would make it possible to minimize casualties and to prevent at least some property damage. There are several approaches to predicting volcanic activity but these succeed only in a limited way.

Active and Extinct Volcanoes

The first step in predicting volcanic hazards is simply to identify the active volcanoes, which is not always as easy

as it might seem. Vesuvius, for example, is obviously an active volcano because it erupts frequently. However, it had been silent for more than 2000 years before the great eruption of A.D. 79 that destroyed Pompeii and Herculaneum. The local people then did not even realize that the mountain is a volcano until it began to erupt. It is likewise possible that some volcanoes may have had their final eruption during historic time and are now dead even though they were recently active. In most cases, the type of volcano and its general appearance probably provide a far better indication of what to expect than does the historic record.

Cinder cones, fissure basalt flows, and plug domes commonly erupt only once from any particular vent. Those types of eruptions typically produce new volcanoes although in most cases there are others like them in the general vicinity. Paricutin, for example, suddenly started erupting from a field in central Mexico in 1943 and remained active (Figure 19–21) for nine years, an extraordinarily long career for a cinder cone. The surrounding region is liberally dotted with older cinder cones, including one

FIGURE 19–21
Paricutin blowing a cloud of ash. The volcano started erupting in a cornfield in central Mexico in 1943 and reached a height of several hundred meters before becoming inactive a few years later. The church tower in the middleground rises above a lava flow that buried a small town. The steep front of the flow is visible in the right foreground just beyond the long building. (K. Segerstrom, U.S. Geological Survey)

FIGURE 19-22
The main peak of Mount Shasta, on the left, is deeply eroded and has obviously been inactive for many thousands of years. Shastina, the satellite peak on the right, is almost uneroded. Some evidence suggests that Shastina erupted in 1785 and no geologist would be surprised if it were to erupt again.

called Jorullo that appeared in 1759, and it seems likely that future eruptions will produce more new volcanoes instead of reactivating one of the old ones. It likewise seems probable that future flood basalt eruptions in the Snake River plain will come from new fissures and that future plug-dome eruptions in the Cascades will quietly add new hills to the landscape.

Composite and shield volcanoes erupt repeatedly from a central vent. Those that are free of deep erosional dissection have obviously been active during the recent geo-

logic past and should be regarded as probably active and likely to erupt again. Deeply eroded volcanoes that have not repaired themselves for many thousands of years are probably extinct. Figure 19–22 illustrates the point.

Signs of an Impending Eruption

Many volcanoes announce their eruptions, as Mount St. Helens did, by producing a series of small to moderate earthquakes during the days or weeks before any lava appears. Volcanic earthquakes are never large, rarely have a magnitude as great as 5.5, and many are peculiar in being felt only within a limited area. That presumably reflects their origin from movements of liquid magma, which cannot propagate shear waves.

If geologists can establish an array of field seismograph stations around a rumbling volcano, they can in many cases watch the magma rise by observing the focal points of the earthquakes become shallower day by day. In a few instances, most notably in Hawaii, eruptions have been forecast accurately by determining the rate at which the magma is rising and projecting the time when it will reach the surface. However, in many other cases, volcanic earthquakes have continued for a while and then ceased without leading to an eruption. Presumably, the magma solidified before it could reach the surface.

Most composite volcanoes begin their eruptive phases, as Mount St. Helens did, by blowing clouds of steam and ash for days or weeks before lava appears. However, that activity may merely mean that that surface water seeping into the hot rocks deep within the volcano is returning as steam, carrying some old volcanic ash along. Once again, an array of field seismographs can help clarify the situation. If the earthquakes are coming from deep beneath the volcano, something must be happening at depth and an eruption may follow. However, if the earthquakes are coming from within the volcano, they probably mean that the mountain is just blowing off steam and is unlikely to produce lava.

The temperatures and compositions of volcanic gases also provide hints. As long as a volcano is blowing steam at a temperature near the boiling point of water, most geologists do not believe that an eruption is imminent. A sharp rise in steam temperature suggests that magma may be approaching the surface and that an eruption is likely. An increase in the amount of sulfur compounds in the volcanic gases also precedes many eruptions.

Many volcanoes swell slightly—as though they were inhaling—as magma rises into them, and then subside—as though they were exhaling—when they erupt. Geologists measure those small movements with sensitive instruments called **tiltmeters,** which detect minute changes in slope. Tiltmeter observations have proved quite helpful in predicting Hawaiian eruptions and they did reflect the movement of a large mass of magma into Mount St. Helens although

that observation did not lead to an accurate prediction of what followed.

Predicting the Type of Eruption

Basalt eruptions generally follow a fairly predictable and reasonably nonviolent course. If the lava contains some steam, it may blow glowing fountains of lava into the air, creating a spectacular fireworks display. However, that is not considered to be violent activity by volcanic standards. When the steam is gone, the remaining lava erupts quietly as a flow, which may cause property damage but poses little threat to human life. On several occasions people have diverted basalt flows into less damaging paths by freezing their fronts with sprays of water.

The course of an eruption from a large composite volcano is impossible to predict. Even minor activity, such as that on Mount Lassen in 1916, is likely to produce devastating floods and mudflows as heat from the eruption melts snow and ice on the mountain. If the water mixes with newly erupted ash, it is likely to create boiling mudflows that may pour dozens of kilometers down stream valleys at speeds measured in tens of kilometers per hour. Volcanic mudflows are extremely dangerous and people who find themselves near an erupting composite volcano should stay out of stream valleys. Relatively minor eruptions may also produce glowing clouds that sweep down the slopes of the volcano with absolutely no advance warning and incinerate everything in their paths. Furthermore, a low and apparently unthreatening level of activity may progress almost instantly and without warning, as Mount St. Helens did, to a violently explosive eruption. An erupting composite volcano may stop at any moment or continue until it has reduced itself to a gaping caldera.

BATHOLITHS

As the slow processes of erosion strip old volcanoes off the landscape and expose their roots, they commonly expose masses of coarse-grained granite called **batholiths** if they cover areas greater than about 100 km², and **stocks** if they cover lesser areas. Only rarely do large masses of gabbro appear in the roots of old volcanoes. It seems that basaltic magmas almost invariably erupt whereas the majority of sialic magmas crystallize at depth without erupting. The explanation for that important difference hinges upon the role of water.

Water, Melting Point, and Pressure

Laboratory experiments with sialic magmas show that addition of water lowers their melting points. That is no surprise because the addition of an extra component invariably lowers the melting point of any mixture. However, it is a surprise to find that the melting points of water-saturated sialic magmas increase, rather than decrease, with relief of pressure. The melting points of all dry rocks and indeed of almost all substances with the conspicuous exception of ordinary ice drop with a decrease in pressure

because they expand as they melt. Evidently, water-saturated sialic magmas must, like water, expand slightly as they crystallize into their solid forms. Figure 19–23 summarizes the relationship between pressure and melting point for dry and water-saturated sialic magmas. Let us consider its geologic implications.

Granitic Plutons

The pressure on a mass of magma decreases as it rises. If the magma contains no water, the drop in pressure will also lower its melting point and thus help maintain it in a molten condition even if it cools somewhat. Such magmas continue to rise until they erupt. That almost certainly explains why basaltic magmas, which are incapable of absorbing much water, generally erupt. It likewise explains

FIGURE 19-23

Curves showing the influence of pressure—plotted here as depth below the surface—on the melting temperatures of dry and water-saturated sialic magmas. The melting temperatures of dry sialic magmas, like those of dry magmas of other compositions, decrease as they approach the surface so they generally remain molten and finally erupt. The melting temperatures of wet sialic magmas, on the other hand, increase as they approach the surface so most freeze at depth without erupting to become plutons. (Adapted from diagrams in "Controls on source and depth of emplacement of granitic magma" by D. W. Hyndman, GEOLOGY, 1981, v. 9, pp. 244–249.)

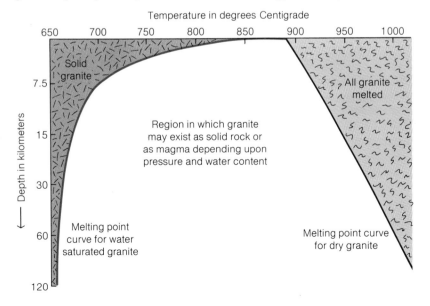

the eruption of sialic plug domes and lava flows, which contain very little water.

However, if a mass of sialic magma absorbs large amounts of water, its melting point will rise as it ascends into higher levels of the earth's crust where the pressure is lower. When its melting point has risen to match the actual temperature of the magma, the mass will freeze where it stands. The magma does not freeze, as lava flows and plug domes do, because it has lost heat and cooled to a temperature that matches its melting point. Instead, it freezes because the drop in pressure has raised the melting point to the temperature of the magma. Most granitic plutons probably crystallize with little or no decrease in temperature and then slowly cool as the solid rock loses heat to its surroundings.

The fate of a rising mass of sialic magma depends upon the accident of whether and where it happens to encounter water on its way to the surface. The majority absorb water at some depth, rise a bit higher, and then crystallize to form plutons deep within the earth's crust. A few remain dry and reach the surface to erupt quietly as plug domes or lava flows. Only the very few that rise almost to the surface before they absorb large amounts of water can generate the cataclysmic eruptions that convert composite volcanoes into calderas. Those eruptions spew a potential granite pluton, or at least a large part of a potential pluton, across the landscape as a sheet of welded ash and pumice.

Batholiths and Batholiths

Although all batholiths resemble each other in being enormous masses of granitic rocks, they form in a variety of geologic settings and vary considerably in their natures.

The vast majority of batholiths appear to have formed beneath volcanic chains and express that origin in their generally linear map outline. One of the best examples of a large mass of granite of that type is the Sierra Nevada batholith of California. It is not a single mass of granite but instead consists of a large number of individual granitic plutons, which vary greatly in size but average perhaps 25 km in diameter. In the southern part of the Sierra Nevada, those plutons pack closely together to form a virtually continuous mass of granite. Farther north, they are more widely scattered in the metamorphosed subduction complex that comprises the basic framework of the range. Radioactive age dates show that plutons of the same age tend to align in north to south trends that shifted back and forth with time. That perhaps reflects the same kind of shift that occurred between the first and second periods of activity in the Cascade volcanoes. Such batholiths consist of swarms of plutons that invade a subduction complex and help convert it into typical continental crust.

Like the Sierra Nevada batholith, the Idaho batholith of central Idaho formed on the landward side of a convergent plate boundary during Mesozoic and early Tertiary time. However, the magmas that formed the Idaho batholith penetrated through old sialic crust that had existed since Precambrian time and therefore did not enlarge the continent.

The Idaho batholith formed in two major stages of significantly different activity. The first, which occurred during late Mesozoic time, involved emplacement of large granite plutons deep within the crust. Very little or no volcanic activity accompanied that first stage. Then much of the overburden of older rocks slid off the batholith while it was still crystallizing and exposed the newly formed granite at the surface. The second stage, which followed after a lull of at least 15 million years, involved emplacement of a second group of large granitic plutons at extremely shallow depth during Eocene time. That stage was accompanied by a large number of extraordinarily violent rhyolite eruptions that blanketed large areas of the batholith beneath a thick cover of volcanic rocks, the *Challis volcanics*. Those eruptions opened calderas that dwarf any formed during historic time. By volcanic standards, the 1980 eruption of Mount St. Helens was a shot from a popgun.

The Boulder batholith of western Montana is an isolated large mass of granite emplaced during late Mesozoic time. It stands well to the east of the Idaho batholith, too far east to be associated in a simple way with the convergent plate boundary that existed off the west coast of the United States when it formed. The Boulder batholith is peculiar in consisting almost entirely of a single, large pluton, which appears to have crystallized beneath a cover of andesitic volcanic rocks erupted from the same body of magma. Perhaps the Boulder batholith formed in an isolated volcanic center somewhat similar to the one now active in Yellowstone Park.

Plutons and Ore Deposits

Prospectors have known for centuries that promising ground generally lies near the margins of granitic plutons, especially in the rocks above them. They systematically search the rocks enclosing granitic plutons, and the igneous rocks themselves, for deposits of gold, silver, copper, lead, molybdenum, and other metals. However, some granitic plutons have ore bodies associated with them whereas others do not. Water seems to play the major role in determining which are mineralized and which remain barren.

Many masses of sialic magma crystallize to form plutons that have neither contact metamorphic rocks nor ore bodies around them. Evidently, the water pressure in the enclosing rocks was high enough to confine steam within the magma and therefore prevent significant transfer of heat and mineral matter into the rocks around the contact. Other plutons, generally those that crystallized at relatively shallow depth, are surrounded by extensive zones of contact metamorphosed rock, which commonly contain ore bodies. Those plutons evidently expelled steam into their enclosing rocks and the steam carried both heat and dissolved mineral matter.

Some plutons also contain ore bodies within the igneous rock. In many such cases, they consist of copper minerals disseminated through granite that shows evidence of having been altered by reaction with steam. Such plutons generally have a porphyritic texture so ore deposits of that type are called **porphyry coppers** although most contain other metals as well.

Many granitic plutons drive a strong circulation of water through the rocks surrounding them (Figure 19–24), which may alter the original minerals and emplace ore deposits. Water heats into steam and dissolves minerals as it rises through the rocks surrounding the crystallizing mass of magma. Then the water deposits those minerals in fractures as it rises into the cooler rocks above the pluton. Steam escaping from the pluton may also deposit minerals in those fractures. In either case, the mineral veins consist mostly of quartz accompanied in many cases by pyrite and in some cases by a variety of other sulfide minerals as well as native elements, such as gold or silver. Presumably, the solubilities of the ore minerals in steam must resemble that of quartz because they could not otherwise occur together. Quartz veins may contain a wide variety of ore mineral assemblages, depending partly upon what happened to be available and partly upon the temperature of vein formation.

Quartz veins are difficult to mine profitably because they tend to contain relatively small bodies of high-grade ore separated by much larger masses of low-grade ore or barren rock. The irregular ore distribution makes it difficult to estimate reserves accurately and therefore difficult to plan capital investments rationally. The history of mining is replete with examples of properties that produced little after having yielded just enough high-grade quartz vein ore to inspire an optimistic investment in capital equipment.

Porphyry copper deposits, on the other hand, contain low-grade ore in enormous tonnages of remarkably uni-

FIGURE 19–24

In many cases, hot water and steam circulate through a crystallizing pluton and its surrounding rocks, altering them and creating a wide variety of metallic mineral deposits. Some of the circulating water may reach the surface as hot springs or geysers, which form potential energy resources. Geologists prospecting for new mineral reserves commonly look first for alteration zones, which may be quite large, and then search within them for ore bodies.

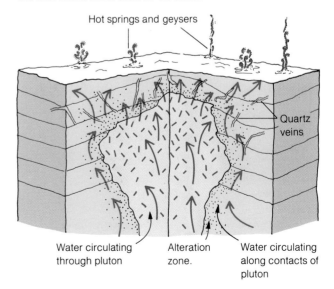

Hot springs and geysers

Quartz veins

Water circulating through pluton

Alteration zone.

Water circulating along contacts of pluton

form quality. That makes it possible for mining operators to estimate their reserves accurately and therefore to plan capital investments with reasonable projections of their probable returns. Most such deposits are mined from large open pits that produce great tonnages of ore at a small but manageable profit. They include such mines as those at Bingham Canyon, Utah, Ajo, Arizona, and Butte, Montana.

SUMMARY

Igneous rocks crystallize from magmas that melt in a variety of geologic situations. The most abundant magma is basalt, which forms where the peridotite of the mantle partially melts. Most basalt magma forms at the crest of the oceanic ridge where separating lithospheric plates relieve pressure on the peridotite beneath, permitting it to partially melt and then erupt to form the oceanic crust.

Much of the oceanic crustal basalt that sinks back into the mantle at convergent plate boundaries melts again as it reaches the asthenosphere and returns to the surface through volcanic chains paralleling the trench. Many of those volcanoes also erupt a wide variety of intermediate and sialic magmas, which apparently form as molten basalt rising from the descending plate melts rocks in the continental crust. Where crustal extension relieves pressure at depth in continental areas, basalt and rhyolite magmas form through partial melting of the upper mantle and lower

part of the continental crust, respectively. Volcanoes in such regions tend to erupt basalt and rhyolite unaccompanied by lavas of intermediate composition.

The viscosities of magmas and their abilities to absorb water both increase with their silica content. Basalt erupts quietly because the small amount of water it contains escapes easily from the fluid lava, which runs out to make thin flows. A long sequence of basalt flows from the same vent builds a broad shield volcano. The more silica-rich sialic magmas also erupt quietly unless they contain large amounts of steam, which expands the erupting lava into a foam that blows off as ash or pumice. Intermediate lavas alternate between erupting as flows or as fragmental material and build large composite volcanoes with steeply conical forms.

The melting points of water-saturated sialic magmas rise with decrease of pressure as they ascend toward the earth's surface. Therefore, most such magmas crystallize within the earth's crust where they reached a level at which the melting point of the magma rose to match its actual temperature. They become granitic plutons and batholiths, which form most abundantly beneath the volcanic chains associated with convergent plate boundaries but also develop in other geologic situations. In some cases, water circulating through and around the crystallizing pluton alters it and the surrounding rocks by removing some minerals and adding others, which may include valuable ores.

KEY TERMS

ash flow	porphyry copper
batholith	pumice
caldera	scoria
cinder cone	shield volcano
Columbia volcanic plateau	stock
composite volcano	tiltmeter
flood basalt	volcanic ash
glowing cloud eruption	volcanic bomb
lava dome	volcanic cinder
plug dome	welded ash
	welded tuff

REVIEW QUESTIONS

1. Under what circumstances will a body of sialic magma erupt explosively, erupt quietly, or crystallize at depth without erupting?

2. Briefly explain the interrelationship between the compositions and viscosities of magmas, their ability to absorb water, and the manner of their eruption.

3. Why does addition of silicon to a magma make it more viscous and also more capable of absorbing water?

4. Explain why you would or would not expect a shield volcano to produce a violent eruption.

5. Why do sialic magmas tend to crystallize at depth to form plutons whereas those with a more basaltic composition generally erupt?

6. Why does relief of pressure on the rocks in the lower crust or mantle permit them to melt?

SUGGESTED READINGS

Bullard, Fred M. *Volcanoes of the Earth.* Austin, Texas: University of Texas Press, 1976.

A beautifully illustrated, eminently readable, and authoritatively comprehensive treatment of volcanoes. Includes careful case histories of numerous historic eruptions.

Green, J. and N. M. Short. *Volcanic Landforms and Surface Features.* New York: Springer-Verlag, 1972.

An elegant photographic atlas illustrating a wide variety of volcanic phenomena.

Herbert, D. and F. Bardossi. *Kilauea: Case History of a Volcano.* New York: Harper & Row, 1968.

A detailed study of the behavior of a large shield volcano.

Hyndman, D. W. *Petrology of Igneous and Metamorphic Rocks.* New York: McGraw-Hill Book Co., 1972.

A standard advanced text. Includes an excellent discussion of batholiths.

McDonald, G. A. *Volcanoes.* Englewood Cliffs, N.J.: Prentice-Hall, 1972.

A comprehensive discussion of volcanoes.

CHAPTER TWENTY

Intensely contorted granitic gneisses typical
of the kinds of rock that form large
portions of the continental basement.
(W. R. Hansen, U.S. Geological Survey)

CONTINENTS

The earth constantly generates new oceanic crust at the oceanic ridges and consumes the old in the trenches so its ocean floors vividly reflect the dynamic processes now at work. However, the ocean floors renew themselves so rapidly that no part of them is old enough to preserve a record of events that occurred more than about 200 million years ago. We must turn to the continents for information about the more remote geologic past. That is where the earth keeps its archives.

Continental crust, once formed, becomes a permanent part of the earth's surface that does not vanish back into its interior. The old rocks preserved in and on the continents are the only surviving witnesses to the earth's distant past and they are not quite so mute as certain old proverbs would have us suppose.

Continents consist basically of rafts of sialic rock approximately 40 km thick. They consist of a mosaic of complexes of granitic and metamorphosed sedimentary and volcanic rocks that formed piecemeal during the long geologic past. Geologists call those coarsely crystalline rocks the **basement** because they lie beneath all the other kinds of rocks and extend continuously downward beyond the range of direct observation.

Most of the continental basement lies buried beneath layers of sedimentary rock, the **sedimentary veneer** (Figure 20–1), which preserves at least a partial record of geologic events since the underlying basement rock formed. Because it rides on top of the continental platform, the sedimentary veneer generally escapes incorporation into subduction complexes. Geologists devote much of their attention to the sedimentary veneer because it contains most of the earth's accessible fossil fuel reserves as well as a large share of its other mineral resources. The same rocks also contain most of the record of life on earth.

Although the continents generally tend to be relatively stable, involvement in plate tectonic events may deform the basement and the sedimentary veneer into mountains quite different from those that form in oceanic trenches. The Rocky Mountains of western North America are an excellent example; their formation involved large-scale breakage of the basement accompanied by deformation of the sedimentary veneer.

Plate tectonic processes also dismember the continents and assemble the pieces in new arrangements. Those events have made the basement into an intricate mosaic of pieces that geologists have only begun to interpret. They also create mountain ranges of which the Appalachians of eastern North America are an outstanding example.

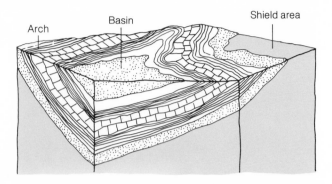

FIGURE 20-1

A schematic diagram showing part of a continent with a shield area, which is devoid of sedimentary cover, and a sedimentary veneer, which varies greatly in depth being thickest in basins and thinnest on arches.

THE CONTINENTAL BASEMENT

Geologists can study the basement directly only in areas where the sedimentary veneer is missing or from occasional drill cores obtained where the sedimentary veneer is relatively thin. Fortunately, each continent contains at least one large region, called a **shield,** in which the basement rocks are exposed. The Canadian shield of North America extends over most of Canada from western Manitoba to the Atlantic Coast and into the northern tier of the United States from Minnesota east. Lesser exposures of basement rock occur in widely scattered areas of the continent, including parts of the Piedmont and Blue Ridge region of the south Atlantic states, smaller areas in the Missouri Ozarks and central Texas, and numerous large and small areas in the Rocky Mountain region of the United States and Canada.

Age Provinces

Radioactive age dating reveals large regions called **age provinces** (Figure 20–2) in which all the basement rocks are nearly the same age. Most age provinces cover regions measurable in many thousands of square kilometers in which the rocks are not only about the same age but also to some extent distinctive in type and in structural style.

For example, the rocks of the Grenville province in Quebec and southeastern Ontario are approximately 1 billion years old and contain large amounts of marble whereas those of the adjacent Superior province to the west are about 2.5 billion years old and contain no marble. The distinction between those provinces and the sharp line separating them were recognized during the last century.

Age provinces conclusively show that the continents have grown. If one part of the continental crust is older than another, then the continent must have become larger when the younger part formed. However, a quite large proportion of the continental crust formed during early Precambrian time, before about 2.5 billion years ago. Geologists call that oldest continental crust the **Archean basement** and find that it differs from the younger parts of the continents.

Archean Basement

The Archean provinces, which formed between about 2.5 and 3.8 billion years ago, resemble the younger basement provinces in consisting mostly of complexly deformed schists and gneisses extensively intruded by granite. However, they differ from the younger age provinces in containing linear belts of less metamorphosed sedimentary and volcanic rocks. Many geologists nickname those rocks ''greenstone'' because of their dominantly green color, which is due to abundant chlorite. Greenstone belts differ greatly in size, ranging from a few to a few hundred kilometers in length and are generally about one-tenth as wide as they are long. Most seem to have the structural form of crudely canoe-shaped synclines (Figure 20–3) folded down into the surrounding granites, gneisses, and schists. Their interpretation is a matter of lively debate.

Most of the rocks in greenstone belts consist of pillow basalts and muddy sediments resembling those on modern ocean floors. Many geologists simply interpret them as remnants of ancient ocean floor incorporated into the continent. However, that interpretation leaves unanswered the question of why typical greenstone belts do not occur in basement rocks formed since Archean time. Therefore, some geologists believe that greenstone belts may record the operation of processes fundamentally different from those that have shaped the earth since Archean time. Remember that the earth still had most of its original inventory of radioactive elements during Archean time and therefore had a larger source of internal heat then than it has had since.

However, there is little reason to doubt that basement rock formed after about 2.5 billion years ago records the operation of plate tectonic processes similar to those that operate today. The younger age provinces tend to be smaller than those formed during Archean time, are more nearly linear in map outline, and consist of the same kinds of rocks that form along modern convergent plate boundaries.

Therefore, it is easy to imagine that the continents should logically consist of ancient cores formed during Archean time surrounded in a sort of giant bullseye pattern by concentric belts of successively younger basement rocks. However, the pattern is actually far more complex because the continents have been moving in various ways as well as growing. The map outlines of the age provinces shows a sort of crazy-quilt pattern, which doubtless records a long history of plate tectonic movement even though most of that story remains to be deciphered.

The Lost Piece of North America

Continental rifting evidently detached a large area of basement rock from the region of the northwestern United States and western Canada sometime between 800 and 1200 million years ago. Sedimentary rocks of that age in the northern Rocky Mountains contain swarms of basalt sills and dikes many of which have yielded radioactive age dates close to 1 billion years. Similar swarms of basalt sills and dikes intruded parts of the east coast of North America and the west coast of Europe about 200 million years ago just as the Atlantic Ocean was beginning to open along the rift that has since become the mid-Atlantic ridge. In the northern Rocky Mountains we also find that the old continental crust and the Precambrian sedimentary veneer both end abruptly along a line that trends north through easternmost Oregon and Washington and on into British Columbia. It looks very much as though there must have been a continental rift along that line about a billion years ago. If so, then where did the detached piece of continental crust go?

Some geologists believe it now forms a large part of eastern Siberia. There are basement rocks there of the appropriate age and their structural style and magnetic

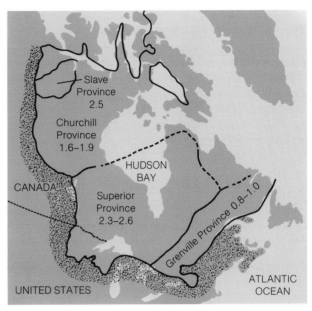

Sedimentary veneer

FIGURE 20-2

The major age provinces in the Precambrian basement rocks of the Canadian shield. Samples obtained from drill cores and more scattered exposures of basement rock have made it possible to extend the province boundaries beneath the sedimentary veneer, although with considerably greater margin for error. (Adapted from J. A. Lowden and others, Geological Survey of Canada, 1963.)

FIGURE 20-3

The greenstone belts in Archean basement provinces consist of tightly folded troughs that contain pillow basalts and sedimentary rocks that are much less metamorphosed than the surrounding granites, gneisses, and schists.

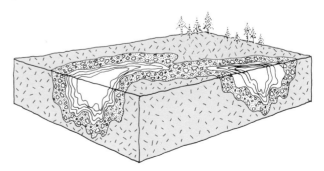

polarization would fit them nicely into the old western margin of North America if they were rotated a bit. Furthermore, there is a Precambrian sedimentary veneer on parts of that eastern Siberian basement closely resembling that in the northern Rocky Mountains. Few geologists would argue that the evidence constitutes conclusive proof that a large part of eastern Siberia was once part of North America. However, most would agree that the resemblances are at least strongly suggestive.

The Western Microcontinents

The same part of North America that lost a large piece of continental crust during late Precambrian time seems to have acquired several new pieces during Mesozoic and early Tertiary time. The lithospheric plate that began to sink beneath western North America during early Mesozoic time evidently contained several small continents that were added to North America as they arrived at the con-

FIGURE 20-4

A scenario for addition of a microcontinent to western North America shown in sections drawn along the line of the border between the United States and Canada. The suggested existence of an established convergent boundary along the western margin of the microcontinent could explain why it joined the North American continent without crushing the suture between them into a mountain range. (a) During Cretaceous time the microcontinent was far offshore, embedded in a plate that was sinking beneath the western edge of the North American continent. Meanwhile, the Farallon plate was sinking beneath the western edge of the microcontinent at another convergent boundary. (b) By early Tertiary time the eastern convergent boundary had swallowed the last of the ocean floor separating the microcontinent from North America and the two joined along a suture marked by a subduction complex. At that point the plate in which the microcontinent had been embedded ceased to exist as a separate entity. The Farallon plate continues to sink at the convergent boundary immediately off the Pacific Coast.

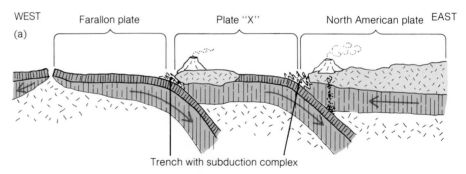

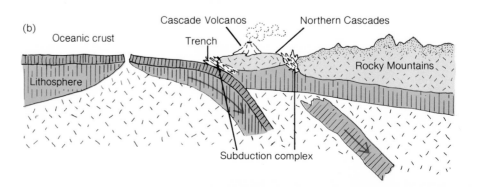

vergent boundary. They now form a large part of British Columbia, the Canadian northwest, and Alaska.

One of the largest of those small continents forms the northern Cascade Range of Washington and extends north through much of the British Columbia Coast Ranges. All of that region is underlain by continental basement rock locally veneered with sedimentary formations quite different from those that cover basement rock farther east. The formations of the western sedimentary veneer also contain distinctive fossil assemblages.

The line of suture between that large microcontinent and North America follows the Okanagon Valley of central Washington and British Columbia, which contains a subduction complex complete with ophiolites, serpentinites, and blueschists. Evidently, the plate collision smoothly transferred the microcontinent to the North American plate without compressing the rocks along the line of suture into a mountain range similar to the Urals. Figure 20–4 outlines a hypothetical plate tectonic scenario that may explain how that happened.

The Appalachians

Pangaea was merely a chance aggregation of continental crust that lasted only briefly and the dispersal of its fragments was just one of the latest in a long series of plate tectonic events. An earlier event involved an ocean that existed where the Atlantic is now and then vanished as the continents assembled to form Pangaea. Figure 20–5 summarizes the sequence of events that left the Appalachians as one of their more obvious souvenirs.

The story of the ancestral Atlantic began to emerge more than a century ago when some of the first geologists to study Cambrian fossils found that the faunas on opposite sides of the Atlantic differ considerably. That was no surprise; animals that live on opposite sides of oceans generally do differ considerably. But those early geologists were surprised to find a few small areas on each side of the Atlantic where the Cambrian rocks contain fossils that clearly belong on the other side of the ocean. For example, the Cambrian rocks in much of eastern New England and some parts of the Maritime Provinces contain fossils that belong to the European faunal realm. Cambrian rocks in the northern British Isles and along the coast of Norway contain fossils such as those generally found in North America. That curious distribution of Cambrian fossils,

along with other observations, eventually inspired some of the boldest thinking in modern geology.

Geologists now believe an ocean did separate North America from Europe during early Paleozoic time, an ocean wide enough to permit the animals living on its opposite shores to evolve independently. That early version of the Atlantic Ocean may have existed for several hundred million years while the trailing continental margins on its opposite sides acquired broad continental shelves and coastal plains. Meanwhile, Africa was in the south polar region, wearing a cover of ice similar to the glacier that blankets Antarctica today. Evidence of that ancient glaciation survives in the Sahara region in the form of striated bedrock surfaces covered by deposits of till now lithified into rock.

That early version of the Atlantic Ocean closed about 300 million years ago, during late Paleozoic time, when the last of its floor disappeared into a trench and the European and North American continents joined. We can perhaps suppose them meeting tentatively at first and then squarely colliding to crush their continental shelves and coastal plains between them as though between the jaws of a vise. Meanwhile, the southern continents were moving north after a long sojourn in the south polar regions. What is now Africa collided with what is now the southeastern seaboard of the United States to complete the assembly of Pangaea.

The oceanic sedimentary rocks crushed between the colliding continents were profoundly metamorphosed and invaded by vast masses of granitic magma that formed as they partially melted. They thus became new continental basement rock joining North America, Europe, and Africa as though it were a welding bead run along the seam between them. By late Paleozoic time that seam may well have been a long range of mountains perhaps resembling the Ural Mountains that mark a similar join between Europe and Asia.

The Appalachian Mountains are among the more visible mementos of that ancient plate collision. They consist essentially of the sedimentary rocks of the inner continental shelf and coastal plain of the ancestral Atlantic now crumpled into a long train of anticlines and synclines cut by reverse faults. The force of the continental collision that closed the ancestral Atlantic telescoped them into those folds and faults.

Pangaea broke up as the pattern of plate movements that had assembled it changed about 200 million years

Continental shelves and
coastal plains on trailing margins

1 Near the end of Precambrian Time, North America separated from Europe and Africa along a divergent plate boundary that opened an ocean between them. Large blankets of continental shelf and coastal plain sediments covered the trailing margins of the separated continents.

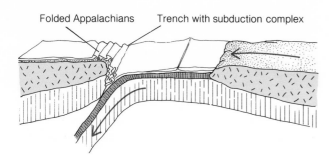

Folded Appalachians   Trench with subduction complex

2 During Ordovician Time, a convergent plate boundary began to consume the ocean floor along its western margin.

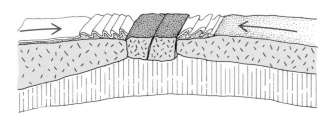

3 By late Paleozoic Time, the ocean closed and the metamorphosed sedimentary rocks caught between the colliding continents formed a band of new sialic continental crust along the suture between them. The collision also crumpled the inner continental shelf and coastal plain sediments into folds, the Appalachians.

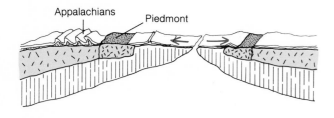

Appalachians   Piedmont

4 During early Mesocoic Time, a new divergent boundary rifts the joined continents apart along a line that approximately follows the suture between them. Basement rocks exposed east of the Appalachians such as those in the Piedmont and Blue Ridge are part of the new continental crust that formed in the suture.

FIGURE 20-5
How the Atlantic Ocean opened, closed, and then opened again, creating the Appalachian Mountains and the Paleozoic basement rock east of them in North America. Comparable souvenirs exist on the other side of the modern Atlantic.

ago, during earliest Mesozoic time. A new divergent plate boundary—the mid-Atlantic ridge—split first Africa and then Europe away from North America and swept those continents apart, opening the present Atlantic Ocean between them. The join that had connected those continents must have remained as a persistent line of weakness because the new rift followed it almost exactly, wavering just enough to switch a few areas that had been on one side of the ancestral Atlantic to the opposite side of the modern ocean. Geologists now interpret the existence of American fossils in the Cambrian rocks of western Massachusetts and European fossils in those of the eastern part of Massachusetts as evidence that those areas were once on opposite sides of the ancestral Atlantic.

THE SEDIMENTARY VENEER

Figure 20–6 shows a fold in the uneven blanket of sedimentary and volcanic rocks that covers most parts of the crystalline basement. Although we see those rocks of the sedimentary veneer nearly everywhere except in the shield areas, they actually comprise only a minuscule proportion of the continent because their thickness is small compared to that of the sialic crust. We are somewhat in the position of ants crawling on a painted board who might reasonably suppose it to be composed of paint unless they happen on a worn spot where the wood is exposed.

Most of the rocks that comprise the sedimentary veneer accumulated on the continents during times when parts of the continents were temporarily submerged below sea level. Hudson Bay, the North Sea, and the Baltic Sea are modern examples of such areas as is a broad expanse of shallow sea floor off the west coast of Florida. All those areas will someday rise above sea level and then the sediments now accumulating on them will remain as formations of sedimentary rock recording their present submergence. Most areas of the continents experienced at least several such episodes of shallow submergence during the past. Each episode added more layers to the continents' sedimentary veneer and more pages to the geologic record of earth history and the evolution of life.

The science—and art—of interpreting sedimentary rocks to extract the historic record they contain and to recon-

struct the ancient environments in which they accumulated is called **stratigraphy.** Stratigraphers resurrect from the rocks maps and cross sections of such structures as shorelines, bays, and lagoons that existed millions of years ago and then vanished, leaving nothing of themselves but layers of sedimentary rocks. The job is not as academic as it may at first seem; energy and mining companies routinely gamble millions of dollars on the validity of their stratigraphic interpretations as they seek new reserves of fossil fuels and other mineral commodities. Interpreting the wreckage of the past is serious business, nowhere more so than in the northern high plains where a large proportion of the remaining North American energy resources exist. Consider the problem of finding that coal.

High Plains Coal

The first white settlers in the northern high plains found thick coal seams, which they mined for local use. However, the generally poor quality of that coal and its remoteness from major markets made large-scale mining economically impossible until recent shortages of other fossil fuels offset those disadvantages. Today, it is vitally important to find and evaluate that region's coal reserves and that is a problem in applied stratigraphy.

Most of the world's large coal deposits, including those on the northern high plains, began as deposits of peat laid down as soggy masses of partially decomposed vegetation in tidewater swamps. Then younger sediments deposited on top of the peat compress it beneath their weight and lithify it into coal. The southwest coast of Florida where the land melts imperceptibly into the sea through a bewildering maze of marshes and mangrove swamps is a good modern example of a coal-forming sedimentary environment. Much of the lush vegetation along that coast grows on thick beds of actively accumulating peat, which will turn into coal if it is eventually buried beneath thick sections of younger sediment. Most ancient coals, like the modern peat beds of southwestern Florida, are closely associated with marine sediments.

Conditions similar to those that now prevail along the southwest coast of Florida must have existed in the northern high plains during Cretaceous and early Tertiary time. Much of the region was flooded by a shallow inland sea and the climate was warm, judging by the plant fossils in

FIGURE 20-6

An air view of a broad fold in the sedimentary veneer near Laramie, Wyoming. Differential erosion has stripped away the less resistant layers to make valleys while leaving the more resistant beds standing high as winding ridges. (J. R. Balsley, U.S. Geological Survey)

the rocks. Lush vegetation flourished in broad tidewater swamps along the coast, laying down thick beds of peat. Meanwhile, soft muds accumulated both on the shallow bottom offshore and in lagoons along the coast. Sand accumulated in channels threading through the coastal marshes and along occasional stretches of beach. A map of those ancient sedimentary environments would show a shoreline fringed by a ribbon of peat that was locally interrupted by patches of mud and sand and bordered on both sides by muddy areas. That pattern shifted back and forth with every slight change of sea level or land elevation. During parts of Cretaceous time the shoreline was as far west as the front of the Rocky Mountains, which were then forming, and by early Tertiary time it had moved into the Dakotas and Saskatchewan.

Every seaward shift in the shoreline moved the peat swamps seaward, which was generally to the east because the inland sea flooded the central part of the continent. When that happened, new deposits of lagoonal muds buried the old coastal peats. Landward shifts in the shoreline moved the peat swamps west and laid new deposits of offshore marine muds over the old peats that had just been flooded. Many changes of land elevation or of sea level shifted the coastline back and forth across the region, slowly weaving a complex pattern of interfingering layers of mud, peat, and sand. Figure 20–7 shows the general scheme.

Each coal seam marks an old shoreline. The problem of sorting them out is basically a matter of identifying the individual coal units and plotting their positions on maps. That is not as easy as it might seem. Geologists tracing coal seams must depend for their solid information upon

FIGURE 20-7

A schematic diagram showing how shifting sedimentary environments lay down a layered sequence of sedimentary rocks.

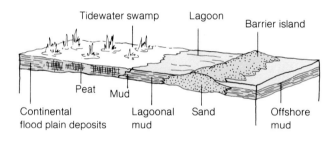

1 Terrestrial flood plain sediments grade seaward into peat accumulating in a tidewater swamp, mud accumulating in a coastal lagoon, a barrier island composed of sand, and marine muds deposited offshore. All constitute a single time rock unit.

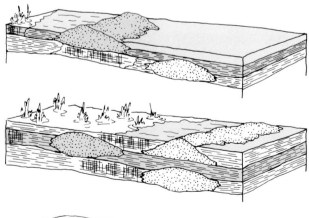

2 A rise in sea level or drop in land level causes landward transgression of the shoreline and another time rock unit similar to the first but shifted in position accumulates. The weight of the younger sediments on the peat compresses it into coal.

3 A drop in sea level or rise in land level causes seaward regression of the shoreline, and yet another time rock unit accumulates, again shifted in position.

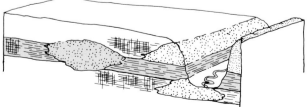

4 A geologist examining the sequence of sedimentary rocks exposed in the walls of a modern river valley has the problem of determining their environment of deposition. In this case, a correct interpretation would establish that the rocks along the river were deposited seaward of any possible coal and indicate the right direction in which to pursue the search.

chance outcrops, roadcuts, and drill holes. One coal seam generally looks very much like the others—as do the beds of mudstone and sandstone associated with them. The problem, then, resembles one of those frustrating jigsaw puzzles made of pieces that look about the same and assemble into an abstract pattern. Nor does it help to discover, as often happens, that crucial pieces of the puzzle are either lost to erosion or deeply buried. All practicing geologists must become adept at assembling scattered items of information into a coherent picture.

The first step in solving such a stratigraphic problem is to match the layers of rock exposed in various places, a procedure called **correlation.** Consider, for example, the hypothetical but nevertheless typical problem posed by a situation such as that shown in Figure 20–8. Do the two coal seams exposed in the river bluff correlate with the two being mined a few kilometers away or are they completely different? If they are different, the region contains four coal seams instead of two. There is no way to count coal seams, plan orderly mining, estimate reserves, or begin the search for new reserves until such questions have been resolved.

Correlating sedimentary rock units from one exposure to another is generally complicated by their persistent tendency to grade laterally into different rock types, a phenomenon known as a **facies change.** Figure 20–9 illustrates how high plains coal seams tend to grade into an offshore mudstone facies in the seaward direction and into a lagoonal mudstone facies landward. However, a coal seam may continue virtually unchanging for long distances if followed along the trend of the old shoreline. Tracing a coal seam laterally into its contemporaneous mudstone or sandstone facies identifies the pattern in which sedimentary environments were arranged during a brief interval of geologic time. That knowledge can be most useful in predicting the lateral extent of sedimentary facies, such as coal seams, in predicting whether a certain coal seam will extend into the neighboring ranch or turn into sandstone before it gets that far.

Any fossils the rocks may contain generally help both in establishing correlations and interpreting sedimentary environments. Most assemblages of fossils include the remains of at least a few creatures that evolved rapidly enough to be useful in establishing time correlations and were also widely enough distributed to appear in rocks deposited in different environments. Other creatures were so closely linked to certain habitats that their fossil remains can help geologists differentiate between otherwise similar sedimentary rocks laid down in different environments.

Geologists put their information together, plot it on a map, and then attempt to infer from their scattered data the situation in which the rocks formed. If they are looking for coal, they plot what they believe to have been the position of an ancient shoreline; in other words, where they think the coal is. To put the matter more directly, those maps show which properties the geologists believe contain coal and which do not.

Mining and energy firms use such geologic interpretations as their chief basis for deciding where to lease mineral rights and how much to offer for them. Then, if drilling reveals that the coal seam is indeed where their geologists expected it, the mineral rights leases will eventually become mines. If the coal is not there, then all the expense of the geologic work, mineral rights leases, legal work, and drilling must be written off as part of the cost of exploring for new reserves. Geologists who can create an effective interpretation on the basis of a small amount of information provide their employers with great competitive advantage.

Coal is only one of many possible examples. The search for oil and gas and many other mineral commodities hidden in the sedimentary veneer involves the use of basically similar techniques with the prospect of similar risks and rewards. However, many of the basic concepts of stratigraphy and structural geology developed along with the early large-scale exploitation of European and North American coal fields at the beginning of the industrial revolution. One of the pioneering American geologists brought his career as a state geologist to a serious crisis by establishing correlations based on fossils that showed large areas of New York to be underlain by rocks older than those in the European coal fields. On that basis, he announced that those areas were most unlikely to contain undiscovered coal thus greatly infuriating many members of the state legislature, who felt that he was not demonstrating a properly positive attitude.

Basins and Arches

The sedimentary veneer is an uneven blanket consisting of scattered, deep **basins** in which the accumulated sediments may be as much as 10 km thick and separated by

The situation

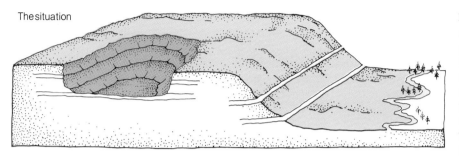

Some possible interpretations

A

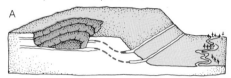

B

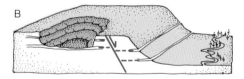

C

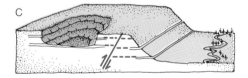

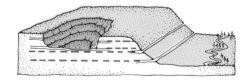

FIGURE 20-8

A typical problem in correlation. Suppose geologists find two horizontal coal seams exposed at different elevations in a mine and stream bank several kilometers apart with no intervening exposures to provide further data. The problem is to determine whether they are the same two coal seams, in which case there must be some structural explanation for the difference in elevation, or different coal seams, in which case the area contains a total of four beds of coal. It may become necessary to drill to solve some of the problems.

FIGURE 20-9

Rock units commonly vary greatly in thickness and grade laterally from one rock type to another as they pass through sedimentary facies deposited in different sedimentary environments.

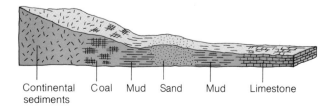

Continental sediments   Coal   Mud   Sand   Mud   Limestone

broad **arches** on which the sedimentary cover is relatively thin and perhaps locally absent. The sedimentary rocks filling the basins generally represent much longer and more continuous lengths of geologic time than those on the arches where unconformities tend to be numerous and to represent long time intervals. Evidently, the basins have persistently tended to subside whereas the arches have tended to remain high.

Sedimentary basins contain most of our fossil fuel reserves in the form of petroleum trapped in their thick, sedimentary fillings and in the coal seams around their margins where the shorelines were. The quest for new fossil fuel reserves begins broadly with efforts to find the sedimentary basins and then focusses on more detailed analysis of the rocks and structures within them.

Sedimentary basins apparently exist for a variety of reasons, which depend in one way or another on the underlying basement rock. Some appear to lie astride sutures in age provinces. Others appear to fill aulacogens that split the continental crust but did not separate the pieces very far—the Anadarko basin of southwestern Oklahoma may be an example of such a structure. The Denver-Julesburg basin of California lies immediately east of the large fault that raised basement rock to the summit of the Front Range. However, in most cases geologists still know too little about the underlying basement rock to explain the existence of sedimentary basins.

SUMMARY

The continental crust consists of a raft of complex granitic and metamorphic rocks extending to an average depth of about 40 km. Vast regions of the continental crust that formed before about 2.5 billion years ago differ somewhat from the younger crustal rocks in ways that may reflect the operation of tectonic processes different from those that now function. Continental crustal rocks formed since about 2.5 billion years ago appear to have originated along convergent plate boundaries similar to those that now exist.

Radioactive age dates show that the continental crust consists of an irregular mosaic of age provinces formed at different times. They provide clear evidence that the continents have grown and also record past plate movements that dismembered continents and assembled the fragments in new arrangements elsewhere. The Appalachian Mountains formed during a plate collision that closed an ancestral version of the Atlantic Ocean. A large region of northwestern North America consists of a patchwork of microcontinents added to the North American continent by convergent plate boundaries during Mesozoic and early Tertiary time.

In most regions the granitic and metamorphic rocks of the continental crust lie beneath an irregular blanket of younger sedimentary rocks. Most of the sedimentary cover accumulated in shallow seas during periods of partial continental submergence. The sedimentary veneer is relatively thin in some areas and much thicker in others which have, for various reasons, shown a persistent tendency to subside. The regions of thicker sedimentary cover are economically important because they contain most of the world's reserves of fossil fuels as well as large amounts of other mineral commodities. Geologists searching for those deposits carefully analyze the sedimentary rocks to reconstruct the ancient environments in which they accumulated and thus predict the locations of rocks likely to contain mineral deposits.

KEY TERMS

age province	correlation
arch	facies change
Archean basement	sedimentary veneer
basement	shield
basin	stratigraphy

REVIEW QUESTIONS

1. Why do the continents retain a more complete record of the earth's past than do the ocean basins?

2. What kinds of plate tectonic events could create boundaries between age provinces?

3. Why do people who drill oil wells customarily give up and stop drilling immediately after they encounter basement rock?

4. Why does petroleum exploration start with the search for sedimentary basins?

5. Briefly suggest several reasons why layers of sedimentary rock generally change in thickness and in rock type from one area to another.

SUGGESTED READINGS

King, P. B. *The Evolution of North America.* Princeton: Princeton University Press, 1959.

For many years a standard reference on the major structural features of the North American continent. However, some of the interpretations are now out of date because they were not based on plate tectonic theory.

LaPorte, L. F. *Ancient Environments.* Englewood Cliffs, N.J.: Prentice-Hall, 1968.

A good general introduction to the interpretation of sedimentary rocks.

McCall, G. J. H., ed. *The Archean: Search for the Beginning.* Stroudsburg, Pa.: Dowden, Hutchinson and Ross, Inc., 1977.

A collection of technical articles on many aspects of the earliest rocks, ranging from the origin of greenstone belts to the earliest signs of life.

Windley, B. F. *The Evolving Continents.* London: John Wiley, 1977.

An outstanding technical review of almost every aspect of continental geology from the origin of the Archean basement to continental movements.

APPENDIX A: Some Common Rock-Forming Minerals

GROUP 1:

Minerals with a metallic or submetallic luster

MINERAL	COLOR	DESCRIPTION	OCCURRENCE
Pyrite (FeS_2)	pale brassy yellow	Generally forms glittering, metallic cubes much harder than a knife blade. The cubes do not cleave. Weathers readily to rusty yellow spots of limonite. Pyrite has a greenish-black streak.	Widely disseminated in virtually all kinds of rocks.
Magnetite (Fe_3O_4)	black	Generally forms small, black grains that are harder than a knife blade and do not cleave. Many magnetite grains have an octahedral shape. The only common mineral that will stick to a magnet.	Widely disseminated in most kinds of rocks except the carbonates. A magnet drawn through sand will generally pick up a few grains of magnetite.

GROUP 2:
Minerals with a nonmetallic luster

Subgroup A: Minerals that are harder than a knife blade and do not cleave

MINERAL	COLOR	DESCRIPTION	OCCURRENCE
Olivine (Fe,Mg isolated tetrahedra silicates)	pale green	Forms small grains that have a glassy appearance and rarely show regular crystal outlines. Weathers readily to form rusty spots of limonite-stained clay and alters readily to form dull green chlorite or serpentine.	Primarily in dark green or black igneous and metamorphic rocks in which it associates with pyroxenes and in some cases calcium-rich plagioclase or garnet.
Garnet (Fe,Mg,Ca,Al isolated tetrahedra silicates)	commonly red; also pink, green, black, or white	Garnets are generally red and almost invariably form perfectly shaped twelve-sided crystals. Any glassy-looking red mineral should be suspected of being garnet.	In virtually all kinds of igneous and metamorphic rocks in association with a wide variety of other minerals. Garnets also occur in small quantity in many sands and sandstones.
Quartz (SiO_2)	typically clear and colorless; may be almost any color	Commonly forms transparent, glassy grains lacking well-developed crystal faces. Also occurs as well-formed crystals, as massive and milky-looking vein fillings, as a fine-grained rock called chert, and in a wide variety of other forms. The most variable in appearance of any mineral.	Common in sialic igneous and metamorphic rocks in which it associates with sodium-rich plagioclase, orthoclase, amphiboles, and micas. Quartz is also abundant in sediments and comprises the bulk of all sand and silt. It is the most widely distributed of all minerals and can be found almost anywhere.

Subgroup B: Minerals that are harder than a knife blade and do cleave

MINERAL	COLOR	DESCRIPTION	OCCURRENCE
Pyroxene (Fe,Mg,Ca single-chain silicates)	black or dull green	Typically form stubby crystals with a dull surface that cleave in two directions meeting at a 90° angle. Many varieties readily alter to dull green chlorite or serpentine. Augite, the commonest pyroxene, is black.	Generally in very dark green or black igneous and metamorphic rocks in which the pyroxenes associate with calcium-rich plagioclase and olivine.
Amphibole (Fe,Mg,Ca double-chain silicates)	black, brown, green, or white	Typically form long, thin crystals with a glossy surface that cleave in two directions meeting at angles of 60° and 120°. The most abundant amphibole, hornblende, is black.	Primarily in light-colored igneous and metamorphic rocks of sialic composition. Amphiboles typically associate with sodium-rich plagioclase, orthoclase, quartz, and micas. However, some schists composed primarily of amphibole are black.
Orthoclase ($KAlSi_3O_8$)	white or salmon pink	Typically form blocky crystals that cleave in two directions meeting at an angle of 90°. Although orthoclase may be perfectly white, most specimens are at least slightly pink. One rare variety of microcline, a mineral closely related to orthoclase, is green. Orthoclase hydrates readily to clay minerals, which impart a milky appearance to most specimens.	Typically in light-colored igneous and metamorphic rocks of sialic composition in which it associates with sodium-rich plagioclase, quartz, amphiboles, and micas. Also occurs in sands and sandstones in which the milky appearance of the grains makes it easy to recognize.
Plagioclase ($NaAlSi_3O_8$ and $CaAl_2Si_2O_8$)	white or greenish	Forms blocky crystals that cleave in two directions meeting at an angle of 90°. The sodium-rich plagioclases are generally white whereas those rich in calcium tend to be greenish. All plagioclases hydrate readily to form clay minerals that commonly make the crystals look milky. Close examination with a lens will reveal fine striations on crystal faces that distinguish plagioclase from orthoclase.	In almost all kinds of igneous and metamorphic rocks. The sodium-rich plagioclases associate with orthoclase, quartz, amphibole, and micas in the light-colored sialic rocks. Calcium-rich plagioclase occurs with pyroxene and olivine in dark-colored mafic rocks.

Subgroup C: Minerals that are softer than a knife blade and do cleave

MINERAL	COLOR	DESCRIPTION	OCCURRENCE
Mica (K,Al,Fe layer silicates)	black, white, or yellow	All micas are easy to recognize because they cleave easily in one direction into thin flakes that have a shiny surface and will bend without breaking.	The black mica, biotite, and the white mica, muscovite, both occur in light-colored igneous and metamorphic rocks of sialic composition. They associate with sodium-rich plagioclase, orthoclase, quartz, and amphibole. The golden yellow mica, phlogopite, occurs in black igneous rocks of mafic composition in which it associates with pyroxene, olivine, and garnet.
Chlorite (Fe, Mg layer silicate)	dull green	Large crystals cleave with difficulty in one direction to form brittle flakes. However, the mineral commonly occurs in fine-grained aggregates in which individual grains are not visible. Any dull green rock should be suspected of containing chlorite.	Typically forms through alteration of augite in dark igneous and metamorphic rocks. Also occurs in many low-grade metamorphic rocks.
Calcite ($CaCO_3$)	white; in many cases with a gray or yellow tint	Cleaves in three directions meeting at angles of 72° and 108° to form rhombohedral fragments. The only abundant mineral that fizzes in weak acid.	Most abundantly in limestone and marble. Calcite is also familiar as the mineral component of most seashells. Calcite also occurs as vein fillings and in certain uncommon igneous rocks.
Dolomite (Ca,Mg,CO_3)	white or pinkish	Closely resembles calcite but can be distinguished from it because it will not fizz unless powdered.	Most abundantly in dolomite and in dolomite marble.

Subgroup D: Minerals that are softer than a knife blade and do not cleave

MINERAL	COLOR	DESCRIPTION	OCCURRENCE
Goethite (Fe_2O_3 with H_2O)	rusty shades of yellow and brown	Goethite may occur as earthy masses or as thin coatings on grains of other minerals. Its presence should be suspected in any rock having a rusty color. Goethite is an excellent pigment; small concentrations can brightly color a rock.	Goethite forms through weathering of iron-bearing minerals and is therefore abundant in many soils and weathered rocks. It also occurs in many sediments, primarily deposits of mud and sand.
Hematite (Fe_2O_3)	various shades of red	Hematite forms primarily through dehydration of goethite and also occurs as earthy masses and as coatings on mineral grains. Any reddish rock is likely to contain hematite. Like goethite, it is an excellent pigment.	Basically similar to that of goethite except that hematite is more abundant in solid rocks than in sediments.

Subgroup E: Minerals that are softer than a fingernail

MINERAL	COLOR	DESCRIPTION	OCCURRENCE
Serpentine (Mg layer silicate)	dull green	Serpentine generally occurs in fine-grained masses that carve easily with a knife and have a soapy feel. Those properties in combination with its green color are distinctive.	In association with dark igneous and metamorphic rocks with a mafic composition. Forms primarily through reaction of augite and olivine with water.
Talc (Mg layer silicate)	white or pale green	Very soft, pearly-looking mineral that cleaves easily in one direction to form brittle flakes. Carves easily with a knife.	Generally in association with serpentine.
Gypsum ($CaSO_4 \cdot 2H_2O$)	generally white; many colors exist.	Large crystals cleave in one direction to make transparent, brittle flakes. More commonly occurs in fine-grained massive rocks that are distinctive in their softness.	Forms large crystals in many arid soils and some evaporite deposits. Also forms a fine-grained evaporite rock called alabaster, which is commonly colorful.

APPENDIX B: Metric Conversion

Length

1 millimeter (mm) = 0.039 inch, about the thickness of a worn dime.

1 centimeter (cm) = 0.39 inch, slightly more than ⅜ of an inch.

1 meter (m) = 39.4 inches, about 3 inches longer than a yard.

1 kilometer (km) = 3,280 feet, slightly more than ⁶⁄₁₀ of a mile.

Area

1 square kilometer (km²) = 0.39 square mile, about 247 acres.

Volume

1 cubic kilometer (km³) = 0.24 cubic mile.

Mass

1 metric ton = 2,205 pounds.

Temperature

$$°C = \frac{°F - 32}{1.8}$$

1 **H** **1.0080** Hydrogen								

Key

6 **C** **12.01** Carbon	← Atomic number ← Symbol ← Atomic weight ← Name ← Elements abundant in the earth's crust

3 **Li** **6.940** Lithium	**4** **Be** **9.013** Berylium							
11 **Na** **22.991** Sodium	**12** **Mg** **24.32** Magnesium							
19 **K** **39.100** Potassium	**20** **Ca** **40.08** Calcium	**21** **Sc** **44.96** Scandium	**22** **Ti** **47.90** Titanium	**23** **V** **50.95** Vanadium	**24** **Cr** **52.01** Chromium	**25** **Mn** **54.94** Manganese	**26** **Fe** **55.85** Iron	**27** **Co** **58.94** Cobalt
37 **Rb** 85.48 Rubidium	**38** **Sr** 87.63 Strontium	**39** **Y** 88.92 Yttrium	**40** **Zr** 91.22 Zirconium	**41** **Nb** 92.91 Niobium	**42** **Mo** 95.95 Molybdenum		**44** **Ru** 101.1 Ruthenium	**45** **Rh** 102.91 Rhodium
55 **Cs** 132.91 Cesium	**56** **Ba** 137.36 Barium	**57** **La** 138.92 Lanthanum	**72** **Hf** 178.50 Hafnium	**73** **Ta** 180.95 Tantalum	**74** **W** 183.86 Wolfram	**75** **Re** 186.22 Rhenium	**76** **Os** 190.2 Osmium	**77** **Ir** 192.2 Iridium
87 **Fr** (223) Francium	**88** **Ra** (226) Radium	**89** **Ac** (227) Actinium	**90** **Th** (232) Thorium	**91** **Pa** (231) Protactinium	**92** **U** 238.07 Uranium			

58 **Ce** 140.13 Cerium	**59** **Pr** 140.92 Praseodymium	**60** **Nd** 144.27 Neodymium	**61** **Pm** (147) Promethium	**62** **Sm** 150.35 Samarium

TABLE OF THE NATURALLY OCCURRING CHEMICAL ELEMENTS

					2 **He** 4.003 Helium
5 **B** 10.82 Boron	6 **C** 12.011 Carbon	7 **N** 14.008 Nitrogen	8 **O** 16.000 Oxygen	9 **F** 19.00 Fluorine	10 **Ne** 20.183 Neon
13 **Al** 26.98 Aluminum	14 **Si** 28.09 Silicon	15 **P** 30.975 Phosphorus	16 **S** 32.066 Sulfur	17 **Cl** 35.457 Chlorine	18 **Ar** 39.944 Argon

28 **Ni** 58.71 Nickel	29 **Cu** 63.54 Copper	30 **Zn** 65.38 Zinc	31 **Ga** 69.72 Gallium	32 **Ge** 72.60 Germanium	33 **As** 74.91 Arsenic	34 **Se** 78.96 Selenium	35 **Br** 79.916 Bromine	36 **Kr** 83.80 Krypton
46 **Pd** 106.4 Palladium	47 **Ag** 107.880 Silver	48 **Cd** 112.41 Cadmium	49 **In** 114.82 Indium	50 **Sn** 118.70 Tin	51 **Sb** 121.76 Antimony	52 **Te** 127.61 Tellurium	53 **I** 126.91 Iodine	54 **Xe** 131.30 Xenon
78 **Pt** 195.09 Platinum	79 **Au** 197.0 Gold	80 **Hg** 200.61 Mercury	81 **Tl** 204.39 Thallium	82 **Pb** 207.21 Lead	83 **Bi** 209.00 Bismuth	84 **Po** (210) Polonium	85 **At** (210) Astatine	86 **Rn** (222) Radon

63 **Eu** 152.0 Europium	64 **Gd** 157.26 Gadolinium	65 **Tb** 158.93 Terbium	66 **Dy** 162.51 Dysprosium	67 **Ho** 164.94 Holmium	68 **Er** 167.27 Erbium	69 **Tm** 168.94 Thulium	70 **Yb** 173.04 Ytterbium	71 **Lu** 174.99 Lutetium

GLOSSARY

A-horizon The uppermost level of the soil in which plant roots, humus, and other organic matter are most abundant.

Abyssal fan A gently sloping apron of sediment deposited seaward of the mouth of a submarine canyon.

Abyssal plain Large areas of the deep ocean floor that are nearly level and quite smooth because they are deeply blanketed by sediments.

Accretion Continental growth either through addition of old sialic crust or formation of a new complex of sialic igneous and metamorphic rocks at a convergent plate boundary.

Accretionary prism Same as a subduction complex.

Active layer A layer of thawed surface soil resting on frozen subsoil.

Aftershock A smaller earthquake that follows after, and evidently results from, a larger one.

Age province A region of basement rock that formed during the same period of time.

Aggradation Depositional buildup of a streambed.

Alluvial fan An accumulation of stream-deposited sediments shaped like a segment of a cone; most commonly found at the mouths of desert canyons.

Alpine glacier Same as valley glacier.

Amphiboles A large group of silicate minerals based on double-chain structures.

Andesite Volcanic rock intermediate between basalt and rhyolite in composition and most other properties.

Angle of repose The steepest angle at which a loose aggregate will stand.

Angular unconformity An unconformity separating two sequences of sedimentary layers that lie at an angle to each other.

Anion An atom that has acquired a negative charge by gaining one or more extra electrons.

Antecedent stream A stream that cuts through an uplifted region because it maintained its course as the uplift proceeded.

Anticline A fold in which the layers of rock bend upward to form an arch.

Aquiclude A body of impermeable rock that functions as a barrier to ground-water movement.

Aquifer A body of rock porous and permeable enough to deliver a useable quantity of water to a well.

Arch A region of the continent in which the sedimentary veneer is relatively thin.

Archean basement Continental crustal rocks formed more than about 2.5 billion years ago.

Arete A narrow ridge of rock separating two glacially gouged valleys.

Artesian aquifer A confined aquifer that contains water under pressure, which causes it to rise in the well.

Ash flow A dense cloud of freshly erupted volcanic ash that pours across the surrounding countryside.

Asthenosphere A zone of weak and probably partially melted rock in the upper mantle that in most regions extends from 100 to about 250 km below the surface.

Atomic number The number of protons in the nucleus of an atom.

Atomic weight The sum of the number of protons and neutrons in an atom of an element.

Augite A common variety of black pyroxene abundant in most mafic rocks.

Aulacogen A failed continental rift filled with sediment.

Axial plane   An imaginary surface that slices a fold along its length by passing through the hinge of each rock layer.

B-horizon   The intermediate level of the soil which contains no humus and may include accumulations of clay washed down from above.

Bajada   A part of a desert plain, a depositional surface graded smooth by running water and floored by sedimentary deposits.

Barrier island   An offshore beach fringing the mainland coast.

Basalt   An abundant volcanic rock composed mostly of augite and calcium-rich plagioclase. It forms most of the oceanic crust and erupts from many terrestrial volcanoes.

Base level   A lower limit to the depth to which a stream can erode; imposed ultimately by sea level and locally by the elevations of trunk streams, lakes, and resistant ledges of rock.

Basement   The complex of sialic igneous and metamorphic rocks that forms the continental crust.

Basin   A term used to describe both a radially symmetrical anticline and a region of the continent in which the sedimentary veneer is abnormally thick.

Basin plain   Extensive smooth and level surfaces on the ocean floor created by deposition of turbidite sediments.

Batholith   A more or less continuous mass of granitic rock covering an area greater than about 100 km^2.

Bauxite   A type of lateritic soil that contains a large proportion of aluminum oxide minerals and very little iron; used as aluminum ore.

Bayou   The same as a slough.

Beach   The ribbon of wave-washed sediment within the surf zone extending from high tide level seaward to the line of breakers.

Bed load   Sediment transported along the bottom of a stream; primarily composed of grains larger than sand.

Benioff zone   A plane surface within the upper mantle defined by the focal points of earthquakes originating from a sinking lithospheric slab.

Biotite   A common variety of black mica.

Blowout   A depression eroded by the wind.

Blueschist   Metamorphic rock formed by recrystallization of sandstone or mudstone under conditions of very high pressure and moderate temperature.

Bolson   An undrained desert basin of internal drainage.

Bowen's reaction series   The sequence in which silicate minerals crystallize within a cooling magma.

Braided channel pattern   A stream that flows in a network of criss-crossing broad channels separated by low bars and islands.

Breaker bar   A wave-built ridge of sediment offshore from the beach and parallel to it that forms where storm waves break.

Brittle deformation   A change in shape effected by breakage; faulting is a good example.

C-horizon   The lowest level of the soil composed largely of partially weathered bedrock.

Calcareous mud   A nonclastic sediment composed largely of calcite that lithifies into limestone.

Calcareous ooze   Sediment consisting largely of the calcium carbonate shells of minute animals. It accumulates on parts of the deep ocean floor beyond the reach of turbidity flows and above the carbonate compensation depth.

Calcite   One of the mineral forms of calcium carbonate.

Caldera   A volcanic crater formed by collapse of the ground surface as an eruption withdraws magma from beneath.

Caliche   Deposits of calcite that accumulate in dry soils.

Capacity   The total load of all kinds of sediment that a stream is capable of carrying.

Capillary fringe   A zone immediately above the water table kept constantly moist by upward seepage from the ground-water reservoir.

Carbonates   Minerals based on the carbonate group $CO_3^=$, and rocks composed dominantly of those minerals.

Carbonate compensation depth   The level in the ocean at which the water becomes capable of dissolving calcite.

Cation   An atom that has acquired a positive charge by losing one or more electrons.

Cementation   A diagenetic process that fills void spaces between grains in a sedimentary deposit with new mineral matter that binds them into solid rock.

Chemical weathering   The complex of processes that convert the original minerals of a rock into new minerals of the soil primarily through reaction with water.

Chert   A sedimentary rock consisting of extremely fine-grained quartz.

Chlorites   A large group of generally greenish layer silicate minerals.

Cinder   A small chunk of lava ejected from a volcano.

Cinder cone   A small volcano composed mostly of small chunks of basalt or andesite lava ejected from a central vent during a single eruption.

Cirque   A broad basin at the head of a glaciated valley.

Clastic load   Stream sediment consisting of fragments eroded from older rocks or soil.

Clastic sediment   Deposited material consisting of fragments derived from erosion of older rocks or soils.

Clay minerals   A group of layer silicate minerals that typically form extremely minute crystals.

Cleavage   The tendency of many crystals to break easily along planes of inherent weakness in their crystal structures.

Coal   A sedimentary rock consisting of lithified peat.

Coastal lagoon   A body of shallow water enclosed between an offshore barrier island and the mainland coast.

Compaction   A diagenetic process in which the volume of a sediment diminishes under pressure.

Competence   Refers both to the largest sediment particles a stream can carry and to rocks that resist plastic deformation.

Composite volcano   A conical edifice consisting of a mixture of andesitic ash, mud flows, and lava flows.

Compressional wave   Earthquake wave motion consisting of a series of compressions and expansions. The rock particles move back and forth in a direction parallel to that of wave advance.

Cone of depression   A funnel-shaped dimple that develops in the water table around a producing well.

Confined aquifer   A body of water-bearing porous and permeable rock that is either beneath layers of impermeable rock or sandwiched between them.

Conglomerate   A type of sedimentary rock formed by lithification of gravel.

Connate water   Old seawater trapped in the pore spaces of sedimentary rocks since their original deposition.

Contact metamorphism   Local recrystallization caused by heat escaping from a pluton.

Continental crust   A thick raft of sialic igneous and metamorphic rocks embedded in the upper part of a lithospheric plate.

Continental drift   The idea that continents move independently about the earth's surface. Now superseded by plate tectonic theory, which holds that continents move as parts of plates.

Continental glacier   A sheet of flowing ice that covers an area of regional extent.

Convergent boundary   A line along which two lithospheric plates collide.

Core   A sphere about 3500 km in diameter that forms the central part of the earth and is believed to be composed of an alloy of iron and nickel.

Coriolis effect   The tendency of all moving objects to drift to the right in the Northern Hemisphere and to the left in the Southern Hemisphere because of the earth's rotation.

Correlation   Matching bodies of rock from one exposure to another.

Covalent bond   A chemical bond in which electrons are shared between two atoms.

Crevasse   An open fracture in the upper, brittle zone of a glacier.

Cross beds   Smaller sedimentary beds enclosed within larger ones to which they are not parallel.

Cross-cutting relationships   The rule that later geologic events overprint the effects of earlier ones.

Crust   The upper part of the lithosphere composed of rocks other than peridotite. The continental crust, which averages about 40 km thick, consists mostly of sialic igneous and metamorphic rocks. The oceanic crust, about 10 km thick, consists of basalt.

Crystal gliding   Deformation of a crystal by internal slippage along planes in its lattice.

Crystallography   The study of the external forms and internal structures of crystals.

Crystal structure   The regular geometric pattern in which the atoms composing a crystal are arranged.

Curie point   The temperature at which a mineral loses its permanent magnetism; that of magnetite is $580°C$.

Deep   A long depression in the ocean floor marking a convergent boundary where a lithospheric plate sinks into the mantle; same as a trench.

Deflation   Wind erosion of sand or dust.

Delta   A deposit of sediment accumulated where a stream empties into the ocean or into a lake.

Density current   A flow of water that sinks because it is heavy and then moves laterally either on the bottom or on a deeper layer of still heavier water.

Desert pavement   An armor of pebbles left covering the ground surface as the wind blows the smaller particles away.

Desert plain   A smoothly graded and gently sloping land surface created by running surface water.

Desert varnish   A dark coating composed mostly of iron and manganese oxides that commonly forms on the upper surfaces of rocks weathering in an arid environment.

Diagenesis   The combination of chemical and mechanical processes that operate within sediments and sedimentary rocks at temperatures and pressures below the metamorphic range.

Dike   A tabular body of igneous rock formed from magma injected into a fracture that cuts across the structure of the enclosing rock.

Dip   The angle at which a rock surface slopes downward from a horizontal plane.

Discharge   The amount of water flowing through a stream as determined by multiplying the cross-sectional area of the channel by the average flow rate.

Dissolved load   Material transported in solution.

Divergent boundary   A line, generally an oceanic ridge, along which two lithospheric plates move away from each other. New oceanic crust forms as the gap opens between them.

Dolomite   The term refers both to a mineral composed of calcium magnesium carbonate—$(Ca,Mg)CO_3$—and a sedimentary rock composed mostly of that mineral.

Dome   A radially symmetrical anticline the flanks of which dip uniformly in all directions from the crest.

Drawdown   The depth of the cone of depression in the water table around a producing well.

Drift   All sediments associated with glaciation, including those deposited from ice and from meltwater.

Dripstone   Travertine formations, such as stalactites or stalagmites, composed of calcite deposited from dripping or trickling water.

Drumlin   A hill composed of glacial till molded by flowing ice into a streamlined form.

Dune   A pile of sand that moves before the wind.

Dust   Operationally defined, dust consists of particles light enough to travel in suspension in the wind; generally particles less than about 0.2 mm in diameter.

Eclogite   An uncommon metamorphic rock type that consists mostly of garnet and pyroxene and forms through recrystallization of basalt under conditions of very high pressure and moderately high temperature.

Effective wave base   The greatest depth at which waves can move sediment on the bottom; depends upon the size of both the waves and the sediment.

Elastic deformation   A temporary change of shape under stress. When the deforming stress is removed, the body regains its original shape.

Entrenchment   The erosional deepening of a streambed.

Epicenter   The point on the ground surface directly above the focus of an earthquake where the fault movement actually occurred.

Erratic   A transported rock that rests on a different kind of bedrock. The term is most commonly applied to glacially transported rocks.

Esker   A winding ridge composed of well-sorted and layered sediments deposited in a glacial stream channel and left when the ice melted.

Estuary   The sea-flooded mouth of a stream valley.

Eugeosyncline   An assemblage of oceanic sediments accumulated in a deep trough.

Evaporite   The term refers both to minerals, such as salt deposited from evaporating water, and to rocks composed of such minerals.

Exfoliation   A weathering process that detaches successive sheets of partially decomposed rock; typically affects coarse-grained igneous and metamorphic rocks.

Facies change   Gradation from one rock type to another within the same body of rock.

Fault   A fracture in the earth's crust, the opposite sides of which have moved relative to each other.

Fault plane   A fracture surface along which the opposite sides have moved.

Fault scarp   A sharp break in topographic slope created by movement along a fault.

Fault zone   A body of rock broken by movement along a fault.

Faunal succession, law of   The observation that assemblages of fossils always occur in the same vertical order in an undisturbed section of sedimentary rocks.

Feldspar   A group of framework silicate minerals in which aluminum partially substitutes for silicon and various combinations of potassium, sodium, and calcium redress the balance of electrical charges.

Flood basalt   A basalt lava flow measuring at least several tens of cubic kilometers in volume.

Floodplain   The area subject to inundation during periods of high stream flow.

Focus   The point within the earth where fault movement generated an earthquake.

Fold   A bend in layers of rock.

Foliation   Rock texture created by parallel alignment of flat mineral grains, most commonly mica or feldspar.

Foot wall   The block of rock beneath an inclined fault surface.

Fossils   The remains of plants or animals preserved in rocks.

Franciscan rocks   The subduction complex of strongly deformed oceanic sediments that forms the Coast Range of California and southern Oregon.

Frost wedging   Mechanical breakage of rocks caused by the expansion of water freezing to ice within them.

Gabbro   A mafic plutonic igneous rock consisting mostly of augite pyroxene and calcium-rich plagioclase accompanied in some cases by olivine.

Garnet   A large group of silicate minerals based on isolated tetrahedra that occur in a wide variety of igneous and metamorphic rocks. Most garnets are red and they typically occur as well-formed crystals.

Geosyncline   A hypothetical deep trough in which thick sections of sedimentary rocks are supposed to accumulate.

Geothermal energy   Energy furnished by the earth's internal heat, generally through natural hot water or steam.

Geothermal gradient   The natural increase in temperature with depth; about 3°C for every 100 m in most regions of the upper crust.

Goethite   The mineral form of hydrated iron oxide, represented by the formula $Fe_2O_3 \cdot nH_2O$. It is essentially the same substance as common rust and is responsible for most of the yellowish colors in rocks.

Glacier   A mass of ice that flows under its own weight.

Glowing cloud eruption   A type of volcanic outburst consisting of a blast of steam filled with small shreds of lava that pours down the slope of a volcano as an incandescent cloud. Often called a *nuee ardente*.

Gneiss A textural term describing foliated metamorphic rocks that show color banding and lack enough mica or amphibole to confer a flaky or splintery quality.

Graded bed A layer of sediment in which the grain size gradually fines upward.

Graded stream A stream in a condition of balance between erosion and deposition.

Gradient The steepness of a slope.

Granite An abundant plutonic igneous rock of sialic composition that contains orthoclase, sodium-rich plagioclase, and quartz; commonly accompanied by mica or hornblende amphibole.

Gravity anomaly A discrepancy between the observed and theoretically calculated values for the strength of the earth's gravitational field. A positive anomaly indicates an excess of mass and a negative anomaly a deficiency.

Gravimeter An instrument that measures local differences in the strength of the earth's gravitational field.

Greenhouse effect Atmospheric warming caused by substances, such as glass or carbon dioxide, that permit entry of the sun's visible radiation but forbid exit of heat radiation from the earth.

Ground moraine A generally formless deposit of glacial till plastered onto the ground surface.

Ground water Water stored in open pore spaces beneath the water table.

Gypsum The mineral form of calcium sulfate—$CaSO_4 \cdot 2H_2O$.

Half life The period during which half the total inventory of a particular kind of radioactive atom decays.

Halite The mineral form of common table salt, NaCl; a common evaporite mineral.

Hanging valley A glaciated tributary valley that enters the larger valley at a level above its floor, commonly over a waterfall or series of steep rapids.

Hanging wall The block of rock above an inclined fault surface.

Hardness The relative ability of minerals to scratch each other.

Hematite The mineral form of iron oxide—Fe_2O_3. Commonly responsible for red or brown colors in rocks.

Hornblende The commonest variety of amphibole; occurs as long, black crystals with a glossy surface.

Horn peak A mountaintop made craggy by glacial erosion from several sides.

Hotspot An isolated local area of volcanic activity remote from a plate boundary.

Humus Partially decayed plant material in soils.

Hydration reaction A class of chemical reactions that combine water and silicate minerals to yield clay and various soluble ions. Most proceed with an increase in volume.

Igneous rock An aggregate of mineral grains crystallized from a melt. Plutonic igneous rocks form from melts that crystallize within the earth's crust and volcanic igneous rocks from those that erupt to the surface.

Incompetent The mechanical behavior of rocks that deform plastically as though they had very little strength.

Initial dip The degree to which sedimentary layers depart from a horizontal position when deposited.

Intensity, of an earthquake The violence of an earthquake assessed in terms of the damage it causes.

Ion An electrically charged atom.

Ionic bond A chemical link based upon the electrical attraction between oppositely charged ions.

Isomorphs Mineral species that have similar crystal structures and different chemical compositions.

Isomorphic substitution Variation in composition of a mineral that occurs when atoms of different elements resembling each other in size and chemical behavior occupy equivalent positions in the crystal structure.

Isostasy The tendency of the lithosphere to remain in floating equilibrium, rising and sinking in response to unloading or loading of the earth's surface.

Isotope One of several forms of a chemical element all of which have the same number of electrons and protons but differ in their numbers of neutrons and therefore in their atomic weights.

Joint A fracture in a rock along which there has been little or no relative movement.

Kame A hill composed of sorted and layered glacial outwash sediment that was in most cases deposited either in a hole in the ice or as a fan at the front of the glacier.

Kaolinite A kind of clay mineral—$Al_4(Si_4O_{10})(OH)_8$—that typically forms in the heavily leached soils of wet regions.

Karst topography A type of landscape created dominantly by solution of carbonate rocks in which sinkholes are abundant and drainage flows largely through caverns.

Kettle A collapse depression formed where a mass of ice incorporated in glacial till or outwash melted.

Landslide A mass of soil and weathered rock that moves as a unit on a slip surface.

Lateral moraine A deposit of till plastered onto a valley wall.

Laterites Soils composed mostly of iron and aluminum oxides and the clay mineral kaolinite; typically form in regions having warm and very wet climates.

Lattice The three-dimensional pattern in which the atoms composing a crystal are arranged.

Lava Molten rock erupted to the surface.

Lava dome   A type of volcano that consists of a large mass of extremely viscous sialic magma extruded onto the earth's surface.

Lava tube   A tunnellike cave within a lava flow, generally basalt, that formed as molten lava drained out from beneath a solid surface.

Layer silicates   Minerals in which the crystal structure is based on layers of silicate tetrahedra, each of which shares three oxygens with its neighbors.

Leaching   Extraction of dissolved substances.

Left-handed fault   Same as left-lateral fault.

Left-lateral fault   A strike-slip fault that moves horizontally in such a way that an observer looking along the line of the fault sees the side on his left coming toward him.

Limestone   A type of sedimentary rock composed largely of calcite.

Lineation   Rock texture created by parallel alignment of needle-shaped crystals, most commonly of amphiboles.

Lithification   The hardening of soft sedimentary deposits into solid rock.

Lithosphere   The relatively cold and therefore rigid outer shell of the earth; about 100 km thick in most regions.

Loess   Soil composed of windblown dust.

Longshore current   A slow, wave-driven movement of water parallel to the coast.

Longshore transport   Surf-driven movement of sediment along a coast.

Low velocity zone   A region of the upper mantle in which seismic waves travel slowly and lose part of their shear wave component; corresponds to the asthenosphere.

Luster   The appearance or shine of a mineral surface in reflected light; generally classified into metallic, submetallic, and a number of nonmetallic categories.

Mafic rocks   Generally dark colored igneous or metamorphic rocks rich in magnesium, iron, and calcium. They generally consist largely of pyroxene, calcium-rich plagioclase, and olivine.

Magma   Molten rock is called magma as long as it remains below the earth's surface but becomes lava if it erupts.

Magnetic anomaly   An area in which the strength of the earth's magnetic field is abnormally high or low.

Magnetic equator   The line around the earth midway between the magnetic poles.

Magnetic latitude   The angle between the earth's surface and its magnetic lines of force. They are parallel to each other at the magnetic equator, at right angles to each other at the magnetic poles, and at intermediate angles at points between.

Magnetic pole   The area in the Arctic or Antarctic region where the earth's magnetic lines of force dip vertically downward.

Magnetite   The mineral form of the iron oxide Fe_3O_4.

Magnetometer   An instrument that measures the strength of the earth's magnetic field.

Magnitude, of an earthquake   The size of an earthquake measured in terms of the amount of energy it releases.

Mantle   The spherical shell of peridotite about 2,900 km thick that surrounds the earth's core and comprises most of its volume.

Marble   A metamorphic rock type composed largely of coarsely crystalline carbonate minerals and formed through recrystallization of limestone or dolomite.

Mass wasting   All the processes that move masses of soil or rock downslope under the simple influence of gravity without the intervention of a fluid medium, such as running water.

Massive texture   A term describing rocks in which the mineral grains are randomly oriented so that specimens lack directionality and look essentially the same from any angle.

Massive sedimentary sulfides   Sediments or sedimentary rocks consisting largely of sulfide minerals.

Meander   A looping bend in the course of a stream channel.

Mechanical weathering   Processes that break rocks without altering them chemically.

Melange   A body of scrambled rocks within a subduction complex.

Mercalli Intensity Scale   A system of classifying earthquakes according to the types of damage they cause.

Metallic bond   A form of chemical combination in which some electrons remain unattached to any particular atom and roam freely throughout the crystal.

Metamorphic rock   Rock formed by recrystallizing igneous or sedimentary rocks within the earth's crust under conditions of high temperature combined in most cases with high pressure and deforming stresses.

Metamorphic grade   The temperature and pressure at which a metamorphic rock recrystallized, as indicated by its mineral assemblage.

Metamorphism   Recrystallization of rock at temperatures above about 300°C to form new mineral assemblages and textures. In most cases, the high temperatures are accompanied by high pressures and deforming stresses.

Micas   A group of minerals based on silicate layer structures.

Migmatite   A type of rock consisting of an intimate mixture of igneous and metamorphic components.

Mineral   A naturally occurring chemical element or compound defined in terms of its composition and internal crystalline structure.

Miogeosyncline   A hypothetical linear trough in which sediments deposited in relatively shallow water accumulate.

Moh's Scale   A set of 10 minerals arranged in order of increasing hardness with talc, the softest, being number 1 and diamond, the hardest, number 10.

Moho discontinuity The surface of contact between the earth's outer crust and the mantle beneath.

Moraine A deposit of glacial till.

Mudflow A thick slurry of soil and rock.

Mudstone A type of sedimentary rock consisting of lithified mud, a mixture of silt and clay.

Muscovite A common variety of colorless mica represented by the formula $KAl_3Si_3O_{10}(OH)_2$.

Natural levee Stream banks raised above the general level of the floodplain by deposition during floods.

Neap tides The minimum tides that occur when the sun and moon are at right angles to each other as seen from the earth.

Negative ion An atom that has gained one or more extra electrons, which give it a net negative charge; same as an anion.

Nonclastic Sedimentary material consisting of substances— such as sea shells, peat, or evaporite minerals—that originated within the sedimentary environment.

Normal fault A fracture in the earth's crust along which the block above the fault surface moved down relative to that below. The term also applies to vertical faults that moved vertically.

Nuee ardente Same as a glowing cloud eruption.

Obsidian A sialic volcanic rock consisting mostly or entirely of glass; in effect, lava that cooled and hardened without crystallizing.

Oceanic crust The layer of rock that underlies the ocean basins. It is approximately 5 km thick, composed mostly of basalt, and forms at the crest of the oceanic ridge system.

Oceanic ridge A linear topographic swell that rises about 600 m above the general level of the ocean floor and winds through all the earth's major ocean basins. It has a rift at its crest that marks a divergent plate boundary.

Olivine Minerals composed of isolated silicate tetrahedra linked together by ions of iron and magnesium or, in rare cases, calcium.

Ophiolite complex A slice of oceanic crust exposed on land; typically incorporated into subduction complexes of rocks swept together at convergent plate boundaries.

Original horizontality, principle of The observation that sedimentary layers commonly form in a nearly level position.

Orthoclase Feldspar with the composition $KAlSi_3O_8$.

Outwash Sediment transported and deposited by glacial meltwater.

Outwash plain A gently sloping and smoothly graded surface underlain by deposits of glacial outwash.

Overthrust fault A fault that moves one slab of rock over others along a surface that dips at an angle less than about $10°$.

Oxbow lake An old stream meander left abandoned on the floodplain when the stream cut across its neck.

Oxidation Chemical weathering involving reaction of elements, such as iron and sulfur, with oxygen.

Oxides A group of minerals based on combinations of various elements with oxygen.

P wave Same as compressional wave.

Paired terraces Sets of stream terraces that match in elevation on opposite sides of the valley.

Pangaea A supercontinent that formed during the latter part of Paleozoic time through assembly of most of the earth's continental crust and then broke up and began to disperse at the beginning of Mesozoic time.

Peat A type of nonclastic sediment composed mostly of partially decayed plant material. It becomes coal upon lithification.

Pedalfer A large class of soils that typically contain high concentrations of iron and aluminum; forms in humid regions.

Pediment A part of a desert plain; a bedrock erosional surface smoothly graded by running water.

Pedocal A large class of soils that typically contain high concentrations of calcium and other soluble materials; forms in dry regions.

Pegmatite Extremely coarse-grained plutonic igneous rock.

Perched water table The upper surface of a local groundwater reservoir that rests on a body of impermeable rock.

Peridotite A plutonic ultramafic rock that consists mostly of pyroxene and olivine.

Permafrost Permanently frozen ground.

Permeability The ability of rocks to transmit fluids—such as water or petroleum—through pores, fractures, and other connected openings.

Petrology The branch of geology that deals with the study of rocks.

Photosynthesis A metabolic process characteristic of green plants that use energy from the sun to extract carbon from atmospheric carbon dioxide.

Phenocryst A large crystal set in a matrix of smaller crystals in a porphyritic igneous rock.

Piedmont glacier A sheet of flowing ice that forms where valley glaciers spread across a lowland.

Piezometric surface The level to which the pressure on water in an artesian aquifer would enable it to rise.

Pillow basalt Basalt lava flow consisting of a pile of cylindrical forms that suggest pillows when seen sectioned in a cliff or roadcut; forms when the lava erupts underwater.

Placer A local concentration of heavy minerals within sediment deposited by streams, waves, or the wind.

Plagioclase group Feldspars with compositions that vary continuously in the range from $NaAlSi_3O_8$ to $CaAlSi_2O_8$.

Plastic deformation   A permanent change in shape that occurs without breakage.

Plate   A segment of the earth's lithosphere.

Plate tectonics   The body of theory that deals with the origin, movement, and destruction of lithospheric plates and other associated phenomena.

Playa   An ephemeral lake in the lowest part of an undrained desert basin.

Plug dome   A craterless volcano consisting of a mass of viscous sialic lava extruded through the ground surface.

Plume   A hypothetical column of hot rock rising through the mantle; proposed as an explanation for volcanic hotspots.

Pluton   A mass of igneous rock that crystallized within the earth's crust.

Plutonic igneous rock   Rock that crystallized from a magma within the earth's crust; typically coarse grained.

Point bar   Sediments deposited on the inside bank of a river meander.

Pole of rotation   The point on the earth's surface about which a moving lithospheric plate can be imagined to rotate.

Polymorphs   Mineral species that have similar chemical compositions and different crystal structures.

Porosity   Open void space within a sediment or a rock.

Porphyritic texture   A term describing igneous rocks in which large crystals exist within a matrix of very much smaller crystals.

Porphyry copper   An ore body consisting of an altered granitic pluton that contains small amounts of ore minerals uniformly disseminated through it.

Positive ion   An atom that has lost one or more electrons, leaving it with a net positive charge; same as a cation.

Pothole   A cylindrical hole drilled in bedrock by pebbles turning beneath a persistent eddy in a stream.

Precambrian   All of geologic time before the beginning of the Cambrian period about 570 million years ago.

Primary sedimentary structures   Features of sediments and sedimentary rocks—such as ripple marks, mudcracks, and layering—that originate in the sedimentary environment.

Pumice   A glassy variety of rhyolite that consists of a mass of tiny gas bubbles and resembles frozen foam.

Pyrite   A mineral form of iron sulfide—FeS_2.

Pyroxene   A group of minerals that consist of silicate single chains linked by ions of iron, magnesium, and calcium.

Quartz   The common mineral form of silica—SiO_2.

Quartzite   A rock composed of quartz sand grains so tightly welded together that fractures break across rather than around them; most commonly forms through metamorphism of quartz sandstone.

Radioactive dating   The technique of determining the age of a mineral or rock by measuring the ratio of undecayed parent atoms to the daughter atoms.

Radioactive decay   The spontaneous transformation of one element into another through disintegration of the atomic nucleus. Heat energy released during radioactive decay drives the earth's internal processes.

Rainsplash erosion   Movement of soil particles by splattering raindrops.

Recessional moraine   A ridge of till deposited by a glacier that had retreated from its position of farthest advance.

Recrystallization   The destruction of old crystals accompanied by corresponding growth of new ones.

Recurrence interval   The statistically estimated probability that a flood of given magnitude will occur within any particular year.

Relief   The local difference in elevation between the high and low parts of a landscape.

Remanent magnetism   The permanent magnetic polarization parallel to the earth's magnetic field that most rocks acquire as they form and retain thereafter.

Residence time   The average length of time an atom of an element remains dissolved in the ocean before being removed.

Reverse fault   Same as thrust fault.

Rhyolite   Sialic volcanic rock that corresponds in composition to granite; generally pale.

Richter Magnitude Scale   A system of classifying earthquakes according to the amount of energy they release; every one-point increase in magnitude represents a 10-fold increase in ground motion.

Ridge and ravine landscape   The typical landscape of well-vegetated regions in which soil creep is the dominant process of hillslope erosion. Consists of a maze of hills and valleys generally drained by perennial streams. The hills all have convex summits, straight midsections, and concave toes.

Right-handed fault   Same as right-lateral fault.

Right-lateral fault   A strike-slip fault that moves so that an observer looking along the line of the fault sees the side on his right coming toward him.

Rip currents   Streams of water that flow from the beach past the line of breakers. They return water beached by the surf to the ocean.

Ripple marks   Waves that form on the surfaces of sediments under transport by waves, running water, or the wind.

Roche moutonnee   Same as a whaleback.

Rock   An aggregate of mineral grains.

Rock avalanche   An extremely mobile mass of shattered rock that moves as though it were fluid.

Rockfall   A slide that moves on weak surfaces within the bedrock.

Rock flour Pulverized rock ground within a glacier.

Rock glacier A stream of rubble that moves slowly through flowage of ice, which fills interstices between blocks.

S wave Same as shear wave.

Saltation A latinate word that means bouncing; often used to describe the motion of sand grains transported by wind or water.

Salt cracking A process of mechanical weathering in which salts crystallizing between the mineral grains in a rock wedge them apart.

Sand Fragments in the size range between about .06 and 2 mm in diameter. Although the vast majority of sand grains are quartz, they may be composed of any mineral.

Sand drift A pile of sand trapped in the wind shadow of an obstacle, such as a boulder.

Sandstone Sedimentary rock consisting of lithified sand.

Schist A textural term used to describe strongly foliated metamorphic rock in which micas or amphiboles are abundant enough to give to rock a flaky or splintery quality.

Scoria Volcanic rock honeycombed with large gas bubbles.

Seafloor spreading The idea that oceanic crust forms at oceanic ridges and moves away from them.

Seamount A submerged volcano.

Sediment Detrital material either clastic or nonclastic that has been transported and deposited by wind, water, ice, or gravity.

Sedimentary rocks Lithified sedimentary material.

Sedimentary structures Features—such as ripple marks, mudcracks, graded bedding, and layering—that formed during accumulation of a sedimentary deposit. They record information about the sedimentary environment.

Sedimentary veneer The blanket of sedimentary rocks that covers the continental basement in most regions.

Seismic sea wave An oceanic wave, also called a *tsunami*, caused by movement of the ocean floor, most commonly due to movement along a fault during an earthquake but in some cases by submarine volcanic eruptions or landslides. Seismic sea waves may rise to catastrophic heights as they enter shallow water and are incorrectly called tidal waves.

Seismograph An instrument that detects and records earthquake wave motion.

Seismology The study of earthquakes.

Serpentine A group of magnesium-rich and iron-poor layer silicate minerals that commonly form through reaction of pyroxenes and olivines with water at low temperatures.

Serpentinite Metamorphic rock composed largely of serpentine minerals; generally forms through reaction of peridotite with water.

Shadow zone A doughnut-shaped dead area on the side of the earth opposite an earthquake in which seismographs record no shock waves because they are blocked by the core.

Shale Sedimentary rock consisting of lithified clay that tends to split into thin flakes in a direction parallel to that of the depositional layering.

Shear wave A kind of earthquake wave motion in which particles of rock move in a direction transverse to that of the wave advance. Somewhat resembles wave motion in a stretched rope.

Shield A large region in which the sedimentary veneer is missing and continental basement rocks are exposed at the earth's surface.

Shield volcano A broad and gently sloping pile of thin basalt lava flows erupted from a central vent.

Sialic rocks Rocks that contain relatively high concentrations of silicon, aluminum, and potassium and relatively small amounts of iron, magnesium, and calcium. Sialic igneous and metamorphic rocks consist largely of sodium-rich plagioclase, orthoclase, quartz, and mica or amphibole. The average chemical composition of the clastic sedimentary rocks is the same as that of sialic igneous and metamorphic rocks.

Silicates All minerals composed basically of silicon and oxygen in combination with various other elements.

Silicate group A silicon atom surrounded by a cluster of four oxygen atoms arranged as though they were at the corners of a tetrahedral pyramid. Silicate groups are also called silicate tetrahedra.

Siliceous ooze Deep ocean sediment consisting mostly of silica shells of minute organisms mixed with fine clay. It accumulates on parts of the ocean floor beyond the reach of turbidity flows and beneath the carbonate compensation depth. Siliceous ooze lithifies into thin beds of brightly colored chert.

Sill A tabular body of igneous rock formed from magma injected parallel to the layering of the enclosing rock.

Siltstone A sedimentary rock composed of lithified silt, sedimentary particles that range from about 0.004 to about 0.06 mm in diameter.

Sinkhole A topographic depression created either by collapse of an underground opening, such as a cave, or by locally concentrated surficial solution of carbonate rocks.

Slate Low-grade metamorphic rock formed by slight recrystallization of shale or mudstone. Slate develops a tendency to split in a direction parallel to an imaginary surface that would bisect the folds in the rock into mirror image halves—a property known as slaty cleavage.

Slough An old stream channel left abandoned on the floodplain when the stream breached its natural levees and formed a new channel.

Smectite A kind of clay mineral that typically forms in the poorly drained soils of dry regions. Smectites absorb water and fertilizer nutrients.

Soil creep A composite of several erosional processes that act together to move soil down hillslopes as though it were an extremely viscous liquid. The upper levels move fastest and the rate of movement diminishes with increasing depth.

Soil flowage Downslope movement of a surface layer of very weak and water-saturated soil; most commonly occurs when thawed surface soil moves on a base of frozen subsoil.

Soil polygons A network of rock-filled cracks that develops in permanently frozen ground.

Spheroidal weathering The tendency of weathering processes, especially hydration, to round the edges and corners of angular rocks.

Spring tides The largest variations in sea level caused by coincidence of the solar and lunar tidal bulges. Spring tides occur during the full moon and the new moon when the earth, moon, and sun are in a straight line.

Stacks Small outposts of rock left standing in the surf as remnants of eroding headlands.

Stalactite A travertine dripstone formation that hangs like an icicle from a cave roof.

Stalagmite A travertine dripstone formation that stands like a pedestal on a cave floor.

Strain energy Energy stored in elastically deformed rocks—as in a bent bow—and released as earthquake waves if the rocks regain their original shape as they suddenly slip along a fault.

Strain gauge An instrument that measures elastic deformation in rocks.

Stratification Layering in sedimentary rocks.

Stratigraphy The branch of geology that deals with the interpretation of sedimentary rocks.

Streak The color of a fine powder of a mineral.

Stream terrace A remnant of an old floodplain left above the modern floodplain as the stream eroded the floor of its valley to a lower level.

Striations Sets of parallel grooves scored in a rock surface.

Strike The compass direction of a horizontal line drawn on the surface of an inclined layer of rock.

Strike-slip fault A fracture in the earth's crust the opposite sides of which move past each other in a horizontal direction.

Structural geology The branch of geology that deals with the interpretation of deformed rocks.

Structural style The pattern of deformation in an area.

Subduction The sinking of a lithospheric plate into the mantle at a convergent boundary.

Subduction complex A scrambled mass of sedimentary and volcanic rocks scraped into a trench from the surface of a plate sinking into the mantle at a convergent boundary.

Sulfides A group of minerals all of which consist of cations bonded to sulfur anions.

Superimposed stream A stream that cuts through a high region or a body of resistant rock that it encountered as it eroded its course deeper.

Superposition, principle of The rule that an undisturbed sequence of sedimentary rocks becomes younger from bottom to top.

Surface runoff erosion Movement of soil particles by water flowing over the ground surface.

Surface sealing Development of an impermeable ground surface through the puddling and hammering effects of rain falling on bare soil.

Surface wave A kind of earthquake wave motion that resembles swells on the open ocean.

Surge A sudden rapid advance of a glacier.

Suspended load Stream-transported silt and clay that do not settle to the bottom.

Suture A line along which formerly separate parts of the continental crust are now joined.

Syncline A fold in which the layers of rock buckle down to form a trough.

Tarn A small lake filling a basin gouged into bedrock by a valley glacier.

Tectonic creep Continuous slow movement along a fault.

Tectonics The study of large movements of the crust and lithosphere.

Terminal moraine A ridge of till that marks the farthest advance of a glacier.

Texture The term refers to the size and shape of the mineral grains in a rock and to the pattern of their arrangement.

Thrust fault An inclined fracture along which the hanging wall moved up relative to the footwall. Movement along thrust faults shortens the earth's crust and probably occurs in response to compressional forces.

Tidal wave Same as seismic sea wave.

Till Glacially deposited sediment that typically consists of an unlayered and unsorted mixture of all sizes of material.

Time rock unit An assemblage of related rocks formed during the same time.

Tombolo A sand spit that ties an island to the shore, converting it into a peninsula.

Trailing margin An edge of a continent that faces toward a divergent plate boundary. The opposite coasts of the Atlantic Ocean are a good example.

Transform boundary A fault along which two lithospheric plates slide past each other without gaining or losing area. Transform boundaries begin and end in divergent and convergent boundaries in any combination.

Transform fault A plate boundary along which one plate slides horizontally past another with no gain or loss in plate area.

Transpiration Return of water to the atmosphere by plants.

Travertine A rock composed of calcite deposited from solution in caves, hot springs, or in streambeds; commonly develops into distinctive dripstone formations, such as stalactites or stalagmites.

Trench A linear depression in the ocean floor marking a convergent boundary where one plate sinks into the mantle; same as a deep.

Tsunami Same as seismic sea wave.

Tundra plain A landscape leveled by soil flowage and drained largely through soil polygons.

Turbidites Sequences of sediments consisting mostly of graded beds deposited from turbidity flows.

Turbulent flow A type of fluid motion in which individual particles follow an irregular and unpredictable path, making swirls and eddies. Air and water both flow turbulently unless the rate of movement is extremely slow.

Turbidity currents A current of moving fluid that flows downslope because its content of suspended matter makes it dense.

Unit cell The smallest portion of a crystal that retains the composition and symmetry of the whole.

Unconformity A surface of erosion or nondeposition separating two sequences of rock of different ages.

Uniformitarianism The assumption that physical and chemical processes operated in the past as they do today; that the present is the key to the past.

Unpaired terraces Sets of stream terraces that do not match in elevation on opposite sides of the valley.

Valley glacier A mass of ice that flows down a valley as though it were an extremely viscous river; often called alpine glaciers.

Ventifact A wind-carved rock. Ventifacts typically have surfaces composed of concave faces that meet in sharp edges and corners.

Vesicle Gas bubble in a volcanic rock.

Volcanic ash Volcanic rock fragments smaller than about 4 mm in diameter that commonly drift long distances in the wind before they settle.

Volcanic bomb A large chunk of volcanic rock that was blown from the vent while still molten; most have an aerodynamically streamlined shape.

Wave refraction The effect that drag on shallow bottoms has in causing approaching waves to change direction so they nearly conform to the outline of the coast.

Water table The upper surface of the zone of water-saturated ground.

Wave-built terrace A gently sloping and smoothly graded bottom surface composed of wave-deposited sediment.

Wave-cut bench A smooth and gently sloping surface cut in bedrock by wave erosion of a sea cliff.

Wave refraction The effect that drag on shallow bottoms has in causing approaching waves to change direction so they nearly conform to the outline of the coast.

Weathering The complex of mechanical and chemical processes that convert rock into soil.

Welded ash A type of volcanic rock composed of fragmental material that fused into a solid mass after having been erupted, typically sialic in composition.

Welded tuff Same as welded ash.

Whaleback A glacially sculptured knob of bedrock polished and striated on its upflow side and roughly quarried by plucking on its downflow side.

Zone of ablation The part of a glacier on which the amount of annual melting exceeds the amount of snow accumulation.

Zone of accumulation The upper part of a glacier on which annual snowfall exceeds melting.

INDEX

Page numbers in boldface indicate definitions, those in italics illustrations.